AF324675

THE MAGNETISM OF AMORPHOUS METALS AND ALLOYS

THE MAGNETISM OF AMORPHOUS METALS AND ALLOYS

Editors

J. A. Fernandez–Baca
Oak Ridge National Laboratory, USA

Wai-Yim Ching
University of Missouri – Kansas City, USA

Published by

World Scientific Publishing Co. Pte. Ltd.

P O Box 128, Farrer Road, Singapore 9128

USA office: Suite 1B, 1060 Main Street, River Edge, NJ 07661

UK office: 57 Shelton Street, Covent Garden, London WC2H 9HE

THE MAGNETISM OF AMORPHOUS METALS AND ALLOYS

ISBN 981-02-1033-7

Printed in Singapore.

2. ELECTRONIC STRUCTURE CALCULATIONS IN MAGNETIC METTALIC GLASSES
by W.Y. Ching

5. NEUTRON SCATTERING STUDIES OF THE SPIN DYNAMICS OF AMORPHOUS ALLOYS
by Jeffrey W. Lynn and Jamie A. Fernandez-Baca

6. NUMERICAL STUDIES OF MAGNONS IN AMORPHOUS MAGNETS
by D. L. Huber

7. MAGNETISM AND MAGNETO-OPTICS OF RARE EARTH-TRANSITION METAL GLASSES AND MULITLAYERS

by D. J. Sellmyer, R.D. Kirby and S. S. Jaswal

FOREWORD

About four years ago we were invited by the editors of World Scientific Publishing Co. Pte. Ltd. to edit a review volume on selected topics of randomness in magnetism. This subject was later narrowed down to the magnetism of amorphous metallic alloys. Of course the first question that came to our minds is "Why bother in editing a volume on a subject that has already excellent and comprehensive review books and articles like those of Moorjani and Coey[1] and J. Chappert.[2]?" The immediate answer was that perhaps it was worthwhile to make an update to the literature in this field, in the form of a book that would gather a selection of topics written by the people that had made the most recent progress. This would offer current reviews from the point of view of the specialists who would concentrate on their most recent progress while not neglecting to refer the reader to the already abundant existing literature. With this in mind we invited a selected group of specialists to review their progress in this subject, some of them accepted this challenge and this book is the result of their hard work. We thank all of them for their dedication and for their commitment. Because of the way that this book came about, this volume is by no means a comprehensive review of all the progress in the subject. Rather it is a collection of selected topics in the recent progress on the study of the magnetism of amorphous metals and alloys. Finally, we recognize the title of this book is very ambitious because while a great deal has been learned about the magnetism of amorphous metallic alloys, the study of the magnetism of the pure amorphous metals is still a subject of intense activity. We hope that the content of this book will help to motivate research in this direction. We trust that the reader will find in this volume a worthwhile update of some of the progress in the field of amorphous magnetism.

J. A. Fernandez-Baca
W. Y. Ching

---

[1] K. Moorjani and J. M. D. Coey, *Magnetic Glasses,* (Elsevier, 1984).
[2] J. Chappert, "Magnetism of Amorphous Metallic Alloys", in *Magnetism of Metals and Alloys*, ed. M. Cyrot (North Holland, 1982)

II.

J. V. Yakhmi
Bombay

2. R. Hasegawa, Amorphous Magnetism, Vol. II, ed. R. A. Levy and R. Hasegawa (Plenum Press, 1977).

CHAPTER ONE

THEORY OF MAGNETISM IN AMORPHOUS TRANSITION METALS AND ALLOYS

Yoshiro KAKEHASHI

Department of Physics, Hokkaido Institute of Technology
Maeda, Teine-ku, Sapporo 006, Japan

Hiroshi TANAKA

IBM Research, Tokyo Research Laboratory,
IBM Japan Ltd., 1623-14, Shimotsuruma, Yamato, Kanagawa 242, Japan

1. Introduction

In the past two decades, a large number of amorphous transition metal alloys showing fascinating magnetic properties have been found with the development of rapid quenching techniques [1-10]. Their physical properties, such as amorphous structure, magnetic properties, electronic states, transport properties, and magneto-optical properties, have been extensively investigated by means of microscopic experimental techniques. A variety of magnetism caused by structural disorder has opened a new field of "amorphous metallic magnetism", in which structural disorder and metallic magnetism are closely related each other *via* electronic structure. This chapter reviews theoretical aspects of recent developments in amorphous metallic magnetism of transition metals (TM) and alloys.

Needless to say, the most essential character of amorphous systems is structural disorder. In the past, experimental data for amorphous magnetic alloys have often prevented us from understanding the effects of structural disorder, because of the nonexistence of amorphous pure transition metals and the limited range of concentration in amorphous alloys. Amorphous transition metal alloys containing considerable amount of metalloids (typically 20 at.% B or P), were first systematically investigated

1

2

by Mizoguchi *et al.* [11-12], Hasegawa *et al.* [13], and O'Handley *et al.* [14-15] in the 1970's. The data showed that the magnetization and Curie temperature (T_C) were uniformly lower than those of crystalline alloys when they were plotted as a function of average d-electron numbers. Amorphous $Fe_{80}B_{10}P_{10}$ alloy [15], for example, shows ferromagnetism with the ground-state magnetization $M=2.1$ μ_B and $T_C=640$ K, while bcc Fe has $M=2.2$ μ_B and $T_C=1040$ K. Amorphous $Co_{80}B_{10}P_{10}$ alloy [15] also shows ferromagnetism with $M=1.1$ μ_B and $T_C=770$ K, which may be compared with $M=1.7$ μ_B and $T_C=1400$ K in fcc Co.

The magnetization vs. concentration curves were analyzed by using a concept of generalized Slater-Pauling curves suitable for the strong ferromagnets [16-19]. This method is based on the following identity:

$$M = Z_m + 2n_{sp\uparrow} \ . \tag{1.1}$$

Here $n_{sp\uparrow}$ is the averaged number of sp-electrons with up spin. Z_m is the averaged magnetic valence defined by $Z_m = \sum_\alpha c_\alpha(2n_{\alpha\uparrow} - Z_\alpha)$. c_α, $n_{\alpha\uparrow}$, and Z_α are the concentration, d-electron number with up spin, and the chemical valence for the constituent atom α, respectively. Since Z_m and $n_{sp\uparrow}$ are linear with respect to the concentration in transition metal alloys with strong ferromagnetism, we have a linear relation between M and Z_m.

The simple behaviors of M and T_C in TM-metalloids alloys were often regarded as inherent properties of amorphous structure rather than as effects of the presence of metalloids. It was concluded that the structural disorder merely introduced such simple behaviors into the magnetic properties [4]. However, the above picture changed with the appearance of amorphous transition metal alloys containing early transition metals or rare-earth (RE) metals in the 1980's [10,20-27]. In particular, Hiroyoshi and Fukamichi [22], Saito *et al.* [23], Coey *et al.* [24-25], Wakabayashi *et al.* [26], and Fukamichi *et al.* [21] found that the ferromagnetism in Fe-rich amorphous alloys collapses completely beyond 90 at.% Fe, and that a new spin-glass (SG) phase appears in which the transition temperatures are 120 K irrespective of the second elements, contradicting the early data on alloys containing metalloids. Furthermore, Fukamichi, Goto, and Mizutani [27] found that the Curie temperatures in Co-rich Co-Y amorphous alloys are enhanced as compared with those in their crystalline counterparts. It was suggested that the Curie temperature extrapolated to amorphous pure Co should reach 1850 K, which is 450 K higher than for fcc Co. These drastic changes of magnetism in the vicinity of amorphous pure metals have revealed the important role of structural disorder in amorphous metallic magnetism. It is our main purpose to describe such a new feature of amorphous metallic magnetism from the theoretical point of view.

There are two directions in the development of solid state physics. One is to include more and more electron correlations in order to achieve microscopic understanding of magnetism, metal-insulator transition, and high-T_C super conductors [28-29]. In particular, theoretical description of the magnetism in transition metals and alloys has been the main part of correlation problems, since the transition metal alloys show both itinerant and localized behaviors in their magnetic properties [30]. The other direction is to take account of more and more disorder [31]. The theory of transition metals and alloys in this direction has been developed in the order of magnetic impurities, substitutional alloys, and amorphous alloys [32-33]. A quantitative theory of amorphous metallic magnetism is, therefore, a goal that these two lines of investigation should reach at the end of the development.

This chapter is organized as follows. We discuss in the following section the theoretical problems and historical developments in amorphous metallic magnetism. There are discussions on the amorphous structure, electronic structure, and magnetism in Sec. 2.1. The most important advances in the recent theories of amorphous metallic magnetism have been the first-principles band calculation of the ground-state amorphous electronic structure and the development of a finite-temperature theory. The former was established by Fujiwara [34-35] through a combination of the linear-muffin-tin-orbital (LMTO) method [36-37] with the recursion method [38-39] in the framework of the band theory, which will be reviewed in Sec. 2.2. The latter was recently proposed by Kakehashi [40-42] on the basis of the functional integral method [43-45] and the distribution-function method [46-47]. The finite-temperature theory is reviewed in Sec. 2.3. Section 3 is devoted to the magnetism in amorphous transition metals, in particular, amorphous Fe, Co, and Ni. Numerical examples for these systems will demonstrate the basic effects of structural disorder. The magnetic properties of transition metal alloys are governed by both structural and configurational disorders. In Sec. 4, we discuss the magnetic properties of TM-TM alloys (Sec. 4.1), TM-RE alloys (Sec. 4.2), and TM-metalloid alloys (Sec. 4.3) on the basis of their electronic structures, which have been obtained in the last decade.

2. Theoretical Approach to Amorphous Metallic Magnetism

2.1. *Basic Problems and Developments*

Amorphous transition metals and alloys are characterized by structural disorder and itinerant magnetism. In the latter, electrons move from site to site with the electron hopping integrals which form electronic band structure. The magnetism is known to be caused by competition between the electron hopping and the electron-electron Coulomb interactions. The magnetic properties are sensitive to the band structure rather than the Coulomb interactions in metals, because the screened Coulomb interactions act mainly between the on-site electrons, and thus depend very little on the structure of solids. It is therefore essential to take into account the effect of the structural disorder on the electronic band structure. This implies that it is difficult to understand the magnetic properties of amorphous transition metals and alloys without taking the following regular steps:

(1) Construction of amorphous structure

4

(2) Calculation of electronic structure

(3) Constructing a theory of amorphous magnetism on the basis of the electronic structure

In what follows, we briefly discuss the theoretical problems in these three steps, as well as historical developments.

2.1.1. Amorphous structure

In crystalline systems, the crystal structure and lattice constants can be determined by analyzing the Bragg peaks in X-ray or neutron diffraction, and thus the electronic structure can be calculated on the basis of the Bloch theory [37,48]. This procedure is no longer available for amorphous metals and alloys, because of the lack of translational symmetry. X-ray experimental techniques [49,50] give us only the pair distribution functions (PDF) $g_{\alpha\gamma}(R)$, which are defined as follows:

$$g_{\alpha\gamma}(R) = \frac{\rho_{\alpha\gamma}(R)}{\rho_\gamma} \ , \tag{1.2}$$

where $\rho_{\alpha\gamma}(R)$ is the density of a γ atom at a distance R from an α atom, and ρ_γ is the density of the γ atom. The distribution function $g_{\alpha\gamma}(R)$ converges to 1 as R approaches infinity. Reduced PDF $G_{\alpha\gamma}(R)$ are also used, and are defined as follows.

$$G_{\alpha\gamma}(R) = 4\pi R^2 \{g_{\alpha\gamma}(R) - 1\} \ . \tag{1.3}$$

Although we can obtain further information on the amorphous structure from EXAFS (extended X-ray absorption fine structure) and neutron measurements [9], it is not possible to determine experimentally all the atomic positions in the amorphous structure, though these are indispensable for electronic-structure calculations of amorphous systems and microscopic understanding of magnetic properties. This difficulty imposes us on the above theoretical problem (1), namely, how to construct a reasonable model for an amorphous structure.

The simplest model for amorphous metals and alloys is the dense random packing of hard spheres (DRPHS) model. The model was first proposed by Bernal [51-52]. He constructed the model by squeezing and kneading rubber bladders filled with ball bearings of the same size. He found that the DRPHS model consists of only the five types of polyhedra (so-called 'Bernal holes') shown in Fig. 1.1. Finney [53] constructed a much larger cluster model by using the same method as Bernal, and obtained a PDF. Bennett [54] constructed a DRPHS model by using a computer. His algorithm for generating the cluster is as follows: (1) make an equilateral triangle consisting of three hard spheres touching each other, (2) list up all the pockets in which new sphere could be added in hard contact with three spheres which already exist in the cluster, (3) add the new sphere to the nearest pockets from the origin of

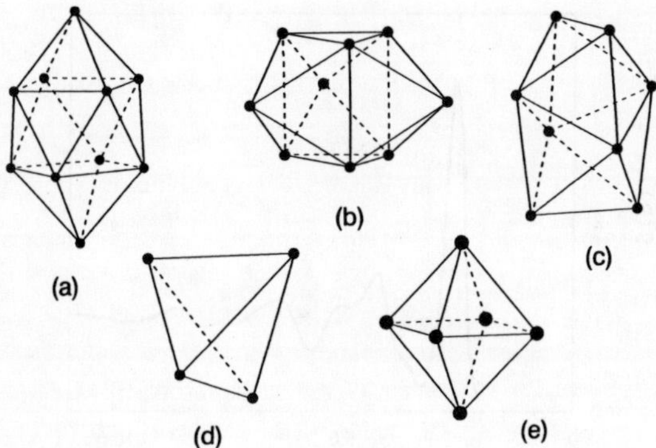

Fig. 1.1: Five types of polyhedra in the DRPHS model: (a) Archimedian antiprism, (b) trigonal prism, (c) tetragonal dodecahedron, (d) tetrahedron, and (e) octahedron

the cluster, and (4) return to the step (3). The PDF calculated from this computer-simulated cluster model was quite similar to that of the ball bearing model. It was however somewhat different from the experimental one. In particular, the splitting of a second peak into two peaks and their relative amplitudes could not be reproduced.

Ichikawa [55] suggested that this splitting was related to the tetrahedral local atomic structure, and improved Bennett's algorithm. He introduced a parameter Λ which measured the perfection of the tetrahedral structure. It is given in the form

$$\Lambda = \frac{r_{ij}^{\max}}{R_i + R_j} \ . \tag{1.4}$$

where each i or j denotes one of the three spheres in the cluster forming a new pocket in step (2) of Bennett's algorithm. r_{ij} is the distance between the spheres i and j. R_i is the radius of the sphere i. Obviously, Λ changes from 1 to 2 for a single-size sphere system in Bennett's algorithm. The deviation from a perfect tetrahedral structure is evaluated as the deviation from the condition $\Lambda = 1$.

Ichikawa constructed a DRPHS model with several Λ values, and found that the PDF was well reproduced when Λ was 1.2. Although the PDF was well reproduced, the packing fraction of the model was smaller than in the experimental data because of porosity of the model structure.

These disadvantages were overcome by a process for structural relaxation of a DRPHS model through the atomic force produced by appropriate pair potentials [56,57]. This method, the so-called relaxed DRPHS model, was proposed by Cargill [49].

6

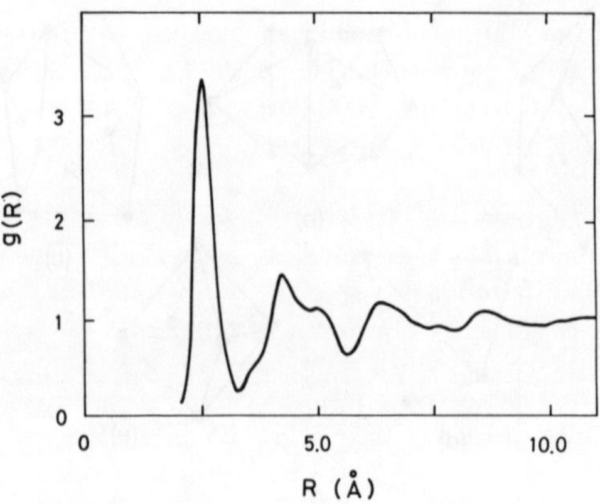

Fig. 1.2: Pair-distribution function of computer-generated amorphous iron [58]

Figure 1.2 shows an example of a pair-distribution function for amorphous Fe calculated by Yamamoto and Doyama [58], using the relaxed DRPHS model. The first peak at $r_1 = 2.54$ Å is sharp, and is clearly separated from the second and third ones. This means that there exists a well-defined nearest-neighbor (NN) shell even in amorphous systems. The ratio of the fluctuation of the NN interatomic distance to the average NN distance is estimated from the width of the first peak to be 0.067, which is in agreement with recent experimental data for Fe-rich amorphous alloys [59]. The second peak, at $r_2 = 1.67r_1$, is considered to originate in the local structures of the rhombi, each of which consists of two regular triangles with a side r_1, and the hexahedra, each of which consists of two tetrahedra with a side r_1. The third peak at $r_3 = 2r_1$ is associated with three contact atoms on a line.

The thermodynamical molecular-dynamics (MD) method [60] is a more sophisticated way of constructing an amorphous structure. In this method, the constituent atoms are distributed in a box with a periodic boundary condition, and the Newton equations of atomic motion are solved by assuming appropriate short-range interatomic pair potentials. Rapid quenching is simulated by reducing the kinetic energy, which is proportional to the temperature, at constant time intervals, under the condition that either the pressure or volume is constant. The method aims to simulate the formation of an amorphous structure according to the physical principle by using computers.

More recently, an *ab-initio* MD method was developed by Car and Parrinello [61]. In this method, interatomic forces are calculated directly from electronic structure and atomic structure without any empirical parameters, so that both the electronic and atomic structure are treated on an equal footing. Car and Parrinello greatly accelerated the *ab-initio* MD calculation by solving the equations of motions for both atoms and electrons at the same time. The method is most efficient when combined with the pseudo-potential technique and plain wave orbitals, and has been applied to amorphous semiconductors [62-63], surfaces of semiconductors, and interdiffusion in semiconductors [64].

Although MD calculations are very effective for constructing an atomic structure model of amorphous materials, there are some limitations at the present stage. The MD calculations are limited to rather small systems (100 $\sim$ 1000 atoms in a box) because of the insufficient efficiency of computers. Moreover, the minimum cooling rate in the MD is about 10^{11} (K/s), which is too large as compared with the experimental rates ($\simeq 10^6$ (K/s)). We do not enter into the details of these problems because they fall outside the scope of our discussion.

2.1.2. *Electronic structure calculations*

Amorphous systems lose translational symmetry and the Bravais lattice. This makes it impossible to calculate their electronic structures by making use of the Bloch theorem. Much theoretical effort, therefore, has been concentrated on this problem in the past twenty years.

The best single-site approximation was first established by Roth [65-66]. She introduced a k-dependent effective selfenergy caused by the structural disorder. The selfconsistent equations for the selfenergy were obtained from a single-site decoupling to the coupled equations for the averaged T-matrices. This is called the effective medium approximation, and is regarded as a natural extension of coherent potential approximation (CPA) in disordered substitutional alloys [67-68].

The single-site approximation does not describe the details of the local environment effects (LEE) on the densities of states and the local magnetic moments, which are in particular important for transition metal alloys. Fujiwara [34] developed a general method going beyond single-site approximation. He combined the tight-binding linear muffin-tin orbital method (LMTO) [36] with the recursion method [38-39] in electronic-structure calculations. The former gives an effective tight-binding Hamiltonian which greatly simplifies the band-structure calculations within the framework of the local-spin density functional theory [69-71]. The latter is powerful for the electronic structure calculations of disordered systems when their Hamiltonians can be described in a tight-binding form. The method provide us with first-principles electronic structure calculations for a given amorphous structure, and is regarded as the best method for calculating the electronic structure of amorphous transition metals and alloys. We will review the theory briefly in Sec. 2.2.

An alternative approach which also aims at the first-principles calculations of amorphous metals and alloys is to simulate the amorphous structure by means of a crystal with a large unit cell. This "supercell" approach [72-76] was made effective by the development of the linear method [36-37] and super computers. It is now possible

8

to perform first-principles calculations for an amorphous "compound" with 50 to 100 atoms in a unit cell.

2.1.3. *Calculating magnetic properties*

When we calculate magnetic properties on the basis of electronic structure, we have to take into account the spin degrees of freedom. This causes two major difficulties in the theoretical investigations of amorphous transition metals and alloys. First, the local magnetic moments (LM) change their directions as well as their amplitudes according to their local environments, as a result of the structural disorder. The question is then how one determines the LM configuration selfconsistently. This is not an easy problem even at the ground state when the ferro- and antiferro-magnetic interactions compete with each other in the disordered system, because many local minima in energy are expected, and finding the lowest energy is beyond the trial-and-error calculations. The problem is essential for the description of spin glasses, in which LM's are spatially disordered with no net magnetization, but are ordered or frozen thermodynamically [77-78].

As far as the ground state is concerned, one possibility may be to use the simulated annealing method [79]. In this method, the total energy E is minimized relative to the spin densities $\{\langle m_i \rangle\}$ by generating a succession of $\{\langle m_i \rangle\}$'s with a Boltzmann-type probability for a fictitious temperature T in a Monte-Carlo method, or by solving the equations of motion for classical particles with a "potential energy" E in the MD method. When $T \to 0$, the state with the lowest energy may be reached. However, these methods have not been applied yet.

The second problem is that transition metals show local-moment as well as itinerant-electron behaviors in their magnetism [80]. The band theory based on the Stoner model explains the non-integer ground-state magnetization, the existence of the Fermi surface, and the T-linear specific heat at low temperatures. But it does not lead to the local-moment behaviors such as a reasonable T_C, the Curie-Weiss susceptibility, and a large specific heat at T_C in Fe, Co, and Ni. Although the difficulty has been a long-standing problem in the theory of itinerant magnetism, it has been clarified in the past decade that it can be solved by taking into account the thermal spin fluctuations missing in the Stoner model [29-30]. Such a spin fluctuation theory has recently been applied to amorphous metallic systems by Kakehashi [40-42]. He took into account thermal spin fluctuations by making use of the functional integral method developed by Cyrot [43], Hubbard [44], and Hasegawa [45], and determined the LM distribution due to the structural disorder using the distribution-function method developed by Matsubara and Katsura [46-47]. This is the only theory which is available at finite temperatures at present. It will be reviewed in detail in Sec. 2.3.

2.2. LMTO-Recursion Method

Recent success in band calculations is founded on the local functional density (LDF) theory [69, 81] and the linear methods developed by Andersen *et al.* [82]. The former has provided the basis for constructing a one electron potential from first principles, while the latter has made band calculations faster by a couple of digit. Among the linear methods, the linear muffin-tin orbital (LMTO) method [37] is the most efficient to calculate the band structure of complicated crystals with many atoms in a unit cell, because it requires only 9 or 16 basis functions (1 for s-, 3 for p-, 5 for d-, and 7 for f-states) per atom. In Sec. 2.2.1, we will explain the basic concept of the linear method, and in Sec. 2.2.2 the tight-binding LMTO method [36], which has been developed on the basis of the conventional LMTO method [37]. Combined with the recursion method developed by Haydock *et al.* [38, 39], the tight-binding LMTO method makes it possible to calculate the electronic structures of amorphous materials without translational symmetry [34, 35]. We will introduce the recursion method in Sec. 2.2.3.

2.2.1. Muffin-tin method and linear method

Before describing the tight-binding LMTO method in detail, we explain the basic concept of muffin-tin (MT) methods such as the augmented plain wave (APW) [83] method and the Korringa, Kohn, and Rostoker (KKR) method [84]. The linear method is a linearized version of these MT methods. In a MT method, the whole space is divided into the muffin-tin part near the nuclei and the interstitial region outside the MT spheres (see Fig. 1.3). In a MT sphere, the wave functions vary rapidly because the one-electron potential is very deep. In this region, the potential can be approximated by a spherically symmetric one, and the Schrödinger equation can be reduced to a radial differential equation. The basis functions are obtained by solving the Schrödinger equation within the MT sphere.

In the interstitial region, the potential is shallow, and slowly varying. Therefore, the basis functions can be approximated by a linear combination of solutions for the following equation, assuming that the kinetic energy k^2 is constant.

$$(\Delta + k^2)\, \psi\,(\boldsymbol{r}) = 0. \tag{1.5}$$

The basis functions thus obtained are called 'augmented wave functions'. The band structure is determined by the condition that the wave functions within MT spheres are connected with those in the interstitial region smoothly (*i.e.* continuously and differentiably) at the MT surface. In this procedure, the information about the potential within an MT is summarized by the logarithmic derivative of the wave function at the MT surface. Therefore, the determination of an energy band by means of the MT method is identical with solving a scattering problem by using a partial-wave scheme, replacing the MT potential with the phase shift determined from the logarithmic derivative at the MT surface.

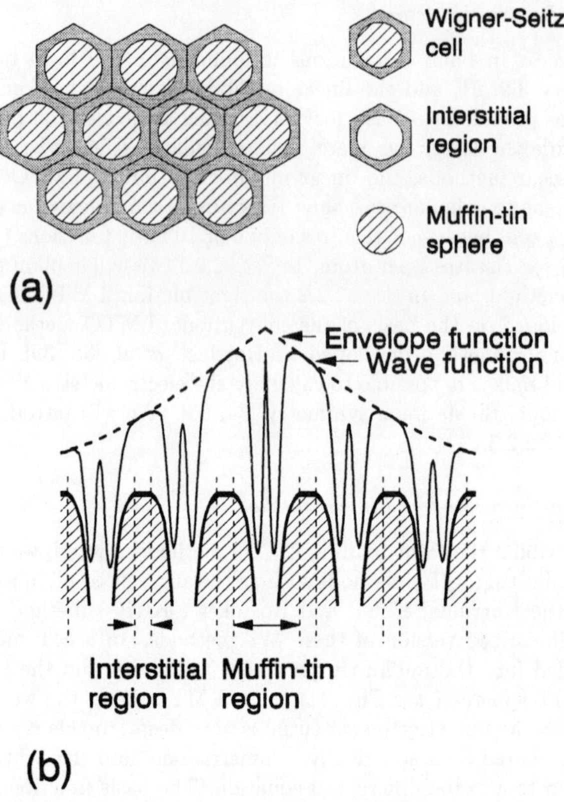

Fig. 1.3: (a) Schematic picture of muffin-tin spheres and the interstitial region. (b) Schematic picture of the muffin-tin potential and envelope function. MT methods assume that the potential is flat in the interstitial region. The wave function is augmented in the interstitial region by the envelope function.

The formula that determines the energy band in the APW (or KKR) method can be described in the following form:

$$\det \left| (\boldsymbol{k}^2 - \epsilon) + \Gamma_{ij}^{\mathrm{APW(KKR)}} \right|. \tag{1.6}$$

Here, $\Gamma_{ij}^{\mathrm{APW(KKR)}}$ is a complex function of the energy ϵ and the logarithmic derivatives of the wave functions, and Eq. (1.6) becomes an eigenvalue problem that is non-linearly dependent on the energy. The linear method is a linearized version of the MT method, where the eigen value equation Eq. (1.6) is reduced to the equation with linear energy dependence

$$\det |\boldsymbol{H} - E\boldsymbol{O}| = 0. \tag{1.7}$$

For this purpose, the wave function is expanded into the Taylor series around a certain energy $E_{\nu l}$, and its energy dependence is taken into account up to the first order. In the next section, we describe the formulation of the LMTO method, and how it can be transferred to the tight-binding form.

2.2.2. *Tight-binding LMTO method*

To give a feeling for the LMTO method, we start from a homo-nuclear diatomic molecule such as H_2. The centers of nuclei are located at the origin and at $\boldsymbol{R}$, and we consider only s-wave. The orbitals of this molecule consist of a bonding ($\phi_B(\boldsymbol{r})$) orbital and an anti-bonding ($\phi_A(\boldsymbol{r})$) orbital. The former is delocalized and has the low eigen value E_B, while the latter is orthogonal to the former and has the high eigen value E_A. They can be approximated by certain atomic wave functions for each atom, as follows:

$$\begin{aligned}
\phi_B(\boldsymbol{r}) &\simeq \phi(\boldsymbol{r}) + \phi(\boldsymbol{r} - \boldsymbol{R}), \\
\phi_A(\boldsymbol{r}) &\simeq \phi(\boldsymbol{r}) - \phi(\boldsymbol{r} - \boldsymbol{R}).
\end{aligned} \tag{1.8}$$

By solving $\phi(\boldsymbol{r})$ and $\phi(\boldsymbol{r} - \boldsymbol{R})$ in terms of $\phi_B(\boldsymbol{r})$ and $\phi_A(\boldsymbol{r})$, we can rewrite Eq. (1.8) as

$$\begin{aligned}
\phi(\boldsymbol{r}) &\simeq \frac{1}{2} (\phi_B(\boldsymbol{r}) + \phi_A(\boldsymbol{r})), \\
\phi(\boldsymbol{r} - \boldsymbol{R}) &\simeq \frac{1}{2} (\phi_B(\boldsymbol{r}) - \phi_A(\boldsymbol{r})).
\end{aligned} \tag{1.9}$$

Introducing a wave function $\phi_{E_0}(\boldsymbol{r})$ which is the eigen function of energy E_0 ($E_B \leq E_0 \leq E_A$), Eq. (1.9) can be approximated in the following form by using $\phi_{E_0}(\boldsymbol{r})$ and its energy-derivative $\dot{\phi}_{E_0}(\boldsymbol{r})$:

$$\phi(\boldsymbol{r}) \simeq \phi_{E_0}(\boldsymbol{r}) + c_1 \dot{\phi}_{E_0}(\boldsymbol{r}) \tag{1.10}$$
$$\phi(\boldsymbol{r} - \boldsymbol{R}) \simeq c_2 \dot{\phi}_{E_0}(\boldsymbol{r}) \tag{1.11}$$

12

Here, Eq. (1.10) corresponds to the wave function centered at the origin, while Eq. (1.11) corresponds to the tail part of the wave function centered at $\boldsymbol{R}$.

The above consideration justifies us in employing the following as a basis function of the bulk system:

$$\chi_{RL}(\boldsymbol{r} - \boldsymbol{R}) = \chi^i_{RL}(\boldsymbol{r} - \boldsymbol{R}) + \phi_{RL}(\boldsymbol{r} - \boldsymbol{R})$$
$$+ \sum_{R'} \sum_{L'} \dot{\phi}_{R'L'}(\boldsymbol{r} - \boldsymbol{R}') h_{R'L',RL}. \tag{1.12}$$

Here $\chi^i_{RL}(\boldsymbol{r})$ is the basis function in the interstitial region. L denotes a set of quantum numbers $\{l, m\}$. $\phi_{RL}(\boldsymbol{r} - \boldsymbol{R})$ is the solution of the Schrödinger equation at a certain energy $E_{\nu l}$,

$$(H - E_{\nu l}) \phi_{RL}(\boldsymbol{r} - \boldsymbol{R}) = 0, \tag{1.13}$$

within the MT centered at $\boldsymbol{R}$, and is described by the spherical harmonic functions in the form

$$\phi_{RL}(\boldsymbol{r} - \boldsymbol{R}) = \phi_{RL}(|\boldsymbol{r} - \boldsymbol{R}|) Y_{lm}(\widehat{\boldsymbol{r} - \boldsymbol{R}}). \tag{1.14}$$

The third term in Eq. (1.12) describes the tail parts of MTOs at neighboring sites centered at $\boldsymbol{R}'$. $h_{R'L',RL}$ will be determined later. $\dot{\phi}_{RL}(\boldsymbol{r} - \boldsymbol{R})$ is the energy derivative of $\phi_{RL}(\boldsymbol{r} - \boldsymbol{R})$ at $E = E_{\nu l}$,

$$\dot{\phi}_{RL}(\boldsymbol{r} - \boldsymbol{R}) = \left. \frac{\mathrm{d}\phi_{RL}(\boldsymbol{r} - \boldsymbol{R})}{\mathrm{d}E} \right|_{E = E_{\nu l}}. \tag{1.15}$$

Both $\phi_{RL}(\boldsymbol{r} - \boldsymbol{R})$ and $\dot{\phi}_{RL}(\boldsymbol{r} - \boldsymbol{R})$ are equal to zero outside the MT centered at $\boldsymbol{R}$.

The basis set $\phi_{RL}(\boldsymbol{r} - \boldsymbol{R})$ is orthogonal to each other, and normalized within the MT as follows:

$$\int_{|r-R| \leq s_R} \phi^*_{RL}(\boldsymbol{r} - \boldsymbol{R}) \phi_{R'L'}(\boldsymbol{r} - \boldsymbol{R}) \mathrm{d}r,$$
$$< \phi_{RL} \mid \phi_{R'L'} > = \delta_{RR'} \delta_{LL'}, \tag{1.16}$$

where s_R is the radius of the muffin-tin centered at $\boldsymbol{R}$. Differentiation of Eq. (1.16) with respect to energy shows that $\phi_{RL}(\boldsymbol{r} - \boldsymbol{R})$ is orthogonal to $\dot{\phi}_{RL}(\boldsymbol{r} - \boldsymbol{R})$:

$$< \phi_{RL} \mid \dot{\phi}_{R'L'} > = 0. \tag{1.17}$$

Here, we introduce a bra-ket notation for convenience in the following derivations. According to this notation, Eq. (1.12) can be written in the form

$$|\chi >^\infty = |\chi >^i + |\phi > + |\dot{\phi} > h, \tag{1.18}$$

where a ket vector with infinity symbol $| >^\infty$ means that the vector function extends over all space $\boldsymbol{r}$, while a simple ket vector $| >$ means that the vector function vanishes outside its own MT.

In order to construct a localized basis function the amplitudes of whose tails decay much faster than $h_{R'L',RL}$ in Eq. (1.12) as a function of $|\boldsymbol{R} - \boldsymbol{R}'|$, we have to prepare a more generalized basis set, defined as follows:

$$|\bar{\chi}>^{\infty} = |\bar{\chi}>^{i} + |\phi> + |\dot{\bar{\phi}}> \bar{h}, \tag{1.19}$$

where

$$|\bar{\phi}> = [1 + (\boldsymbol{E} - \boldsymbol{E}_\nu)\bar{o}]|\phi>, \tag{1.20}$$

so that

$$|\dot{\bar{\phi}}> = |\dot{\phi}> + |\phi> \bar{o}. \tag{1.21}$$

This definition requires that any energy derivative function is a linear combination of the orthogonal function sets $|\phi>$ and $|\dot{\phi}>$. $\bar{o}$ and $\bar{h}$ will be determined later. There exist the following relations:

$$\begin{aligned} <\phi|\dot{\bar{\phi}}> &= \bar{o}, \\ <\dot{\bar{\phi}}|\dot{\bar{\phi}}> &\equiv \bar{p} = \bar{o}^2 + p. \end{aligned} \tag{1.22}$$

It should be noted that the basis functions in Eq. (1.18) or Eq. (1.19) are expected to form a minimal basis set because they are constructed from solutions of the Schrödinger equation and their energy-derivatives at the energy levels in which we are interested.

The coefficient $\boldsymbol{h}$ in Eq. (1.18) or the coefficients $\bar{h}$ and $\bar{o}$ in Eq. (1.19) are determined according to the requirement that the basis function in Eq. (1.18) or Eq. (1.19) should be connected to the envelope function, which is a solution of the Schrödinger equation (1.5) in the interstitial region, continuously and differentiably (see Fig. 1.3 (b)). Note that the basis function in Eq. (1.18) is a special case of the basis function in Eq. (1.19) with $\bar{o} = 0$.

In the LMTO method, the linear combination of solutions for Eq. (1.5) with $\boldsymbol{k}^2 = 0$ (i.e. the Laplace equation $\Delta\psi = 0$), is chosen as the envelope function. Let us consider the following function:

$$K_{RL}(\boldsymbol{r} - \boldsymbol{R}) = \left(\frac{|\boldsymbol{r} - \boldsymbol{R}|}{s_w}\right)^{(-l-1)} Y_{lm}(\widehat{\boldsymbol{r} - \boldsymbol{R}}). \tag{1.23}$$

This is a solution of Eq. (1.5), and can be expanded around another atomic site $\boldsymbol{R}'$ in the region $|\boldsymbol{r} - \boldsymbol{R}'| \leq |\boldsymbol{R} - \boldsymbol{R}'|/2$ as follows:

$$\begin{aligned} K_{RL}(\boldsymbol{r} - \boldsymbol{R}) &= -\sum_{L'} \left(\frac{|\boldsymbol{r} - \boldsymbol{R}'|}{s_w}\right)^{l'} \frac{Y_{l'm'}(\widehat{\boldsymbol{r} - \boldsymbol{R}'})}{2(2l' + 1)} S_{R'L',RL} \\ &= -\sum_{L'} J_{R'L}(\boldsymbol{r} - \boldsymbol{R}') S_{R'L',RL} \end{aligned} \tag{1.24}$$

14

with

$$S_{R'L',RL} = \sqrt{4\pi}\, g_{l'm',lm} \left(\frac{|\boldsymbol{r} - \boldsymbol{R}'|}{s_w}\right)^{-l-l'-1} Y^*_{l+l'm'-m}(\widehat{\boldsymbol{R} - \boldsymbol{R}'}), \qquad (1.25)$$

$$g_{l'm',lm} = (-)^{l+1}\sqrt{4\pi}\,\frac{2(2l''-1)!!}{(2l-1)!!(2l'-1)!!}C_{lm,l'm',l''m''}, \qquad (1.26)$$

where $C_{lm,l'm',l''m''}$ is the Gaunt's coefficient, and s_w is the average Wigner-Seitz sphere radius. $S_{R'L',RL}$ is the so-called canonical structure constant, and is determined solely by the atomic structure. Therefore, we need to calculate it only once for the same atomic structure. Eq. (1.23) can be rewritten in the form

$$K^\infty_{RL} = K_{RL} - \sum_{R'L'} J_{R'L'}S_{R'L',RL}, \qquad (1.27)$$

or in the matrix form

$$|\boldsymbol{K}>^\infty = |\boldsymbol{K}> -|\boldsymbol{J}> \boldsymbol{S}, \qquad (1.28)$$

where $|\boldsymbol{K}>^\infty$ extends over all space $\{\boldsymbol{r}\}$, whereas $|\boldsymbol{K}>$ and $|\boldsymbol{J}>$ vanish outside its own Wigner-Seitz cell. K_{RL} is a irregular solution of the Laplace equation with the pole at $\boldsymbol{R}$, which decays proportionally to $|\boldsymbol{r} - \boldsymbol{R}|^{-l-1}$. On the other hand, J_{RL} is a regular solution of the Laplace equation with the pole at infinity, and is proportional to $|\boldsymbol{r} - \boldsymbol{R}|^l$. In order to obtain a localized basis, we introduce a more generalized envelope function by adding a certain factor of irregular solution $|\boldsymbol{K}>$ to the regular solution $|\boldsymbol{J}>$ in Eq. (1.28):

$$|\bar{\boldsymbol{J}}> = |\boldsymbol{J}> -|\boldsymbol{K}> \bar{\boldsymbol{Q}}, \qquad (1.29)$$

where $\bar{\boldsymbol{Q}}$ is an arbitrary diagonal matrix.

By surrounding a irregular solution K_{RL} centered at $\boldsymbol{R}$ with irregular solutions $K_{R'L'}$ centered at neighboring atoms, we can obtain a short-range envelope function, and thus a localized basis function. The new envelope function is described in the form:

$$\begin{aligned}
|\bar{\boldsymbol{K}}>^\infty &= |\boldsymbol{K}>^\infty (\boldsymbol{I} - \bar{\boldsymbol{Q}}\boldsymbol{S})^{-1} \\
&= |\boldsymbol{K}> -|\bar{\boldsymbol{J}}> \bar{\boldsymbol{S}},
\end{aligned} \qquad (1.30)$$

where $\bar{\boldsymbol{S}}$ is given as

$$\bar{\boldsymbol{S}} = \boldsymbol{S}(\boldsymbol{I} - \bar{\boldsymbol{Q}}\boldsymbol{S})^{-1}, \qquad (1.31)$$

or in a Dyson type equation

$$\bar{\boldsymbol{S}} = \boldsymbol{S} + \bar{\boldsymbol{Q}}\bar{\boldsymbol{S}}. \qquad (1.32)$$

Now, we can determine $\bar{o}$, $\bar{h}$, and $\bar{\boldsymbol{Q}}$ from the condition that the envelope function $|\bar{\boldsymbol{K}}>^\infty$ is connected smoothly with $|\bar{\chi}>^\infty$ on the MT surface. Only the results are listed below (see Ref. 36 for derivation):

$$\bar{o}_{RL} = -\frac{w_{RL}\{\bar{J},\dot{\phi}\}}{w_{RL}\{\bar{J},\phi\}} = -\frac{w_{RL}\{J,\dot{\phi}\} - w_{RL}\{K,\dot{\phi}\}\bar{Q}_l}{w_{RL}\{J,\phi\} - w_{RL}\{K,\phi\}\bar{Q}_l}, \tag{1.33}$$

$$\bar{h}_{R'L',RL} = -\frac{w_{RL}\{K,\phi\}}{w_{RL}\{K,\dot{\phi}\}}\delta_{RR'}\delta_{LL'}$$

$$+\sqrt{\frac{2}{s_w}}w_{R'L'}\{\bar{J},\phi\}\bar{S}_{RL,R'L'}w_{RL}\{\bar{J},\phi\}\sqrt{\frac{2}{s_w}}, \tag{1.34}$$

$$\bar{Q}_l = \frac{w_{RL}\{J,\dot{\phi}\}}{w_{RL}\{K,\dot{\phi}\}} = \frac{(s_R/s_w)^{2l+1}}{2(2l+1)}\frac{D\{\dot{\phi}\} - l}{D\{\dot{\phi}\} + l + 1}. \tag{1.35}$$

Here $D\{a\}$ is the logarithmic derivative of the function a, and $w_{RL}\{a,b\}$ is the Wronskian. They are defined as follows:

$$D\{a\} = s_R a'(s_R)/a(s_R),$$
$$w_{RL}\{a,b\} = s_R a_{RL}(s_R)b_{RL}(s_R)[D\{b_{RL}\} - D\{a_{RL}\}]. \tag{1.36}$$

Note that we can choose an arbitrary value for either $\bar{o}$ or $\bar{Q}$ in the above derivation. If we choose $\bar{o} = 0$, Eq. (1.34) and Eq. (1.35) give the h, and the complete expression of the basis function (Eq. (1.18)). On the other hand, we can construct a short-range structure constant $\bar{S}_{R'L',RL}$, the so-called screened structure constant, by choosing appropriate $\bar{Q}_l$ values, because $\bar{S}_{R'L',RL}$ is given by the Dyson-type equation (1.32). As a result, the basis function (1.19) is localized through Eq. (1.34). To localize the basis set, the following $\bar{Q}_l$ values have been obtained numerically:

$$\bar{Q}_{s,p,d,l\geq3} = 0.34850, 0.05303, 0.01071, 0.0. \tag{1.37}$$

The screened structure constant matrix $\bar{S}$ can be calculated from the bare structure constant matrix S directly in the real space by using Eq. (1.31). An important feature of each screened structure constant $\bar{S}_{l'lm}$ in the unit of Wigner-Seitz radius is that it lies on a universal curve, regardless of the type of structure. The corresponding interpolation formula is given by

$$\bar{S}_{ll'm} = \bar{A}_{ll'm}\exp(-\bar{\lambda}_{ll'm}d/s_w), \tag{1.38}$$

where d denotes an interatomic distance, and the list of coefficients $\bar{A}_{ll'm}$ and $\bar{\lambda}_{ll'm}$ is given in Ref. 35. This fact is convenient for calculating the electronic structure of an amorphous material. The diagonal elements of the screened structure constant matrix can be obtained from the off-diagonal element through Eq. (1.32). Thus obtained screened structure constants, and therefore the basis function defined by Eq. (1.19), decay exponentially, as shown in Eq. (1.38). They vanish before the second or third nearest neighbor distances.

We have now succeeded in obtaining a localized basis function. In the next step, we have to construct the Hamiltonian. Hereafter, we employ the atomic sphere (ASA) approximation, in which the MT sphere centered at R has the same volume as that of the Wigner-Seitz cell. As a result, MT spheres overlap each other to some extent. However, this approximation is valid provided that the overlap of the MT spheres does not exceed 20 % ($i.e.$ $s_R + s_{R'} \leq 1.20 \cdot |R + R'|$) [5] or the system is closely packed. Therefore, the first terms in Eq. (1.18) and Eq. (1.19) are neglected, and these equations are rewritten in the form

$$|\chi >^{\infty} = |\phi> + |\dot{\phi}> h, \tag{1.39}$$

$$|\bar{\chi} >^{\infty} = |\phi> + |\dot{\phi}> \bar{h}. \tag{1.40}$$

By using the basis set Eq. (1.39), we can describe the matrix elements of the Hamiltonian $\tilde{H}$ and overlap integral $\tilde{O}$ as follows:

$$
\begin{aligned}
\tilde{H}_{RL,R'L'} &= {}^{\infty}< \chi_{RL}| - \Delta + v|\chi_{R'L'} >^{\infty} \\
&= h_{RL,R'L'} + E_{\nu L}\delta_{RR'}\delta_{LL'} \\
&\quad + \sum_{R''L''} h_{RL,R''L''} E_{\nu l''} p_{R''L''} h_{R''L'',R'L'},
\end{aligned}
\tag{1.41}
$$

$$
\begin{aligned}
\tilde{O}_{RL,R'L'} &= {}^{\infty}< \chi_{RL}|\chi_{R'L'} >^{\infty} \\
&= \delta_{RR'}\delta_{LL'} + \sum_{R''L''} h_{RL,R''L''} p_{R''L''} h_{R''L'',R'L'},
\end{aligned}
\tag{1.42}
$$

$$p_{RL} = < \dot{\phi}^2_{RL} >. \tag{1.43}$$

In matrix form, Eqs. (1.41) and (1.42) can be written as

$$\tilde{H} = h + E_\nu + hE_\nu ph, \tag{1.44}$$

$$\tilde{O} = I + hph. \tag{1.45}$$

The basis set Eq. (1.39) is the so-called *nearly Löwdin-orthonormalized* set, which is orthonormal to the first order in $h \approx H - E_\nu$ as can be seen from Eq. (1.42).

By using the basis set in Eq. (1.40), we can also describe the Hamiltonian and overlap matrices as follows:

$$
\begin{aligned}
\bar{H} &\equiv {}^{\infty}< \bar{\chi}| - \Delta + v|\bar{\chi} >^{\infty} \\
&= (I + \bar{h}\bar{o})\bar{h} + (I + \bar{h}\bar{o})E_\nu(I + \bar{o}\bar{h}) + \bar{h}E_\nu p\bar{h},
\end{aligned}
\tag{1.46}
$$

$$
\begin{aligned}
\bar{O} &\equiv {}^{\infty}< \bar{\chi}|\bar{\chi} >^{\infty} \\
&= (I + \bar{h}\bar{o})(I + \bar{o}\bar{h}) + \bar{h}p\bar{h}.
\end{aligned}
\tag{1.47}
$$

Note that the Löwdin-orthonormalized basis set is a special case of the present new basis set with $\bar{o} = 0$. We can transform the latter basis set into the former through the equation

$$|\chi >^{\infty} = |\bar{\chi} >^{\infty} (I + \bar{o}\bar{h})^{-1}, \tag{1.48}$$

and obtain the following relation defining $\bar{h}$:

$$h = (I + \bar{h}\bar{o})^{-1}\bar{h} = \bar{h}(I + \bar{o}\bar{h})^{-1}. \tag{1.49}$$

This relation is expressed by the power series of $\bar{h}$ as

$$h = \bar{h} - \bar{h}\bar{o}h = \bar{h} - h\bar{o}\bar{h} = \bar{h} - \bar{h}\bar{o}\bar{h} + \cdots. \tag{1.50}$$

The Hamiltonian in the completely orthonormal representation is given by

$$H = \tilde{O}^{-1/2}\tilde{H}\tilde{O}^{-1/2} = \bar{O}^{-1/2}\bar{H}\bar{O}^{-1/2}, \tag{1.51}$$

and expressed by h and p as follows:

$$H = E_\nu + h + (hE_\nu ph - \frac{1}{2}hphE_\nu - \frac{1}{2}E_\nu hph) - (\frac{1}{2}hphh + \frac{1}{2}hhph) + \cdots. \tag{1.52}$$

The last two terms are corrections of the eigenvalue E of the third order in $(E - E_\nu)$. Therefore, $E_\nu + h$ equals the Hamiltonian H up to the second order in $(E - E_\nu)$. By substituting Eq. (1.50) into Eq. (1.52), the following second-order Hamiltonian can be obtained as a power series of $\bar{h}$ in the two-center tight-binding form:

$$H = E_\nu + h = E_\nu + \bar{h} + \bar{h}\bar{o}\bar{h}. \tag{1.53}$$

Now, we have a linear eigenvalue equation expressed in the form of Eq. (1.7), which determines the band structures. By substituting Eq. (1.33-1.35) into Eq. (1.53), the complete expression of the Hamiltonian is obtained in the two center tight-binding form. The above LMTO method is referred to as the tight-binding LMTO method.

2.2.3. Recursion method

In the previous section, we constructed the short-range tight-binding Hamiltonian on the basis of the LMTO method. The recursion method enables us to calculate the DOS without translational symmetry [38, 39], if the Hamiltonian can be described in the tight-binding form. Therefore, we can calculate the electronic structure of an amorphous material by using the tight-binding LMTO method combined with the recursion method, if its atomic structure model is given [34, 35].

The recursion method is based on the Lanczos method which transforms a symmetric matrix into a tridiagonal matrix, generating a new orthogonal basis set $\{u_i\}$. The procedure is as follows:

$$\begin{aligned}
b_1 u_2 &= H u_1 - a_1 u_1, \\
b_n u_{n+1} &= H u_n - a_n u_n - b_{n-1} u_{n-1}.
\end{aligned} \tag{1.54}$$

18

The starting vector $\boldsymbol{u}_1$ is set to a unit vector such that only the i th component has non zero value $(\boldsymbol{u}_1)_i = 1$. Here, i denotes the orbital whose projected DOS $\rho_i(\epsilon)$ is calculated. The coefficients a_n and b_n are defined as

$$\begin{aligned} a_n &= \boldsymbol{u}_n^\dagger \boldsymbol{H} \boldsymbol{u}_n, \\ b_n &= \boldsymbol{u}_{n+1}^\dagger \boldsymbol{H} \boldsymbol{u}_n, \end{aligned} \tag{1.55}$$

so that $\boldsymbol{u}_l$ with $l \leq n - 2$ does not appear in Eq. (1.54). Then the Hamiltonian $\boldsymbol{H}$ is tridiagonalized after N th recursion in the form

$$\boldsymbol{H}' = (\boldsymbol{u}_m\dagger\boldsymbol{H}\boldsymbol{u}_n) = \begin{pmatrix} a_1 & b_1 & & & & \\ b_1 & a_2 & b_2 & & & \text{\Large 0} \\ & b_2 & a_3 & b_3 & & \\ & & \ddots & \ddots & \ddots & \\ & & & b_{N-2} & a_{N-1} & b_{N-1} \\ \text{\Large 0} & & & & b_{N-1} & a_N \end{pmatrix}. \tag{1.56}$$

The Green's function matrix $\boldsymbol{G}'(z)$ for this Hamiltonian is given by

$$\boldsymbol{G}'(z) = (\boldsymbol{u}_m\dagger\boldsymbol{G}(z)\boldsymbol{u}_n) = \frac{1}{z - \boldsymbol{H}'}, \tag{1.57}$$

where $\boldsymbol{G}(z)$ denotes the Green's function for the Hamiltonian $\boldsymbol{H}$. The diagonal element is obtained in a continued fraction form

$$G_{ii}(z) = G'_{11}(z) = \cfrac{1}{z - a_1 - \cfrac{b_1^2}{z - a_2 - \cfrac{b_2^2}{z - a_3 - \cfrac{b_3^2}{\ddots \atop z - a_n - T^{(n)}(\epsilon)}}}}, \tag{1.58}$$

where $T^{(n)}(\epsilon)$ is a terminator at n-th order recursion. The square-root terminator with the asymptotic recursion coefficients determined by the Beer and Pettifor's method [85] is usually used. Then, the projected DOS is given by

$$\rho_i(\epsilon) = -\frac{1}{\pi}\mathrm{Im}G'_{11}(\epsilon + i0^+). \tag{1.59}$$

In practice, it is most important to determine the appropriate order of the recursion process. As can be seen from Eq. (1.58), high-order recursion coefficients are needed to obtain high-energy resolution in the DOS. On the other hand, a recursion process of too high an order results in non-physical oscillation in the DOS spectrum because of the finite-size effect of the model atomic structure. Therefore, the order of recursion should be determined by a compromise between the energy resolution and the size of the model atomic structure (or computational time).

2.3. *Finite-Temperature Theory of Amorphous Metallic Magnetism*

Finite-temperature theory of amorphous metallic magnetism has recently been proposed by Kakehashi [40-42]. We first consider amorphous metallic systems for brevity. The amorphous metals form a structure close to the dense random packing one, and have a well-defined NN shell, as has been discussed in Sec.2.1 (see Fig. 1.2). The magnetic interactions between the LM's in amorphous metals are expected to be short-range, since the electron scatterings due to the structural disorder cause a strong damping of the interaction strength as a function of the interatomic distance [86-87]. Thus, we might have a physical picture that the central LM is directly influenced by the LM's on the NN shell, but the effect of more distant atoms and their LM's may be treated as an effective medium as shown in Fig. 1.4. In the followings, we formulate a finite-temperature theory of amorphous metallic magnetism, keeping in mind the above physical picture.

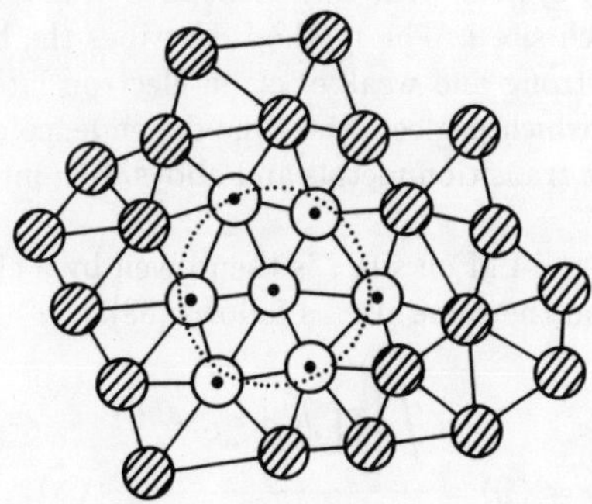

Fig. 1.4: Physical picture for amorphous structure. The central atom has a well-defined nearest-neighbor (NN) shell (dotted circle). More distant atoms (hatched circles) are regarded as an effective medium.

2.3.1. *Functional integral method*

Conduction electrons in transition metals are well-known to consist of a broad *sp* band and a narrow *d* band. The magnetism is carried by *d*-electrons because of their localized nature, even in the metallic state. The degenerate-bands Hubbard model for *d*-electrons with intraatomic Coulomb (U) and exchange (J) interactions is suitable for the description of the present system.

$$\hat{H} = H_0 + H_1 \,, \tag{1.60}$$

20

$$H_0 = \sum_{i,\nu,\sigma} (\epsilon_i^0 - h_i\sigma)\, n_{i\nu\sigma} + \sum_{i,\nu,j,\nu',\sigma} t_{i\nu j\nu'}\, a_{i\nu\sigma}^\dagger a_{j\nu'\sigma} \ , \tag{1.61}$$

$$H_1 = \frac{1}{4} U \sum_i n_i^2 - J \sum_i \mathbf{S}_i^2 . \tag{1.62}$$

Here ϵ_i^0 and h_i in the noninteracting Hamiltonian H_0 are the atomic level and external magnetic field on site i, respectively. $t_{i\nu j\nu'}$ denotes the transfer integral between the orbitals ν on site i and ν' on site j. $a_{i\nu\sigma}^\dagger (a_{i\nu\sigma})$ is the creation (annihilation) operator for electrons with spin σ and orbital ν on site i, and $n_{i\nu\sigma} = a_{i\nu\sigma}^\dagger a_{i\nu\sigma}$. The charge-density and spin-density operators on site i are denoted by n_i and $\mathbf{S}_i$ in the interaction part H_1.

The thermal spin fluctuations are taken into account by means of the functional integral method [43-45]. In this method, the interacting Hamiltonian is exactly transformed into a one-electron system with time-dependent random exchange fictitious fields $\{\xi_i(\tau)\}$ acting on each site i. The method describes the basic behaviors of the spin fluctuations in both strong and weak electron-electron interaction limits within the static approximation, which neglects the time dependence of the fictitious fields, so that it is suitable for the transition metals and alloys with intermediate interaction strength.

The thermal average of the LM on site i is then given by a classical average of the fictitious field variable ξ_i on the same site as follows [88].

$$\langle m_i \rangle = \langle \xi_i \rangle = \frac{\int \left[\prod_j d\xi_j \right] \xi_i e^{-\beta E(\xi)}}{\int \left[\prod_j d\xi_j \right] e^{-\beta E(\xi)}} \ , \tag{1.63}$$

$$E(\xi) = -\beta^{-1} \ln \mathrm{tr}\left(e^{-\beta H(\xi)} \right) + \sum_i \left(-n_i w_i(\xi) + \frac{1}{4}\tilde{J}_i \xi_i^2 \right) . \tag{1.64}$$

From Eq. (1.63) the fictitious-field variable ξ_i is interpreted as a flexible LM on site i. It thermally fluctuates according to the energy functional $E(\xi)$. The first term in the rhs of Eq. (1.64) is the free energy for independent electrons with the Hamiltonian defined by

$$\begin{aligned}
H(\xi) &= \sum_{i,\nu,\sigma} (\epsilon_i^0 - \mu + w_i(\xi) - \frac{1}{2}\tilde{J}\xi_i\sigma - h_i\sigma)\, n_{i\nu\sigma} \\
&\quad + \sum_{i,j,\nu,\sigma} t_{ij} a_{i\nu\sigma}^\dagger a_{j\nu\sigma} \ .
\end{aligned} \tag{1.65}$$

Here μ is the chemical potential. We have adopted the equivalent-bands model for brevity, so that the transfer integrals t_{ij} do not depend on the orbital. The charge potentials $\{w_i(\xi)\}$ in Eq. (1.64) are determined so that each site has a neutral charge n_i [44]. This implies that a strong Coulomb interaction limit ($U \to \infty$) is considered here. The effective exchange energy parameter $\tilde{J}$ is defined by

$$\tilde{J} = \frac{1}{2D}U + \left(1 + \frac{1}{2D}\right)J \,, \tag{1.66}$$

D being the degeneracy of the d bands (*i.e.* $D = 5$).

The last term in the rhs of Eq. (1.64) describes a Gaussian random distribution of the static fictitious fields.

It is well-known that the one-electron free energy is written in logarithmic form, so Eq. (1.64) is written as follows.

$$
\begin{aligned}
E(\xi) &= \int d\omega f(\omega)\frac{D}{\pi}\mathrm{Im}\,\mathrm{tr}\left[\ln(L^{-1} - t)\right] \\
&\quad + \sum_i \left(-n_i w_i(\xi) + \frac{1}{4}\tilde{J}\xi_i^2\right) \,,
\end{aligned}
\tag{1.67}
$$

where $f(\omega)$ is the Fermi distribution function, t denotes the matrix t_{ij}, and the locator L is defined by

$$\left(L^{-1}\right)_{ij\sigma} = L_{i\sigma}^{-1}\delta_{ij} = \left(\omega + i\delta - \epsilon_i^0 - w_i(\xi) + \frac{1}{2}\tilde{J}\xi_i\sigma + \mu\right)\delta_{ij} \,. \tag{1.68}$$

Here δ is an infinitesimal positive number, and external magnetic field was omitted for brevity.

It should be noted that Eq. (1.63) reduces to the Hartree-Fock approximation at the ground state. In fact, the thermal average at low temperatures in Eq. (1.63) is replaced by the saddle-point values $\{\xi_i^*\}$ defined by $\partial E(\xi^*)/\partial \xi_i^* = 0$, from which we obtain

$$\langle m_i \rangle = \xi_i^* = \frac{\mathrm{tr}\left(m_i\,e^{-\beta H(\xi^*)}\right)}{\mathrm{tr}\left(e^{-\beta H(\xi^*)}\right)} \,. \tag{1.69}$$

The result reduces to the Stoner model with the Stoner parameter $\tilde{J}$ for a simple crystalline ferromagnet [89-91].

$$\xi^* = \int d\omega f(\omega) \sum_\sigma \sigma \rho^0 \left(\omega - \epsilon^0 + \mu - w(\xi^*) + \frac{1}{2}\tilde{J}\xi^*\sigma\right) \,. \tag{1.70}$$

Here ρ^0 denotes the density of states (DOS) for the noninteracting system.

At this point, we note some basic problems in the functional integral method. First, the transverse spin fluctuations were neglected in Eq. (1.63). Samson [92] and Kakehashi [88] showed that inclusion of the transverse spin fluctuations, which is called the vector field method, leads to a negative specific heat within the static approximation because of the violation of the commutation relations between the spin density operators on the same site. It is thus necessary to go beyond the static approximation in the treatment of the transverse components, which is left for future investigations. Second, the static approximation leads to the Hartree-Fock one at the ground state, thus it does not take into account the ground-state electron correlations as discussed by many investigators [29,93-94]. This means that the exchange energy parameter $\tilde{J}$ defined by Eq. (1.66) should be regarded as a renormalized or reduced one. Kakehashi and Fulde [95-96], and Hasegawa [97] showed that the static approximation with use of the reduced $\tilde{J}$ can describe the principal part of the local electron correlations in the magnetization vs. temperature curves.

2.3.2. Local moment in an effective medium

The energy functional $E(\xi)$ in Eq. (1.67) determines the central LM. The energy contains two types of disorder. One is the spin disorder which is caused by the thermal spin fluctuations *via* random exchange potentials $\{ \tilde{J}\xi_i\sigma/2 \}$ in Eq. (1.68). Another is the structural disorder characteristic of amorphous metallic systems. The latter appears in the atomic potentials $\{ \epsilon_i^0 - \mu + w_i(\xi) \}$ and transfer integrals $\{ t_{ij} \}$. Consequently, the diagonal disorder in the locator includes both types of disorder, while the off-diagonal disorder in the transfer matrix is caused by the structural disorder only.

The diagonal disorder is treated by introducing the inverse effective locator $\mathcal{L}_\sigma^{-1}(\omega + i\delta)$ into the first term in Eq. (1.67). The deviation from the medium is expanded with respect to the sites [98]. The zeroth order is described by the effective medium only. The first-order correction consists of the sum of single-site energy functionals $E_i(\xi_i)$.

$$E_i(\xi_i) = \int d\omega f(\omega)\frac{D}{\pi}\mathrm{Im}\sum_\sigma \ln\left(L_{i\sigma}^{-1} - \mathcal{L}_\sigma^{-1} + F_{ii\sigma}^{-1}\right)$$
$$+ \sum_i \left(-n_i w_i(\xi_i) + \frac{1}{4}\tilde{J}\xi_i^2\right) , \qquad (1.71)$$

Here the coherent Green function $F_{ij\sigma}$ is defined by

$$F_{ij\sigma} = \left[\left(\mathcal{L}^{-1} - t\right)^{-1}\right]_{ij\sigma} . \qquad (1.72)$$

The second-order correction is the pair-interaction terms $\sum_{(i,j)} \Phi_{ij}(\xi_i, \xi_j)$. $\Phi_{ij}(\xi_i, \xi_j)$ denotes the pair energy between sites i and j:

$$\Phi_{ij}(\xi_i, \xi_j) = \int d\omega f(\omega)\frac{D}{\pi}\mathrm{Im}\sum_\sigma \ln\left[1 - F_{ij\sigma}F_{ji\sigma}\tilde{t}_{i\sigma}(\xi_i)\tilde{t}_{j\sigma}(\xi_j)\right] . \qquad (1.73)$$

Here $\tilde{t}_{i\sigma}(\xi_i)$ is the single-site t matrix defined by

$$\tilde{t}_{i\sigma}(\xi_i) = \frac{L_{i\sigma}^{-1} - \mathcal{L}_{\sigma}^{-1}}{1 + \left(L_{i\sigma}^{-1} - \mathcal{L}_{\sigma}^{-1}\right) F_{ii\sigma}} \; . \tag{1.74}$$

The t matrix describes the impurity scattering when the impurity potential $\epsilon_i^0 - \mu + w_i(\xi) - \tilde{J}\xi_i\sigma/2$ is embedded in the effective medium $\mathcal{L}_\sigma^{-1}$. Higher-order terms are neglected by assuming a small deviation from the effective medium. The energy functional $E(\xi)$ is then written as follows.

$$E(\xi) = \sum_i E_i(\xi_i) + \sum_{(i,j)} \Phi_{ij}(\xi_i, \xi_j) \; . \tag{1.75}$$

Here the zeroth term is dropped, since it does not make any contribution to the thermal average (1.63).

Next, one neglects the direct pair interactions between the central LM and the LM's outside the NN shell according to the physical picture discussed at the beginning. By making use of a molecular-field approximation for the LM's on the NN shell [99], Eq.1.63 reduces to

$$\langle m_0 \rangle = \frac{\displaystyle\int d\xi\, \xi\, e^{-\beta\Psi(\xi)}}{\displaystyle\int d\xi\, e^{-\beta\Psi(\xi)}} \; , \tag{1.76}$$

$$\Psi(\xi) = E_0(\xi) + \sum_{j=1}^{z} \Phi_{0j}^{(a)}(\xi) - \sum_{j=1}^{z} \Phi_{0j}^{(e)}(\xi)\frac{\langle m_j \rangle}{x_j} \; . \tag{1.77}$$

Here z is the number of atoms on the NN shell, and is usually taken to be 12 [49]. The atomic and exchange pair energies $\Phi_{0j}^{(a)}(\xi)$ and $\Phi_{0j}^{(e)}(\xi)$ are defined as follows.

$$\begin{bmatrix} \Phi_{0j}^{(a)}(\xi) \\ \Phi_{0j}^{(e)}(\xi) \end{bmatrix} = \frac{1}{2}\sum_{\nu=\pm} \begin{bmatrix} 1 \\ -\nu \end{bmatrix} \Phi_{0j}(\xi, \nu x_j) \; . \tag{1.78}$$

The quantity x_j in Eqs. (1.77) and (1.78) is an amplitude in the single-site approximation, which is defined by

$$x_j^2 = \frac{\displaystyle\int d\xi\, \xi^2\, e^{-\beta E_j(\xi)}}{\displaystyle\int d\xi\, e^{-\beta E_j(\xi)}} \; . \tag{1.79}$$

Equations (1.76) and (1.77) show up that a flexible central LM ξ feels directly the molecular fields from the LM's $\{ \langle m_j \rangle \}$ on the NN shell (the third term in Eq. (1.77))

24

and indirectly the average molecular field from the LM's outside the NN shell *via* the spin-dependent effective medium $\mathcal{L}_\sigma^{-1}$ which appears in $E_0(\xi)$, $\Phi_{0j}^{(a)}(\xi)$, and $\Phi_{0j}^{(e)}(\xi)$.

The effective medium $\mathcal{L}_\sigma^{-1}$ is chosen so that the correction to the single-site energies in $E(\xi)$ becomes as small as possible. This is called the CPA [67-68], and the condition is that the averaged single-site t matrix vanishes [45].

$$\left[\langle \tilde{t}_{i\sigma}(\xi)\rangle\right]_{\mathrm{s}} = 0 \; . \tag{1.80}$$

Here $\langle \; \rangle$ ($[\;\;]_{\mathrm{s}}$) denotes the thermal (structural) average.

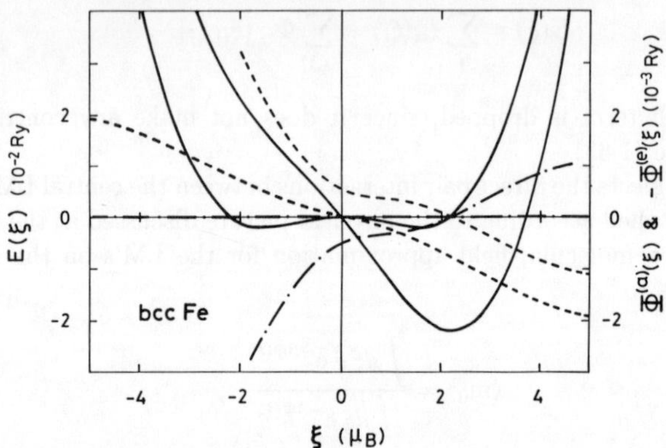

Fig. 1.5: Single-site energy functional $E(\xi)$ (solid curves), atomic pair energy $\Phi^{(a)}(\xi)$ (dot-dashed curves), and exchange pair energy $-\Phi^{(e)}(\xi)$ (dashed curves) of bcc Fe in the ferromagnetic ($T/T_{\mathrm{C}} = 0.074$) and paramagnetic ($T/T_{\mathrm{C}} = 1.11$) states [100]. Note that $E(\xi)$ and $\Phi^{(a)}(\xi)$ are symmetric, while $-\Phi^{(e)}(\xi)$ is antisymmetric in the paramagnetic states.

Figure 1.5 shows an example of energy functionals [100]. At the ground state, the single-site energy $E(\xi)$ has a minimum at $\xi^* = 2.216\,\mu_{\mathrm{B}}$, and rapidly increases for a large $|\xi|$ because of the quadratic term in Eq. (1.71). The pair energies $\Phi^{(a)}(\xi)$ and $\Phi^{(e)}(\xi)$ vanish at $\xi = \xi^*$ because there is no scattering due to thermal spin fluctuations and other disorder. Above the Curie temperature, the uniform molecular magnetic field vanishes, and therefore the effective medium $\mathcal{L}_\sigma^{-1}$ becomes spin independent. The energies $E_0(\xi)$ and $\Phi^{(a)}(\xi)$ then show symmetric behavior, while the exchange pair energy $-\Phi^{(e)}(\xi)$ behaves as an odd function. The latter reduces to $-\mathcal{J}\xi \cdot x_j$ in the local-moment limit. Here $\mathcal{J}$ denotes Anderson's super exchange interaction [101].

2.3.3. *Structural disorder and electronic states*

The central LM (1.76) still depends on the structural disorder outside the NN shell *via* the coherent Green functions $F_{00\sigma}$, $F_{0j\sigma}(= F_{j0\sigma})$, and $F_{jj\sigma}$ in Eq. (1.77). These

Green functions are treated with the Bethe approximation [102-103].

The approximation is based on the locator expansion for coherent Green functions as follows.

$$F_{00} = \mathcal{L} + \mathcal{L} \sum_{j \neq 0} t_{0j} F_{j0} \; , \tag{1.81}$$

$$F_{j0} = \mathcal{L} t_{j0} F_{00} + \mathcal{L} S_j F_{j0} + \mathcal{L} \sum_{i \neq j,0} T_{ji} F_{i0} \; . \tag{1.82}$$

Here we have omitted the spin suffix σ for brevity and neglected the transfer integrals between the central atom and the atoms outside the NN shell. The selfenergy S_j (T_{ji}) consists of the sum of all the paths which start from site j and end at site j (i) without returning to the cluster on the way (see Fig. 1.6). It should be noted that all the information outside the cluster is included in S_j and T_{ji}.

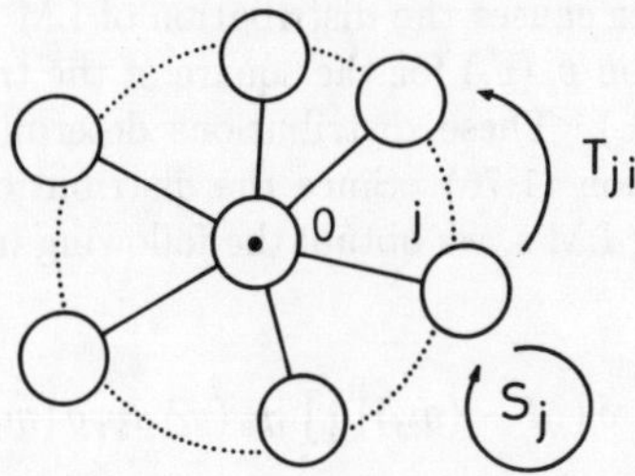

Fig. 1.6: Schematic representation of the irreducible path S_j and T_{ji}.

When we take the structural average outside the cluster, we neglect the last term in the rhs of Eq. (1.82), and replace S_j with $\mathcal{S}$, an effective medium for structural disorder. We then obtain

$$F_{00\sigma} = \left(\mathcal{L}_\sigma^{-1} + \sum_{j=1}^{z} \frac{t_{j0}^2}{\mathcal{L}_\sigma^{-1} - \mathcal{S}_\sigma} \right)^{-1} , \tag{1.83}$$

$$F_{j0\sigma} = \frac{t_{j0}}{\mathcal{L}_\sigma^{-1} - \mathcal{S}_\sigma} F_{00\sigma} \; . \tag{1.84}$$

The diagonal Green functions $F_{jj\sigma}$ in the NN shell, on the other hand, are approximated by the averaged one.

$$F_\sigma = [F_{jj\sigma}]_{\mathrm{s}} = \int \frac{[\rho(\epsilon)]_{\mathrm{s}} \, d\epsilon}{\mathcal{L}_\sigma^{-1} - \epsilon} \; . \tag{1.85}$$

Note that the averaged DOS $[\rho(\epsilon)]_\mathrm{s}$ for noninteracting systems can be calculated by using the ground-state theories, as has been described in Sec. 2.1.

The effective medium $\mathcal{S}_\sigma$ is determined from the condition that the structural average of the central coherent Green function $F_{00\sigma}$ should be identical with the neighboring one.

$$[F_{00\sigma}]_\mathrm{s} = \int \frac{[\rho(\epsilon)]_\mathrm{s}\, d\epsilon}{\mathcal{L}_\sigma^{-1} - \epsilon} \ . \tag{1.86}$$

2.3.4. Distribution-function method

It is obvious now that the central LM in Eq. (1.76) is determined by the neighboring LM's $\{\langle m_j \rangle\}$ on the NN shell, the transfer integrals $y_j = t_{j0}^2$ between the central atom and the neighboring atoms, the effective medium $\mathcal{L}_\sigma^{-1}$ due to the spin fluctuations, and the effective medium $\mathcal{S}_\sigma$ due to the structural disorder outside the NN shell.

The structural disorder causes the distribution of LM $g(\langle m_j \rangle)$ at the neighboring site j, and the distribution $p_\mathrm{s}(y_j)$ for the square of the transfer integral. (The latter is assumed to be known.) These distributions determine the LM distribution at the central site *via* relation (1.76). Since the distribution should be identical with those for the surrounding LM's, we obtain the following integral equation for the LM distribution.

$$g(M) = \int \delta(M - \langle m_j \rangle) \prod_{i=1}^{z} [p_\mathrm{s}(y_i)\, dy_i\, g(m_i)\, dm_i] \ . \tag{1.87}$$

This method, which is called the distribution function method, was first adopted in disordered problems in insulator systems, by Matsubara and Katsura [46-47].

The effective mediums $\mathcal{L}_\sigma^{-1}$ and $\mathcal{S}_\sigma$ are selfconsistently determined from Eqs. (1.80) and (1.86):

$$\int \langle \tilde{t}_{0\sigma} \rangle \prod_{i=1}^{z} [p_\mathrm{s}(y_i)\, dy_i\, g(m_i)\, dm_i] = 0 \ . \tag{1.88}$$

$$\int F_{00\sigma} \prod_{i=1}^{z} [p_\mathrm{s}(y_i)\, dy_i] = \int \frac{[\rho(\epsilon)]_\mathrm{s}\, d\epsilon}{\mathcal{L}_\sigma^{-1} - \epsilon} \ . \tag{1.89}$$

Finally, the LM distribution $g(M)$, and the effective mediums $\mathcal{L}_\sigma^{-1}$ and $\mathcal{S}_\sigma$ are selfconsistently determined by solving Eqs. (1.87), (1.88), and (1.89). The average magnetization $[\langle m \rangle]_\mathrm{s}$ and the SG order parameter $[\langle m \rangle^2]_\mathrm{s}$ are obtained from the distribution $g(M)$ as follows.

$$[\langle m \rangle]_\mathrm{s} = \int M g(M)\, dM \ , \tag{1.90}$$

$$\left[\langle m\rangle^2\right]_{\mathrm{s}} = \int M^2 g\left(M\right) dM \ . \tag{1.91}$$

Since Eqs. (1.87), (1.88), and (1.89) include $2z$-fold integrals, it is impossible without making further approximations to solve the equations. We adopt the following decoupling approximation in Eqs. 1.87-1.89, which is correct up to the second moment:

$$\int M^{2n+k} g\left(M\right) dM \approx \left[\langle m\rangle^2\right]_{\mathrm{s}}^n \left[\langle m\rangle^k\right]_{\mathrm{s}} \ , \tag{1.92}$$

$$\int \left(y - [y]_{\mathrm{s}}\right)^{2n+k} p_{\mathrm{s}}(y) dy \approx \left[(\delta y)^2\right]_{\mathrm{s}}^n 0^k \ . \tag{1.93}$$

Here $k = 0$ or 1. These are the lowest approximation which takes into account the fluctuations. $[y]_{\mathrm{s}}$ is a mean square of a transfer integral, and $\left[(\delta y)^2\right]_{\mathrm{s}}$ is the fluctuation around $[y]_{\mathrm{s}}$, which is calculated from the fluctuations of the NN interatomic distance R as follows:

$$\frac{\left[(\delta y)^2\right]_{\mathrm{s}}^{\frac{1}{2}}}{[y]_{\mathrm{s}}} = 2\kappa \frac{\left[(\delta R)^2\right]_{\mathrm{s}}^{\frac{1}{2}}}{[R]_{\mathrm{s}}} \ . \tag{1.94}$$

Here we adopted Heine's law $t\left(R\right) \equiv t_{j0} \propto R^{-\kappa}$ [104-105]. $[R]_{\mathrm{s}}$ and $[(\delta R)^2]_{\mathrm{s}}^{1/2}$ denote the average interatomic distance and its fluctuation.

Substituting Eqs. (1.87) after the decoupling approximation into Eqs. (1.90) and (1.91), we obtain the selfconsistent equations for $[\langle m\rangle]_{\mathrm{s}}$ and $[\langle m\rangle^2]_{\mathrm{s}}$ as follows.

$$\begin{bmatrix} [\langle m\rangle]_{\mathrm{s}} \\ [\langle m\rangle^2]_{\mathrm{s}} \end{bmatrix} = \sum_{n=0}^{z} \Gamma\left(n, z, \frac{1}{2}\right) \begin{bmatrix} [\langle m\rangle_n]_{\mathrm{s}} \\ [\langle m\rangle_n^2]_{\mathrm{s}} \end{bmatrix} \ , \tag{1.95}$$

$$\begin{bmatrix} [\langle m\rangle_n]_{\mathrm{s}} \\ [\langle m\rangle_n^2]_{\mathrm{s}} \end{bmatrix} = \sum_{k=0}^{n} \sum_{l=0}^{z-n} \Gamma\left(k, n, q\right) \Gamma\left(l, z-n, q\right) \begin{bmatrix} \langle \xi\rangle\left(n, k, l\right) \\ \langle \xi\rangle^2\left(n, k, l\right) \end{bmatrix} \ , \tag{1.96}$$

$$\langle \xi\rangle\left(n, k, l\right) = \frac{\int d\xi\, \xi\, \mathrm{e}^{-\beta\Psi(\xi, n, k, l)}}{\int d\xi\, \mathrm{e}^{-\beta\Psi(\xi, n, k, l)}} \ , \tag{1.97}$$

$$\begin{aligned}
\Psi\left(\xi, n, k, l\right) = {} & E\left(\xi, n\right) + n\Phi_+^{(\mathrm{a})}\left(\xi, n\right) + \left(z - n\right)\Phi_-^{(\mathrm{a})}\left(\xi, n\right) \\
& - \left[\left(2k - n\right)\Phi_+^{(\mathrm{e})}\left(\xi, n\right) + \left(2l - z + n\right)\Phi_-^{(\mathrm{e})}\left(\xi, n\right)\right] \\
& \times \frac{\left[\langle m\rangle^2\right]_{\mathrm{s}}^{\frac{1}{2}}}{x} \ ,
\end{aligned} \tag{1.98}$$

$$q = \frac{1}{2}\left(1 + \frac{[\langle m \rangle]_{\mathrm{s}}}{\left[\langle m \rangle^2\right]_{\mathrm{s}}^{\frac{1}{2}}}\right).$$ (1.99)

Here $\Gamma(n, z, p)$ is the binomial distribution function defined by $[z!/n!(z-n)!]\,p^n (1-p)^{z-n}$.

In the present approximation, the local environments inside the NN shell are described *via* the NN transfer integrals by the contraction $(-[(\delta R)^2]_{\mathrm{s}}^{1/2})$ of the NN interatomic distance R from the average value $[R]_{\mathrm{s}}$ and the stretch $([(\delta R)^2]_{\mathrm{s}}^{1/2})$ of the distance R. Thus, the local structure is specified by means of the number of contracted pairs (n) between the central atom and the atoms on the NN shell. Since a local structure is realized with the probability $\Gamma(n, z, 1/2)$, the averaged LM's are given by Eq. (1.95). The single-site energy and pair energies are also characterized by n, so that the notations $E(\xi, n)$, $\Phi_{\pm}^{(a)}(\xi, n)$, and $\Phi_{\pm}^{(e)}(\xi, n)$ are used in Eq. (1.98). Here the subscript $+(-)$ denotes the contracted (stretched) pair.

The parameter q defined by Eq. (1.99) is interpreted as the probability that the fictitious spin $[\langle m \rangle^2]_{\mathrm{s}}^{1/2}$ points up on a site of the NN shell. The probability of finding up spins of k among n contracted atoms on the NN shell is then given by $\Gamma(k, n, q)$, and the probability of finding up spins of l among $z - n$ stretched atoms is given by $\Gamma(l, z - n, q)$. Therefore, the average central LM in the local environment n is obtained by averaging $\langle \xi \rangle (n, k, l)$ over the spin configurations (See Eq. (1.96)). Here, $\langle \xi \rangle (n, k, l)$ denotes the central LM in the (n, k, l) configuration. Note that the LM's $\langle \xi \rangle (n, k, l)$ are strongly influenced by their local environments (n, k, l) in transition metals and alloys. Such effects are called local environment effects (LEE).

In the same way, one obtains a simplified CPA equation from Eq. (1.88),

$$\sum_{\nu=\pm} \frac{1}{2}\left(1 + \nu\frac{[\langle \xi \rangle]_{\mathrm{s}}}{\left[\langle \xi \rangle^2\right]_{\mathrm{s}}^{\frac{1}{2}}}\right)\left\{L_\sigma^{-1}\left(\nu\left[\langle \xi \rangle^2\right]_{\mathrm{s}}^{\frac{1}{2}}\right) - \mathcal{L}_\sigma^{-1} + F_\sigma^{-1}\right\}^{-1} = F_\sigma,$$ (1.100)

and the following equation from Eq. (1.89).

$$\sum_{\nu=\pm} \frac{1}{2}\left\{\mathcal{L}_\sigma^{-1} - z^*[y]_{\mathrm{s}}\left(1 + \nu\frac{\left[(\delta y)^2\right]_{\mathrm{s}}^{\frac{1}{2}}}{\sqrt{z^*}\,[y]_{\mathrm{s}}}\right)\left(\mathcal{L}_\sigma^{-1} - \mathcal{S}_\sigma\right)^{-1}\right\}^{-1} = F_\sigma,$$ (1.101)

z^* being an effective coordination number.

Finally, the input parameters are the d-electron number N, the effective exchange energy parameter $\tilde{J}$, the DOS $[\rho(\epsilon)]_{\mathrm{s}}$, and the fluctuation of the interatomic distance $[(\delta R)^2]_{\mathrm{s}}^{1/2}/[R]_{\mathrm{s}}$. One determines $[\langle m \rangle]_{\mathrm{s}}$, $[\langle m \rangle^2]_{\mathrm{s}}$, $\mathcal{L}_\sigma^{-1}$ and $\mathcal{S}_\sigma$, solving Eqs. (1.95), (1.100), and 1.101 selfconsistently.

2.3.5. Extension to amorphous alloys

In the case of amorphous transition metal alloys, we have to take into consideration two factors. One is the configurational disorder as seen in the substitutional alloys. To describe such a disorder in a binary alloy, one needs to introduce the configurational average ($[\quad]_c$), the concentration (c_α), and the probability of finding a γ atom at the neighboring site of an α atom ($p^{\alpha\gamma}$). The other is the difference in atomic size. One might draw the following features from the DRPHS model with different size of atoms: (1) a systematic change in the interatomic distance which strongly reflects the atomic size, (2) an average coordination number z_α^* that depends on the type of the central atom α, and (3) alloying effects on the fluctuations of the interatomic distance and the coordination number.

Although one could take into account these effects on the electronic structure along the lines mentioned in Sec. 2.1, it is too laborious to perform the calculations for each concentration. Moreover, one needs the experimental data for pair distribution functions, which are often missing in the literature. It is therefore convenient for the purpose of semiquantitative calculations to introduce the geometrical mean model [106] which satisfies the following conditions: (1) the transfer integrals $t^{AB}(R_{ij}^{AB})$ are given by $[\, t^{AA}(R_{ij}^{AB})\, t^{BB}(R_{ij}^{AB})\,]^{1/2}$, (2) the interatomic distance R_{ij}^{AB} between the atom A on site i and the atom B on site j is given by the geometrical mean $(R_{ij}^{AA} R_{ij}^{BB})^{1/2}$, (3) transfer integrals follow the same power law with respect to the interatomic distance $t^{\alpha\gamma}(R) \approx R^{-\kappa}$, and (4) constituent amorphous pure metals A and B have the same structure ($i.e..$, $R_{ij}^{AA}/R_{ij}^{BB} =$ constant). These conditions lead to the relation

$$t^{\alpha\gamma}(R_{ij}^{\alpha\gamma}) = r_\alpha^*\, \hat{t}\,(R_{ij})\, r_\gamma \ . \tag{1.102}$$

Here $\hat{t}(R_{ij})$ are the transfer integrals for amorphous pure metal, and r_α denotes a factor being independent of the amorphous structure. Equation (1.102) reduces the off-diagonal configurational disorder to a diagonal disorder, thus allows us to calculate the electronic structures of amorphous alloys from those of amorphous pure metals, as it is well known in the substitutional alloys [107].

The geometrical mean model describes the DOS qualitatively or semi-quantitatively, and is therefore regarded as a good reference system. Figure 1.7 shows an example for amorphous $Cu_{35}Zr_{65}$ alloys. The single-site theory using the geometrical-mean model reproduces an overall structure that has a narrow Cu band with a peak and a broad Zr band.

Adopting the geometrical-mean model outside the NN shell, it is easy to obtain the coherent Green functions $F_{00\sigma}$, $F_{j0\sigma}$, and $F_{jj\sigma}$ within the Bethe approximation, so that one can calculate the energy functional (1.77) for alloy systems. It turns out that the local environment surrounding the central LM of type of atom α is specified by the coordination number (z), the number of α atoms on the NN shell (n), the number of contracted atoms among n atoms of type α on the shell (i), and the number of contracted atoms among the remaining $z - n$ atoms of type $\bar{\alpha}$ (j). The

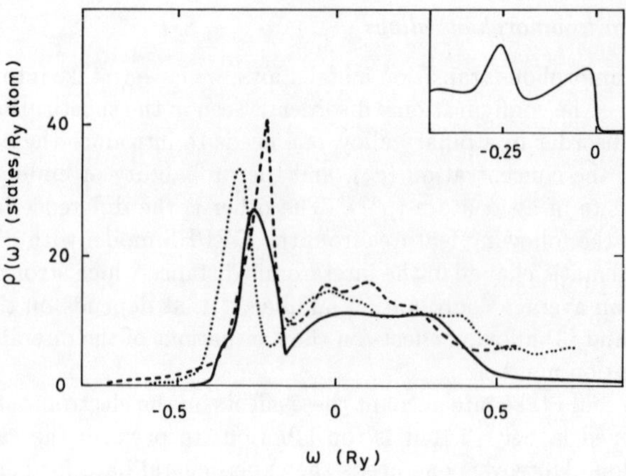

Fig. 1.7: Densities of states (DOS) for amorphous $Cu_{35}Zr_{65}$ alloys obtained by the geometrical-mean model (solid curve) [106], the tight-binding LMTO-Recursion method (dashed curves) [34], and the orthogonalized LCAO method (dotted curves) [76]. The inset shows the experimental data of the UPS by Oelhafen *et al.* [108].

selfconsistent equations for $[\,[\,\langle m_\alpha\rangle\,]_s\,]_c$ and $[\,[\,\langle m_\alpha\rangle^2\,]_s\,]_c$ are obtained by averaging over these configurations (z, n, i, j).

$$
\begin{bmatrix} \left[\left[\langle m_\alpha\rangle\right]_s\right]_c \\ \left[\left[\langle m_\alpha\rangle^2\right]_s\right]_c \end{bmatrix} = \sum_z p_\alpha(z) \sum_{n=0}^{z} \Gamma(n, z, p^{\alpha\alpha}) \sum_{i=0}^{n} \Gamma\left(i, n, \frac{1}{2}\right) \times
$$

$$
\sum_{j=0}^{z-n} \Gamma\left(j, z-n, \frac{1}{2}\right) \begin{bmatrix} \left[\left[\langle m_\alpha\rangle(z,n,i,j)\right]_s\right]_c \\ \left[\left[\langle m_\alpha\rangle^2(z,n,i,j)\right]_s\right]_c \end{bmatrix} . \tag{1.103}
$$

The function $p_\alpha(z)$ in Eq. (1.103) denotes the distribution of the coordination number z for atom α. It is given in its simplest form by

$$
p_\alpha(z) = ([z_\alpha^*] + 1 - z_\alpha^*)\,\delta_{z\,[z_\alpha^*]} + (z_\alpha^* - [z_\alpha^*])\,\delta_{z\,[z_\alpha^*]+1} , \tag{1.104}
$$

where [   ] denotes Gauss' notation.

The probability $p^{\alpha\alpha}$ of finding atom α at the neighboring site of atom α is given by the atomic short-range order parameter τ_α as

$$
p^{\alpha\alpha} = c_\alpha + c_{\bar\alpha}\tau_\alpha . \tag{1.105}
$$

Note that $\tau_{\bar\alpha}$ is obtained from τ_α as

$$
\tau_{\bar\alpha} = 1 - \frac{z_\alpha^*}{z_{\bar\alpha}^*}(1 - \tau_\alpha) , \tag{1.106}
$$

because of the sum rule for A-B atomic pairs.

The LM's $[\, [\, \langle m_\alpha \rangle^k(z,n,i,j) \,]_{\mathrm{s}} \,]_{\mathrm{c}}$ $(k = 1,2)$ in Eq. (1.103), which correspond to Eq. (1.97), denote the averaged LM's in the local environment (z,n,i,j). They are calculated by taking the average over the configurations of the fictitious spins $\{[[\langle m_\alpha \rangle^2]_{\mathrm{s}}]_{\mathrm{c}}^{1/2}\}$ on the NN shell with the probability q_α of finding an up spin, which is given by

$$q_\alpha = \frac{1}{2}\left(1 + \frac{[\,[\,\langle m_\alpha \rangle]_{\mathrm{s}}]_{\mathrm{c}}}{\left[\,[\,\langle m_\alpha \rangle^2]_{\mathrm{s}}\right]_{\mathrm{c}}^{\frac{1}{2}}}\right). \tag{1.107}$$

The theory covers a wide range of amorphous and liquid magnetic alloys, from metals to insulators within a molecular-field approximation, because it is based on the functional integral method which interpolates between the strong and weak Coulomb interaction limits. The theory reduces to the previous theory of local environmental effects (LEE) developed by Kakehashi in the case of substitutional alloys [99,109]. It has been shown that the theory of LEE explains overall feature of the Slater Pauling curves, the Curie-temperature Slater Pauling curves, and the systematic variations of the effective Bohr magneton numbers in $3d$ transition metal alloys on the same footing [33]. Numerical investigations revealed the following points supporting the validity of the theory: (1) the magnetizations at low temperatures are semi-quantitatively described with use of the effective exchange energy parameters $\tilde{J}_\alpha$; (2) the concentration dependence of the Curie temperatures is reproduced, but the absolute values are overestimated by a factor of $1.5 \sim 2.0$ because of the molecular-field approximation and the neglect of the transverse spin fluctuations; and (3) the effective Bohr magneton numbers obtained from the Curie-Weiss susceptibilities are generally underestimated by about 25 % because of the overestimated itinerant character due to the use of the reduced $\tilde{J}_\alpha$.

The most important characteristic of the theory is that it takes into account $[[(\delta\langle m_\alpha\rangle)^2]_{\mathrm{s}}]_{\mathrm{c}}$, the fluctuations of the LM's with respect to the structural and configurational disorders. This leads to the spin-glass (SG) solution ($[[\langle m_\alpha\rangle]_{\mathrm{s}}]_{\mathrm{c}} = 0$ and $[[\langle m_\alpha\rangle^2]_{\mathrm{s}}]_{\mathrm{c}} \neq 0$), in addition to the ferromagnetic ($[[\langle m_\alpha\rangle]_{\mathrm{s}}]_{\mathrm{c}} \neq 0$ and $[[\langle m_\alpha\rangle^2]_{\mathrm{s}}]_{\mathrm{c}} \neq 0$) and paramagnetic ($[[\langle m_\alpha\rangle]_{\mathrm{s}}]_{\mathrm{c}} = [[\langle m_\alpha\rangle^2]_{\mathrm{s}}]_{\mathrm{c}} = 0$) solutions.

The SG transition temperature is shown to be obtained analytically in the local-moment limit as

$$T_{\mathrm{g}}^2 = \frac{1}{2}z^*\{c_{\mathrm{A}}\mathcal{J}_{\mathrm{AA}}^2 + c_{\mathrm{B}}\mathcal{J}_{\mathrm{BB}}^2 + \left[\left(c_{\mathrm{A}}\mathcal{J}_{\mathrm{AA}}^2 - c_{\mathrm{B}}\mathcal{J}_{\mathrm{BB}}^2\right)^2 + 4c_{\mathrm{A}}c_{\mathrm{B}}\mathcal{J}_{\mathrm{AB}}^4\right]^{\frac{1}{2}}\}, \tag{1.108}$$

for substitutional alloys [110], and

$$T_{\mathrm{g}}^2 = \frac{1}{2}z^*\left(\mathcal{J}_+^2 + \mathcal{J}_-^2\right) \tag{1.109}$$

32

for amorphous pure systems [40]. Here z^* is the average coordination number, $\mathcal{J}_{\alpha\gamma}$ is the exchange coupling energy between α and γ atoms, and $\mathcal{J}_+(\mathcal{J}_-)$ is the exchange coupling for a contracted (stretched) pair. Eq. (1.108) is in agreement with the well-known SG temperature T_g for an insulator model in the molecular field approximation [6,47,77]. Equation (1.109) reduces to T_g for the $\pm\mathcal{J}$ model, i.e.. $T_g = \sqrt{z^*}\,|\,\mathcal{J}\,|$ when $\mathcal{J} = \mathcal{J}_+ = -\mathcal{J}_-$ [78]. These facts show that the theory describes both types of SG on the same footing, i.e.. the SG due to configurational disorder and the SG due to structural disorder. Needless to say, the latter are essential for amorphous magnetism. The possibility of such a SG will be discussed in the next section.

3. Magnetic Properties of Amorphous Transition Metals

3.1. General Survey

In theoretical investigations of transition metals and alloys, an important role is played by noninteracting DOS, which are approximately common to all the $3d$ elements because of the similar bonding nature. Figure 1.8 shows the DOS for the amorphous, bcc, and fcc structures calculated by Fujiwara [111] and Moruzzi [112].

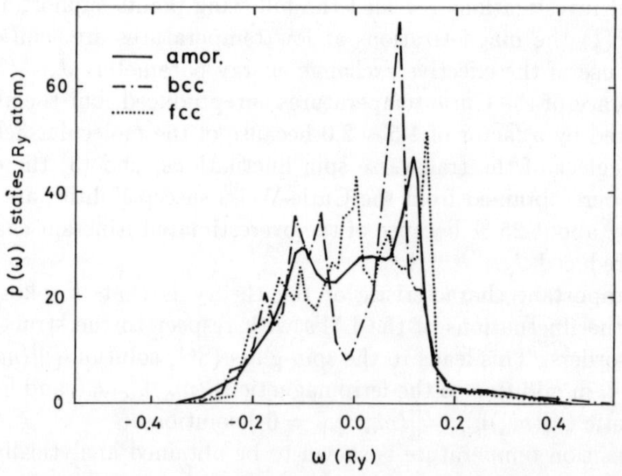

Fig. 1.8: Input DOS for the amorphous (solid curve) [111], bcc (dot-dashed curve) [111], and fcc (dotted curve) [112] structures.

As it is well-known, the appearance of ferromagnetism with uniform magnetization is determined by the Stoner condition $\rho(0)\tilde{J}/2 > 1$, where $\rho(0)$ denotes the noninteracting DOS at the Fermi level. The ferromagnetism of bcc Fe, for example, is stabilized by the main peak near the Fermi level. Fcc Ni shows strong ferromagnetism for the same reason.

The most important characteristic of DOS for amorphous transition metals is that the main peak is located just between the bcc and fcc one. The d-electron numbers N^* with the Fermi level at each main peak are $N^*(\text{bcc}) = 7.440$ for the bcc, $N^*(\text{amor}) = 8.352$ for amorphous structure, and $N^*(\text{fcc}) = 9.048$ for the fcc. They indicate strong ferromagnetic region in each structure.

Fujiwara [34] estimated the Stoner criterion for amorphous Fe to be $\rho(0)\tilde{J}/2 = 0.96 < 1$ with use of $\tilde{J}=0.068$ Ry, which was obtained by Janak [113]. He concluded that amorphous Fe does not develop a uniform magnetization. In the same way, it can be inferred that the ferromagnetism in amorphous Co is enhanced because the main peak is near the Fermi level, while amorphous Ni may not maintain strong ferromagnetism.

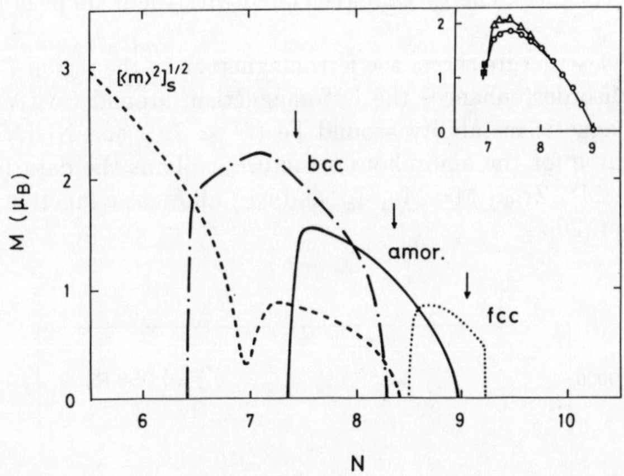

Fig. 1.9: Calculated magnetization vs. d-electron number curves at 75 K for the amorphous (solid curve), bcc (dot-dashed curves), and fcc (dotted curve) structures [42]. The effective exchange energy parameter is fixed to the amorphous Fe ($\tilde{J} = 0.059$ Ry). The dashed curve shows the spin-glass (SG) order parameter. The d-electron numbers N^* integrated up to the main peak of each DOS in Fig. 1.8 are indicated by the arrows. The inset shows the experimental data for amorphous $(\text{Fe} - \text{M})_{90}\text{Zr}_{10}$ alloys [20]. Here, M = Mn ($\blacksquare$), Co ($\triangle$), and Ni ($\circ$).

The second characteristic of the DOS for amorphous transition metals is that they are similar to the fcc DOS rather than the bcc ones, in particular, near the top of the d bands, presumably because the amorphous structure is close to the close-packed structure. Thus the amorphous transition metals and alloys are expected to show anomalous magnetic properties similar to those seen in fcc ones. For example, γFe alloys are known to have nonlinear magnetic couplings that show ferromagnetic couplings when the amplitudes of Fe LM's are large, but that show antiferromagnetic

34

couplings when the amplitudes are small [114-115]. Such a peculiarity may explain
the SG states in Fe-rich amorphous alloys, as will be discussed in the next subsection.
The Invar anomalies [116] as found in Fe-base amorphous alloys [21] are also expected
because of the similarity of the DOS (see Sec. 4.1).

The Stoner model is based on the assumption of the uniform magnetization which
is not generally satisfied in amorphous metallic systems. To examine the existence
of the ferromagnetism, one needs to perform numerical calculations using the finite-
temperature theory presented in Sec. 2.3. Figure 1.9 shows an example of magnetization vs. d-electron number curves. (Note that the input parameters are set as those of
amorphous iron except for the d-electron number (N) to see the qualitative feature.)
It is found that N^*, indicated by arrows, correspond well to the magnetic region of
the d-electron number for each structure. This verifies that the $3d$ ferromagnetism is
stabilized by the magnetic energy gain associated with the main peak in the DOS for
each structure.

The amorphous structure shows the ferromagnetism in the region $7.35 \lesssim N \lesssim 9.0$.
The structural disorder enhances the ferromagnetism around Co ($N \approx 8.0$), while
it causes ferromagnetic instability around Fe ($N \approx 7.0$) and Ni ($N \approx 9.0$). The
magnetization curve for the amorphous structure explains the data in the inset for
amorphous $(Fe - M)_{90}Zr_{10}$ (M= Mn, Co, and Ni) alloys as a function of the average
d-electron number [20].

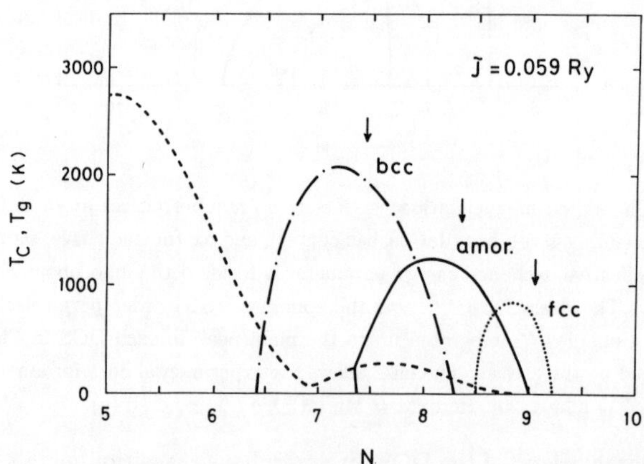

Fig. 1.10: Curie temperature vs. d-electron number curves calculated with use of
the same input parameters as in Fig. 1.9 [42]. The SG temperatures are shown by
the dashed curve. The d-electron numbers N^* are indicated by arrows.

The Curie temperature maxima are also characterized by N^*, as shown in Fig. 1.10.
According to thermodynamic considerations, the Curie temperature T_C is roughly

given by $T_C \sim E_{\text{mag}}/S_{\text{mag}}$ [117]. Here E_{mag} is the magnetic energy defined by the energy difference between the ground state and the paramagnetic state, and S_{mag} is the magnetic entropy. The result in Fig 1.10 indicates that the Curie temperatures for transition metals showing strong ferromagnetism are dominated mainly by the magnetic energy gain associated with the main peak for each structure. This explains why the Curie temperature of amorphous Co is enhanced [27]. A more detailed analysis will be given in Sec. 3.3.

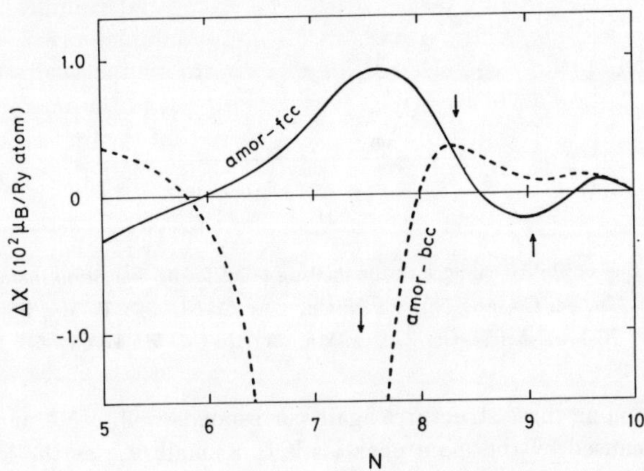

Fig. 1.11: Relative paramagnetic susceptibilities of amorphous and bcc structures (dashed curve), and of amorphous and fcc structures (solid curves) calculated at 2400 K [118]. The arrows indicate the d-electron numbers N^*.

The characteristics of the electronic structure also influence the finite-temperature properties. Figure 1.11 shows systematic variations of the paramagnetic susceptibilities at high temperatures [118]. The susceptibility in the fcc structure shows the highest value among the three structures in the region $5.0 \lesssim N \lesssim 5.85$ and $8.55 \lesssim N \lesssim 9.43$. The bcc susceptibility shows the highest value in the region $5.85 \lesssim N \lesssim 7.98$, and the amorphous one in the regions $7.98 \lesssim N \lesssim 8.55$ and $9.43 \lesssim N \lesssim 10.0$. These results clearly reflect the DOS at the Fermi level in Fig. 1.8 , in particular, the main peaks corresponding to $N^*(\text{bcc}) = 7.440$, $N^*(\text{amor}) = 8.352$, and $N^*(\text{fcc}) = 9.048$. This implies that the susceptibility enhancement due to the magnetic energy gain associated with the DOS at the Fermi level remains even above T_C, in spite of large thermal spin fluctuations. The experimental data [119] for liquid transition metal alloys in Fig. 1.12 are regarded as supporting evidence for such a physical picture.

The susceptibilities follow the Curie-Weiss law at high temperatures. The effective Bohr magneton numbers (m_{eff}) obtained from the susceptibility are presented in Fig. 1.13 [118]. It should be noted that the region for a structure whose m_{eff} takes the

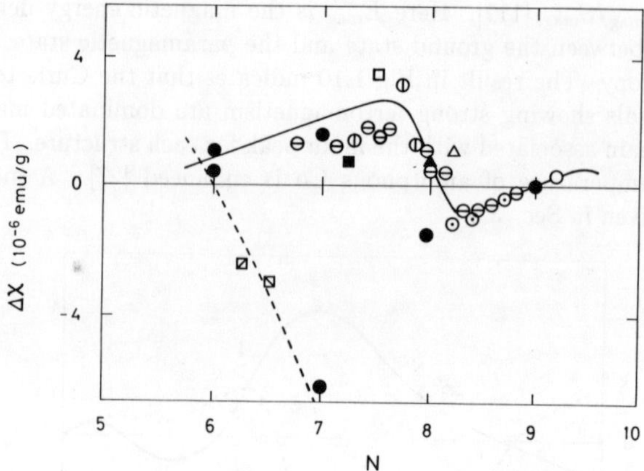

Fig. 1.12: Susceptibility changes at the melting point for $3d$ transition metal alloys [119]. $\bullet$: Mn, Fe, Co, and Ni, $\circ$: Ni-Cu, $\odot$: Co-Ni, $\ominus$: Fe-Mn, Fe-Ni, $\oplus$: Fe-Co, $\triangle$: Ni-Mn, $\blacktriangle$: Ni-Cr, $\square$: Co-Mn, $\blacksquare$: Co-Cr, $\boxslash$: Fe-Cr, $\boxminus$: Fe-V.

minimum value among three structures again corresponds well to N^* since the strong ferromagnetism caused by the main peak leads to a small m_{eff} as the local-moment limit. When the d-electron number decreases from $N^*(\text{fcc})$, the effective Bohr magneton numbers increase gradually in the order of the fcc, amorphous, and bcc structures according to the inequality $N^*(\text{bcc}) < N^*(\text{amor}) < N^*(\text{fcc})$. This behavior is understood as a gradual change of magnetism from strong to weak ferromagnetism [120]. It is found that a rapid increase of the fcc m_{eff} as compared with the amorphous one in the region $N \lesssim 8.5$ corresponds well to the experimental data for m_{eff} in the fcc and liquid transition metal alloys. The systematic change in $\partial M/\partial P$ and $\partial T_{\text{C}}/\partial P$ is also found to be characterized by N^* for each structure [118].

One of the most important features of the magnetism in amorphous transition metals is that the SG solution ($[\langle m \rangle^2]_s \neq 0$ and $[\langle m \rangle]_s = 0$) appears, as shown in Figs. 1.9 and 1.10. Although the SG solutions exist even for Co ($N \approx 8.0$), they are not realized because the ferromagnetic state is more stable. The SG state is expected to be stabilized in the region $N \lesssim 7.35$, where the ferromagnetism disappears. In the region $N \lesssim 6.7$ we find antiferromagnetic NN interactions, irrespective of the local environment. Therefore, it might be expected that the antiferromagnetic order is realized in the region $N \lesssim 6.7$. It should be emphasized, however that the "antiferromagnetic spin structure" can not be distinguished from the "spin-glass spin structure" in amorphous systems [28], particularly, in amorphous metallic systems in which both the direction and the amplitude of LM's can change on the random network, depending on their local environments. The key to understanding of their magnetic properties may be the spin frustrations which dominates the low energy excitations. The "antiferromagnetism" in amorphous metallic magnetism is left as a

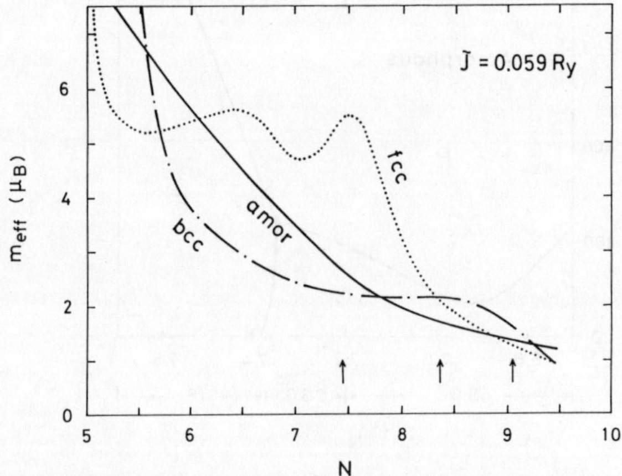

Fig. 1.13: Effective Bohr magneton numbers calculated around $T = 1800$ K [118]. Arrows indicate N^*.

problem for future research.

3.2. Amorphous Fe

The magnetism in the vicinity of amorphous Fe has a complicated structure. Figure 1.14 shows the magnetic phase diagram obtained by the finite-temperature theory outlined in Sec. 2.2. The SG states with $T_g = 100 \sim 200$ K are extended to a reasonable range of amorphous Fe ($6.7 \lesssim N \lesssim 7.35$). It is shown that the existence of the SG around $N \approx 7.0$ is not sensitive to the degree of the fluctuation $[(\delta R)^2]_s^{1/2}/[R]_s$. $T_g = 117$ K is obtained if the value of the fluctuation is taken to be $[(\delta R)^2]_s^{1/2}/[R]_s = 0.067$, which was obtained from an X-ray experiment [59] and by computer simulation [58]. The transition temperature is consistent with the experimental value 120 K [23-26].

The calculated inverse susceptibility follows the Curie-Weiss law with $m_{\text{eff}} = 3.5\,\mu_B$ above 1000 K, but it shows upward convexity below 1000 K and a cusp at T_g, as shown in Fig. 1.15. It should be noted that the deviation from the inverse susceptibilities below 1000 K is also seen in fcc Fe. Thus, it is not due to structural disorder, but due to competition between short-range and long-range magnetic couplings in close-packed Fe [118].

Magnetic behavior below 25 K has not yet been calculated by using the finite-temperature theory, because of numerical difficulties. The ground-state calculations have recently been performed by Krey, Krauss, and Krompiewski [121] using a realistic tight-binding Hartree-Fock Hamiltonian which allows for arbitrary configurations of vector LM's on 54 atoms in a box with a periodic boundary condition. They found the Hartree-Fock spin-glass solution (i.e. random isotropic distribution of LM's)

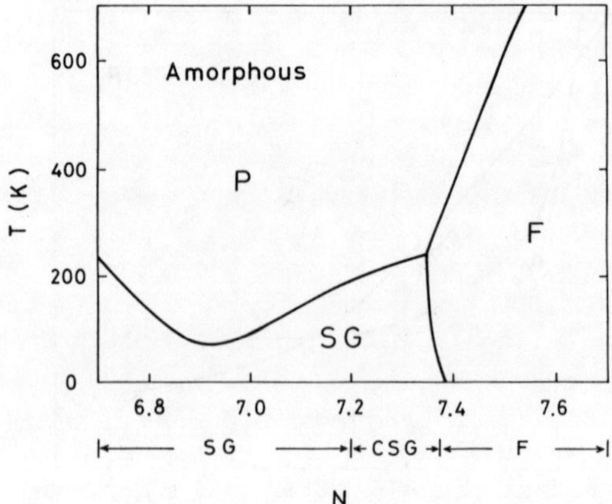

Fig. 1.14: Magnetic phase diagram around amorphous Fe showing the ferromagnetic (F), paramagnetic (P), and spin-glass (SG) states [42]. The cluster SG (CSG) phase is expected to fall in the region $7.2 < N < 7.385$.

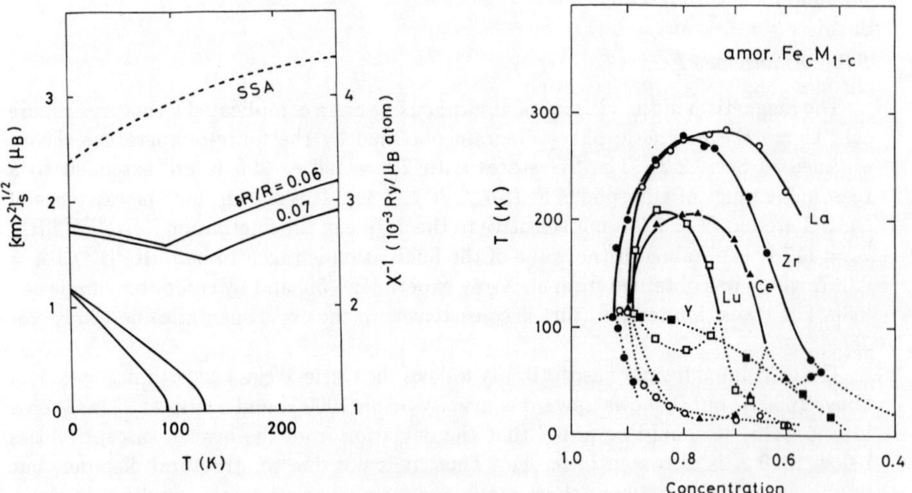

Fig. 1.15: (left) The SG order parameter ($[\langle m \rangle^2]_s^{1/2}$) and inverse susceptibilities as a function of temperature for $[(\delta R)^2]_s^{1/2}/[R]_s = 0.06$ and 0.07 [41]. The dashed curve denotes the inverse susceptibility in the single-site approximation.

Fig. 1.16: (right) Magnetic phase diagram showing the Curie temperatures (solid curves) and spin-glass temperatures (dotted curves) of amorphous Fe$_x$M$_{1-x}$ alloys (M= La, Zr, Ce, Lu, and Y) [21].

with 0.8 μ_B as the average magnitudes of LM's , which lowers the energy by about 0.001 eV as compared with the ferromagnetic solution.

Experimentally, Hiroyoshi and Fukamichi [22] first suggested the existence of an SG in amorphous Fe_xZr_{1-x} alloys with $x \approx 0.9$. Saito and Nakagawa [23], and Coey *et al.* [24-25] determined the magnetic phase diagram for amorphous Fe_xZr_{1-x} alloys up to 93 at.% Fe. They speculated that amorphous pure iron would be the SG with $T_g \approx 150$ K. Subsequently, Fukamichi *et al.* [21,26] performed systematic investigations on amorphous Fe_xM_{1-x} (M= Y, Zr, La, Ce, and Lu) alloys. The important point is that beyond 90 at.% Fe these alloys have the same SG transition temperature irrespective of their second element M (see Fig. 1.16). This implies that the SG's for amorphous alloys with 90 at.% Fe or more are caused not by the configurational disorder of Fe and M atoms, but rather by the structural disorder of amorphous pure Fe.

Kakehashi clarified the mechanism for the formation of SG's. He examined the nature of magnetic couplings. It is determined by the exchange pair energy $-\Phi_{0j}^{(e)}(\xi)$ in $\Psi(\xi)$ in the case of no polarization of the medium (see Eq. (1.77)). The energy $-\Phi_{j0}^{(e)}(\xi)$ is interpreted as the magnetic pair-energy gain for the flexible central LM ξ when the neighboring LM with average amplitude x_j points up, as seen from Eq. (1.77).

Figure 1.17 shows the exchange pair energies $-\Phi_{0j}^{(e)}(\xi)$ in various environments in the SG state. In contrast to bcc Fe (see Fig. 1.5), neighboring Fe LM's show nonlinear magnetic couplings for the environments $3 \lesssim n \lesssim 12$; the Fe LM's with large amplitudes couple ferromagnetically with the neighboring LM's, while the Fe LM's with small amplitudes couple antiferromagnetically with the neighboring ones. This behavior has also been found in γFe crystalline alloys [115]. Since the amplitudes of LM depend strongly on the surrounding environments *via* the single-site energy $E(\xi, n)$ (see the dotted curves in Fig. 1.17), the sign of the magnetic couplings changes with the local environments. The picture obtained for Fe LM's in various environments is shown in Fig. 1.18. The competition between these ferro- and antiferro-magnetic couplings produces the itinerant-electron SG in amorphous iron.

The mechanism mentioned above is characteristic of the amorphous metallic magnetism, since neither the LEE on the amplitude of LM nor the non-linearity of the magnetic couplings are seen in insulators.

Numerical calculations revealed that anomalous nonlinear magnetic couplings appear in the region $6.7 \lesssim N \lesssim 7.2$. When the d-electron number is increased, the NN couplings do not show any non-linearity irrespective of the local environments at $7.2 \lesssim N$. Nevertheless, the SG state is found in the region $7.2 \lesssim N \lesssim 7.385$. This is explained as follows. The LM's form local ferromagnetic orders according to the NN ferromagnetic couplings. If these ferromagnetic orders developed a long-range ferromagnetic order, the effective medium would have a polarization consistent with the ferromagnetic clusters. It was found, however, that antiferromagnetic couplings between the central LM and the medium occur when the medium is polarized. These couplings reverse the central LM's, so that a long-range ferromagnetic order

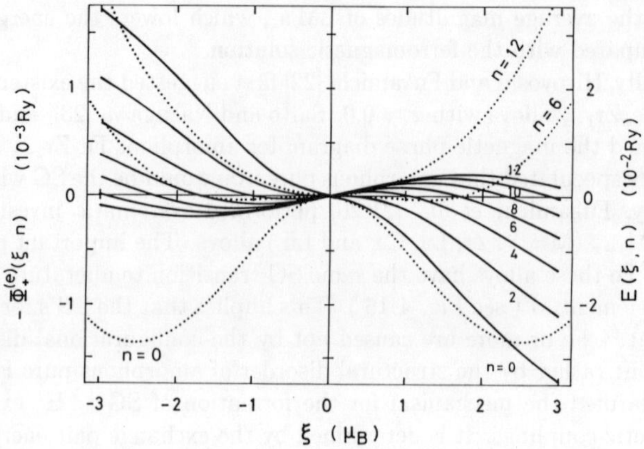

Fig. 1.17: Single-site energy $E(\xi, n)$ (dotted curves) and exchange pair energy $-\Phi_+^e(\xi, n)$ (solid curves) of amorphous iron in various environments (n) at $T = 35$ K [42].

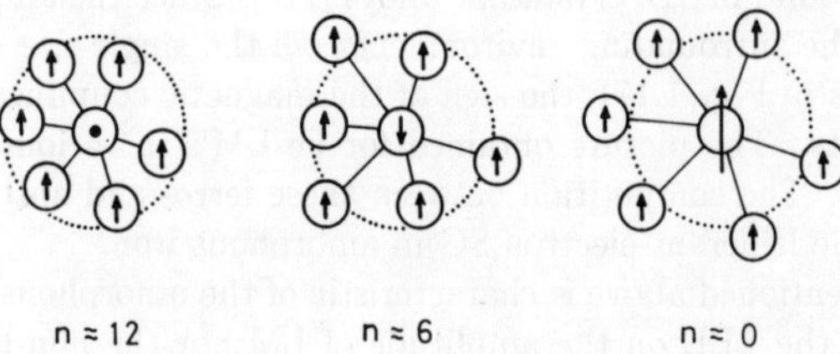

Fig. 1.18: Schematic representation showing the local environment effect on the central local moment (LM) in amorphous iron. n denotes the number of contracted atoms on the NN shell.

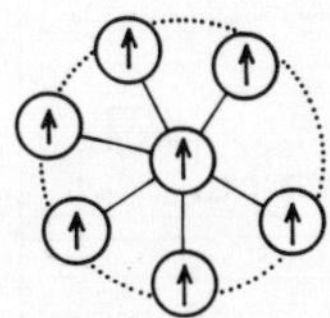 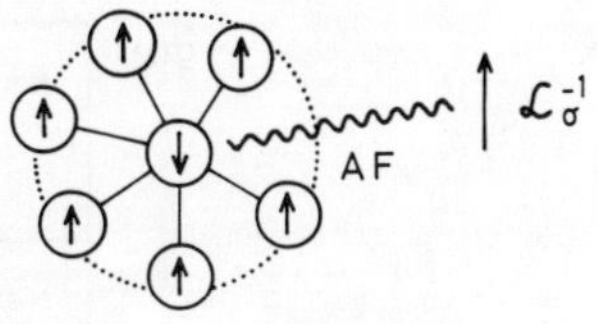

Fig. 1.19: Schematic representation showing the formation of a cluster SG. The short-range ferromagnetic couplings form ferromagnetic clusters. But the long-range antiferromagnetic couplings between the central LM and the polarized medium ($\mathcal{L}_\sigma^{-1}$) suppress ferromagnetism.

can not be developed. (see Fig. 1.19). This implies that the competition between the short-range ferromagnetic couplings and the long-range antiferromagnetic couplings produces the SG in the region $7.2 \lesssim N \lesssim 7.35$. Since the SG is accompanied by the ferromagnetic clusters, it is called a cluster SG. As the d-electron number is increased, the size of the clusters is expected to increase more and more. Finally, a long-range ferromagnetic order is realized in the region $7.35 \lesssim N$.

A recent neutron scattering experiment on $Fe_{90-x}Ni-xZr_{10}$ (x=1, 5, 10, 20) amorphous alloys was reported to show a coexistence of propagating spin-wave excitations and spin-freezing phenomena for x=1 [122]. The data seem to support the physical picture mentioned above.

Theoretical calculations show the broad distributions of LM's in the vicinity of the ferromagnetic instability and the SG region, as shown in Fig. 1.20. They are caused by the LEE on both the amplitudes and the directions of LM's. The results are consistent with the broad distributions of the hyperfine field found in amorphous $Fe_{93}Zr_7$ [25], $Fe_{92}La_8$ [26], and $Fe_{92}Hf_8$ [123] alloys. But, quantitative comparison with the data has not been made.

The most intriguing feature in the magnetic phase diagram in Fig. 1.14 is the reentrant SG in the narrow region $7.350 \lesssim N \lesssim 7.385$. The magnetic states in this region are determined by a detailed balance between the short-range ferromagnetic interactions and the long-range antiferromagnetic interactions. Theoretical calculations showed that the reentrant SG behavior is caused by the temperature-induced enhancement of the short-range ferromagnetic couplings.

The reentrant SG produced by the structural disorder is considered to be realized in Fe-rich amorphous alloys around 90 at.% Fe. In fact, the calculated T-P

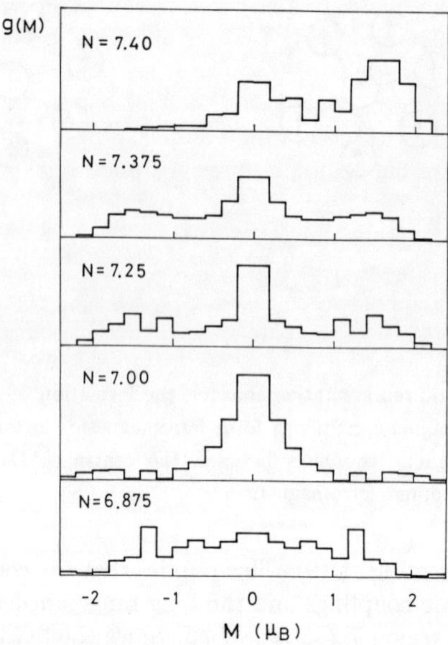

Fig. 1.20: The LM distributions ($g(M)$) around amorphous Fe at 35 K [42]. N denotes the d-electron number.

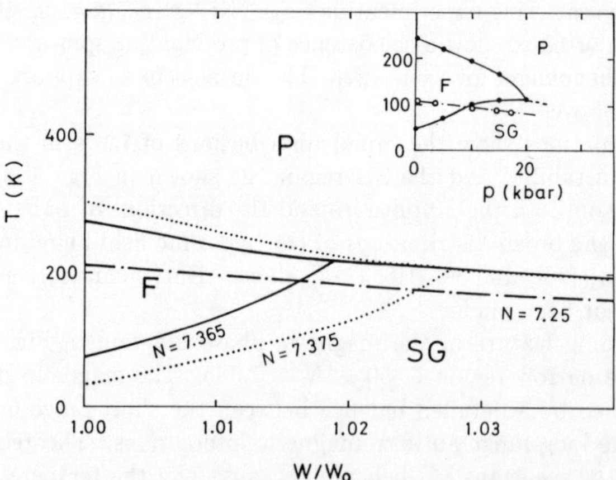

Fig. 1.21: T-W magnetic phase diagram showing ferromagnetic (F), spin-glass (SG), and paramagnetic (P) phases near the reentrant SG region [118]. W_0 denotes the d band width at zero pressure. The inset shows the experimental data for amorphous $Fe_{87.5}La_{12.5}$ (•) and $Fe_{92.5}La_{7.5}$ (○) alloys [124].

phase diagram explains the qualitative features of the experimental data for amorphous Fe_xLa_{1-x} ($x = 0.925$ and 0.875) alloys [124], as shown in Fig. 1.21. The transition from the ferromagnetic to the SG with increasing pressure is explained by a relative change in the ratio of short-range ferromagnetic couplings to long-range antiferromagnetic couplings with decreasing volume [118].

The critical pressure for the complete disappearance of the ferromagnetism was obtained to be 12 [kbar] for $N = 7.365$ and 18 [kbar] for $N = 7.375$. These values are consistent with the experimental value of 20 [kbar] [124]. The calculated $\partial T_C / \partial P$ are -5.3 [K/kbar] for $N = 7.365$ and -6.4 [K/kbar] for $N = 7.375$, which are in good agreement with the experimental value of -5.1 [K/kbar] for amorphous $Fe_{87.5}La_{12.5}$ [124]. The pressure dependence of the Curie temperatures, $\partial T_g / \partial P$, were also calculated as -1.5 [K/kbar] for $N = 7.0$, and -2.1 [K/kbar] for $N = 7.25$. These values are consistent with the experimental value of -1.4 [K/kbar] [124].

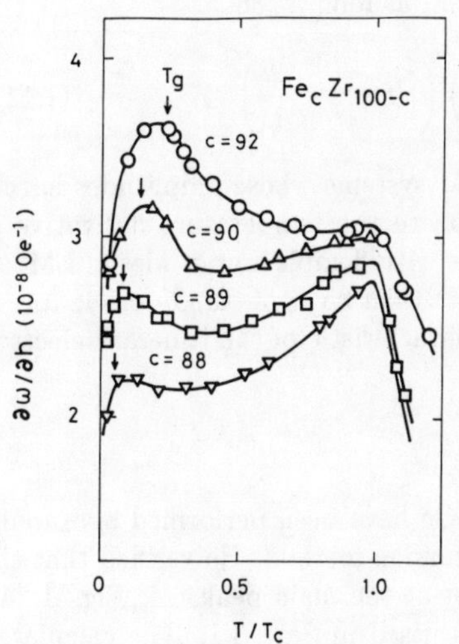
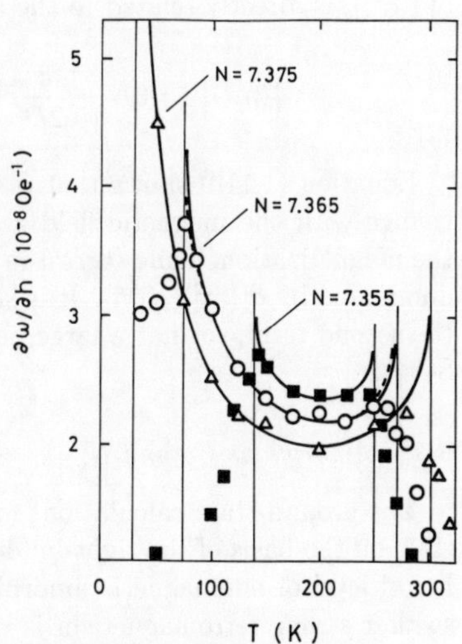

Fig. 1.22: (left) Temperature dependence of the forced volume magnetostriction $\partial \omega / \partial h$ for amorphous Fe-Zr alloys [125]. The arrows indicate the reentrant SG temperatures obtained from the ac susceptibilities.

Fig. 1.23: (right) Calculated $\partial \omega / \partial h$ in the reentrant SG regime [126]. ■, ○, and △ denote the numerical results for a finite field $h \approx 6 \times 10^4$ G, while the solid curves are those obtained from an extrapolation to $h = 0$. The dashed curve for $N = 7.365$ shows the contribution of the second term in the rhs of Eq. (1.110).

Recently, Tange, Tanaka, Goto, and Fukamichi [125] found an anomaly in large forced volume magnetostriction ($\partial\omega/\partial h$) at T_g in $\mathrm{Fe}_x\mathrm{Zr}_{1-x}$ ($0.88 \leq x \leq 0.92$) alloys, as shown in Fig. 1.22. This anomaly is not seen in the insulator SG's. Kakehashi [126] pointed out that it is explained by the itinerant-electron SG found in the region $7.350 \lesssim N \lesssim 7.385$. Figure 1.23 shows some numerical examples of $\partial\omega/\partial h$, as well as the experimental data obtained by Tange et $al.$ The calculated value of $\partial\omega/\partial h$ are very large in comparison with those in bcc Fe (0.05 $[10^{-8}\mathrm{Oe}^{-1}]$) and fcc Ni (0.01 $[10^{-8}\mathrm{Oe}^{-1}]$) [127], and show divergence at both T_g and T_C, in agreement with the experimental data.

According to the theory of magnetovolume effect [118,128-130], $\partial\omega/\partial h$ in transition metals is given by

$$\frac{\partial\omega}{\partial h} = \frac{\dot{D}}{3BV}\left(T\frac{\partial\left[\langle m\rangle\right]_s}{\partial T} + \frac{1}{4}\tilde{J}\,\frac{\partial\left[\langle\xi^2\rangle\right]_s}{\partial h}\right) . \tag{1.110}$$

Here $\dot{D}$ (=3.55 for Fe [131]) is a proportionality constant. Note that the amplitude of $\left[\langle\xi^2\rangle\right]_s$ is directly related to the amplitude of LM as follows [88].

$$\left[\langle m^2\rangle\right]_s = 3N - \frac{3}{2D}N^2 + \left(1 + \frac{3}{2D}\right)\left(\left[\langle\xi^2\rangle\right]_s - \frac{2}{\beta\tilde{J}}\right) . \tag{1.111}$$

Equation (1.110) shows that $\partial\omega/\partial h$ in the LM systems whose amplitudes hardly change with the magnetic field are in proportion to the temperature derivative of the magnetization, while $\partial\omega/\partial h$ in the Invar type with flexible amplitudes of LM are dominated by $\partial[\langle m^2\rangle]_s/\partial h$. It is shown that more than 95 % of $\partial\omega/\partial h$ originates in the second term, so that a large peak at T_g is characteristic of the itinerant-electron SG [126].

3.3. Amorphous Co and Ni

The ground-state calculations for amorphous Co have been performed by Tanaka [132] on the basis of the tight-binding LMTO-recursion method. He verified that the Fermi level of nonmagnetic amorphous Co is just at the main peak (see Fig. 1.24), so that strong ferromagnetism is realized, as discussed in Sec. 3.1. The calculated ground-state magnetizations are 1.63 μ_B for amorphous Co [132], 1.58 μ_B for fcc Co [133-134], and 1.55 μ_B for hcp Co [133,135], and the experimental values are reported to be 1.72 μ_B for both amorphous and hcp Co [27]. Subtracting the orbital contribution 0.15 μ_B [136], we obtained the spin contribution 1.57 μ_B, which is in good agreement with the theoretical results.

The most exciting feature of amorphous Co is that the Curie temperature of this material obtained by extrapolation of the data for amorphous Co-Y alloys amounts to 1850 K, which is 450 K higher than that of crystalline Co [27] (see Fig. 1.25). Kakehashi [42] calculated the magnetization vs. temperature curves for amorphous and fcc Co by means of the finite-temperature theory. The results are shown in Fig.

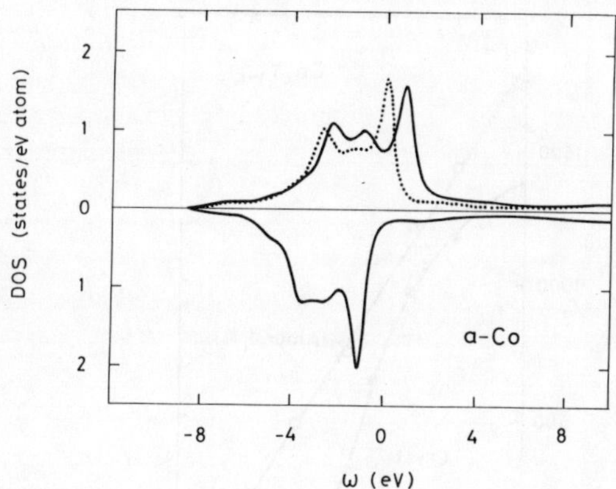

Fig. 1.24: The up and down DOS for amorphous Co at the ground state [132]. The dotted curve shows the down DOS in the nonmagnetic state.

1.26. The calculated T_C are overestimated by a factor of 1.8 because of the molecular-field approximation. The results of T_C for amorphous Co is 490 K higher than that for the fcc Co. This is because the magnetic energy gain due to the main peak at the Fermi level enhances T_C in amorphous Co, while the ground-state magnetization is approximately same as for fcc Co because of the strong ferromagnetism in both amorphous and fcc Co. It is shown that the enhancement of T_C occurs in the range $7.9 \lesssim N \lesssim 8.5$ in accordance with $N^* = 8.352$.

Experimentally, it has been shown by extrapolation that the spin wave stiffness constant is also enhanced by 330 [meV·Å^2] as compared with the hcp Co (510 [meV·Å^2]) [27]. Theoretical investigations for the stiffness constant, however, have not been done yet.

The strong ferromagnetism in Ni is expected to be reduced with the introduction of structural disorder, since the main peak near the top of the d band is broadened and shifts down to the lower energy so that the Fermi level is located above the peak. It is, however, not conclusive whether or not the amorphous Ni shows weak ferromagnetism, because it is sensitive to the choice of parameters in the model. Figure 1.27 shows a diagram for the Stoner ferromagnetism near Ni as a function of N [137], which is obtained from the DOS in Fig. 1.8 after a scaling of the band width by a factor of $W(\mathrm{Fe})/W(\mathrm{Ni})=0.393/0.364$ [112]. It is seen that a weak ferromagnetism ($[\langle m \rangle]_\mathrm{s} \approx 0.5 \ \mu_\mathrm{B}$) may be obtained if one assumes $N = 9.0$ and Janak's value $\tilde{J}=0.074$ Ry [113], but paramagnetism may also be obtained if we choose $\tilde{J} = 0.060$ Ry so as to reproduce the ground-state magnetization (0.615 μ_B) in fcc Ni at $N = 9.1$. Here, the fluctuation of the interatomic distance $[(\delta R)^2]_\mathrm{s}^{1/2}/[R]_\mathrm{s}$ in amorphous Ni was assumed to be the same as in amorphous Fe in Fig. 1.8. Fujita and Fukamichi [138]

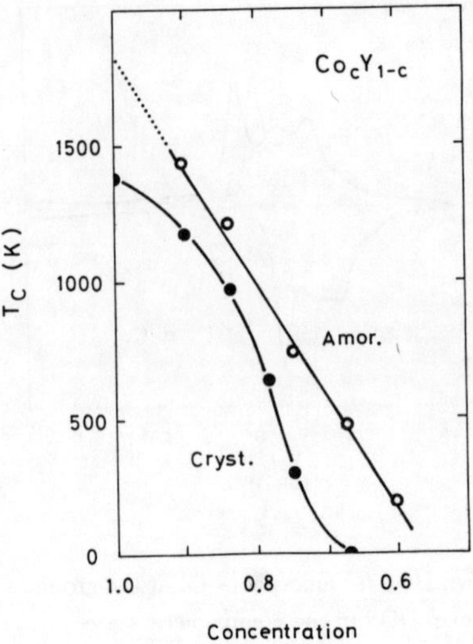

Fig. 1.25: Curie temperatures in crystalline (solid circles) and amorphous (open circles) Co$_x$Y$_{1-x}$ alloys [27]. The dotted line is an extrapolation to amorphous pure Co.

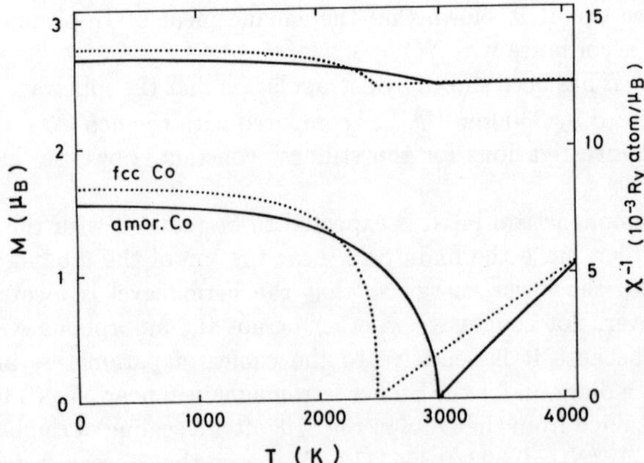

Fig. 1.26: Calculated magnetizations, inverse susceptibilities, and amplitudes of LM's for amorphous Co (solid curves) and fcc (dotted curves) Co ($N = 8.1$ and $\tilde{J} = 0.100$ Ry) as a function of temperature [42].

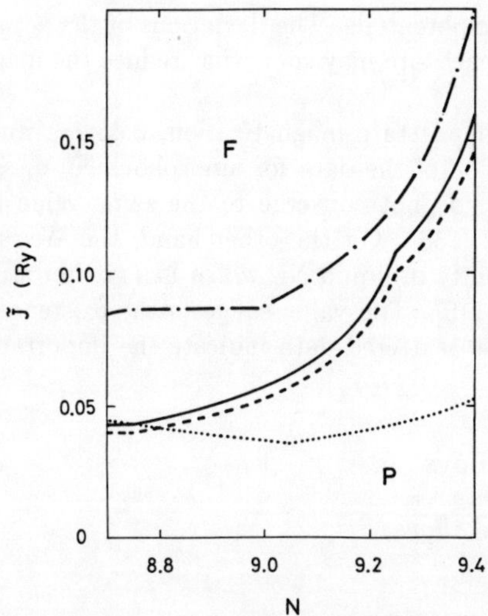

Fig. 1.27: Ferro- (F) and para-magnetic (P) phase boundaries for the amorphous (solid curve), bcc (dot-dashed curve), and fcc (dotted curve) structures on the $\tilde{J}$-N plain, which are obtained by making use of the Stoner theory [137]. The dashed curve shows the boundary for the amorphous structure when the atomic volume is changed from 72.2 a.u. to 76.5 a.u.

measured the pair distribution in amorphous Ni_xY_{1-x} ($0.85 \leq x \leq 0.92$) alloys. They suggested that the fluctuation $[(\delta R)^2]_s^{1/2}/[R]_s$ in amorphous Ni is smaller than that of amorphous Fe by about 10 %. Such a reduction of the structural disorder is favorable to ferromagnetism.

Recently, Ching *et al.* [75,76] calculated the DOS for amorphous Ni by using orthogonalized linear combinations of atomic orbitals (OLCAO). The obtained DOS are broader than that expected from Fujiwara's DOS for amorphous Fe, and show a simple structure with a peak at 0.13 Ry below the Fermi level. The DOS at the Fermi level was estimated to be 6.4 (states/Ry atom), which apparently leads to nonmagnetic amorphous Ni.

The result seems to overestimate the effect of the distribution of atomic levels, since the charge neutrality on each site is often extremely violated in contradiction to the recent selfconsistent calculations for amorphous pure Zr [139], which are based on the LMTO super-cell approach.

Quite recently, Tanaka *et al.* [132] performed spin-polarized tight-binding LMTO calculations for amorphous Ni. They obtained a value of 0.6 μ_B for the ground-state magnetization . The Wigner-Seitz radius $r_{WS} = 2.677$ Å used in the calculation is slightly larger than the experimental data 2.633 Å [140]. Self-consistency was achieved within the averaged atomic potentials. The deviations of the actual atomic potential from the averaged one on each site may somewhat reduce the magnetization because of the band broadening.

Experimentally, the ground state magnetization, 0.45 μ_B, and $T_C = 480$ K are obtained by an extrapolation of the data for amorphous $Ni_x Y_{1-x}$ ($0.75 \lesssim x \lesssim 0.97$) [141]. The data, however, do not converge to the same value beyond 90 at.% Ni when Y is replaced by La [138]. On the other hand, the Weiss constant obtained from the inverse susceptibility in liquid Ni, which has greater structural disorder, is estimated to be -435 K [119]. The value suggests the existence of paramagnetism at the ground state. These scattered data indicate the importance of the degree of structural disorder.

4. TM Amorphous Alloys

4.1. TM-TM Amorphous Alloys

4.1.1. General aspects

The simplest system among actual amorphous magnetic alloys may be transition metal-transition metal (TM-TM) binary alloys. In particular, $3d$-$3d$ amorphous alloys are useful from the theoretical point of view because they enable us to clarify the effect of structural disorder by comparing their magnetic properties with those of substitutional alloys. However, only Fe-Sc [142], Fe-Ti [143], Fe-V [145], and Fe-Ni [146] amorphous alloys have so far been made, in a narrow range of concentrations. Most of TM-TM magnetic alloys are formed by adding $4d$ or $5d$ transition metals (ET = Y, Zr, Nb, Mo, La, Hf, Ta, and W) to $3d$ magnetic transition metals (TM = Fe, Co, and Ni) [3, 5, 10]. These alloys are characterized by a large difference in atomic size between the magnetic $3d$ and non-magnetic $4d$ (or $5d$) elements. It is expected that the NN atomic distance depends strongly on the type of NN pair, and that the coordination number depends strongly on the size of the central atom.

Fe-based ET-TM amorphous alloys rapidly regain ferromagnetism when ET atoms are added, as has been shown in Fig. 2.16, and have a maximum Curie temperature around 80 at.% Fe. Reentrant SG states are found at low temperatures over a wide range of Fe concentrations, $0.6 \lesssim x \lesssim 0.9$ [21-26]. Although it has been explained in Sec. 3.2 that the RSG states around $x = 0.9$ are caused by structural disorder, the mechanism for their formation at lower Fe concentrations has not yet been clarified.

It is worth noting in considering the problem of the structure vs. magnetism that Laves-phase $Fe_x Ti_{1-x}$ ($0.6 \lesssim x \lesssim 0.7$) crystalline alloys show ferromagnetic-antiferromagnetic transition around the $Fe_2 Ti$ composition with increasing Ti concentration [147], while $Fe_x Ti_{1-x}$ ($0.55 \lesssim x \lesssim 0.8$) amorphous alloys show a simple ferromagnetism in the same concentration region [143]. The same type of behavior has been reported in quasi-binary $Fe_2(Hf_x Ta_{1-x})$ ($0 \leq x \leq 1$) amorphous alloys [149].

The Co-based ET-TM (ET = Ti, Y, Zr, Nb, Mo, La, Hf, Ta, and W) amorphous alloys show ferromagnetism with a simple linear concentration dependence of magnetization, as shown in Fig. 1.28. In particular, Co_xY_{1-x}, Co_xZr_{1-x}, and Co_xLa_{1-x}

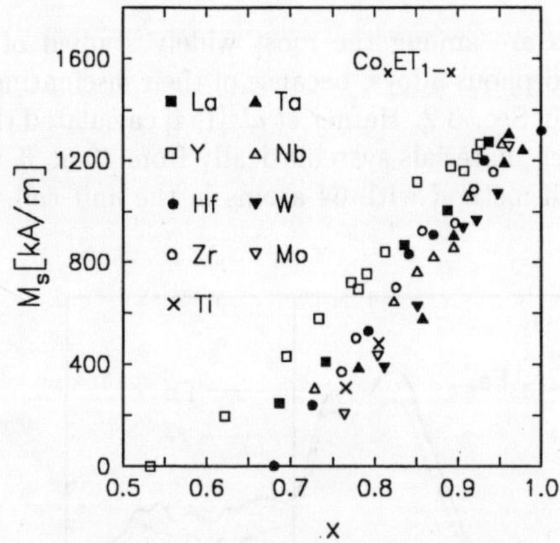

Fig. 1.28: Room temperature magnetization for early TM-Co amorphous alloys [10]

alloys show strong ferromagnetism, which disappears at low critical concentration ($x_c = 0.55$ for Y, 0.63 for La, and 0.68 for Zr). They were analyzed by the generalized Slater-Pauling curve [17, 18]. When the d-electron numbers of an ET atom are increased, the critical concentrations shift to higher levels. This is probably because stronger d-d mixing delocalizes the $3d$ magnetic electrons more.

The Ni-based ET-TM alloys show weak ferromagnetism in the range of Ni concentration $0.85 \lesssim x \lesssim 0.95$. When magnetic $3d$-electrons are changed from Fe to Ni, the critical concentration x_c moves to a higher level, such as $x_c = 0.4$ for Fe_xZr_{1-x} [150], 0.65 for Co_xZr_{1-x} [151], and 0.8 for Ni_xZr_{1-x} [5] amorphous alloys. This trend is explained by a change in the DOS at the Fermi level, with a band filling of $3d$ subbands.

A remarkable change in magnetism due to structural disorder is found in Heusler Cu_2MnZ amorphous alloys with Z = Al, In, and Sn [152]. The crystalline counterparts show strong ferromagnetism with T_C = 600, 520, and 530 K, while amorphous Heusler alloys show the SG state with T_g = 65, 60, and 59 K, respectively. The appearance of the SG was explained in terms of a localized model as a result of a competition between the anti-ferromagnetic NN Mn-Mn interactions and ferromagnetic Mn-Mn interactions at a distance of several angstrom. At the opposite limit of weakly magnetic metals, MnSi amorphous alloys are reported to show a SG state with T_g = 22 K [153], while their crystalline counterparts show a weak helimagnetism. Although competition between long-range ferro and anti-ferromagnetic interactions is expected in this system, the detailed mechanism for the formation of the SG has not yet been clarified.

50

In the following, we discuss in more detail the magnetism and electronic structure of amorphous Fe-Zr, Fe-Ni, and Co-Y alloys.

4.1.2. Fe-Zr amorphous alloys

Fe-Zr amorphous alloys are among the most widely studied of early transition metal-transition metal amorphous alloys, because of their fascinating magnetic properties, which we discussed in Sec. 3.2. Hafner *et al.* [154] calculated the spin-polarized electronic structures of such materials systematically from 40 at.% to 90 at.% Fe by using the LMTO super cell method with 64 atoms in the unit cell. Fig. 1.29 shows

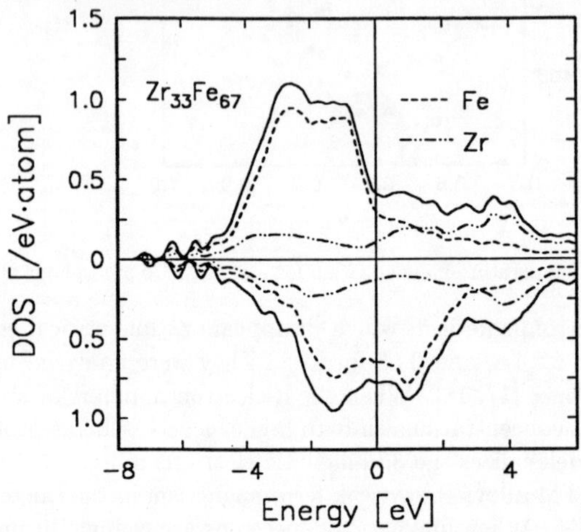

Fig. 1.29: Total and projected DOS's for an amorphous Fe$_{67}$Zr$_{33}$ alloy obtained by the LMTO super-cell method [154]

the calculated total DOS of an Fe$_{67}$Zr$_{33}$ amorphous alloy. Note that Zr atoms also carry magnetic moments anti-parallel to the magnetization. The shape of the obtained DOS is far from the picture of rigid band splitting, because of ferrimagnetism. The calculated magnetic moments decrease with decreasing Fe concentration.

A significant result of the above work is that, in the Fe$_{90}$Zr$_{10}$ system, some Fe moments that are coupled anti-ferromagnetically with their surrounding Fe moments. A similar result was also reported by Krey *et al.* [121], who used a tight-binding vector-spin model Hamiltonian. These facts support a phase transition to a SG state, as we have already explained in Sec. 3.2.

On the other hand, the changes in the coordination number due to the difference of atomic size are important for the development of ferromagnetism at lower Fe concentrations. It is expected that the average coordination number around an Fe atom becomes smaller than 12 with increasing Zr concentration, because the solid angle of a Zr atom is larger than that of an Fe atom. Recently, Kakehashi *et al.* [155] developed a Bethe-type theory for calculating the electronic structure of amorphous alloys on the basis of the geometrical mean model (see Sec. 2.3.5), and calculated the DOS of nonmagnetic $Fe_{65}Zr_{35}$ amorphous alloy, varying the average coordination number z_{Fe}^{*} around Fe atom. Their results are shown in Fig. 1.30 for $z_{Fe}^{*} = 8$ to 12. The total

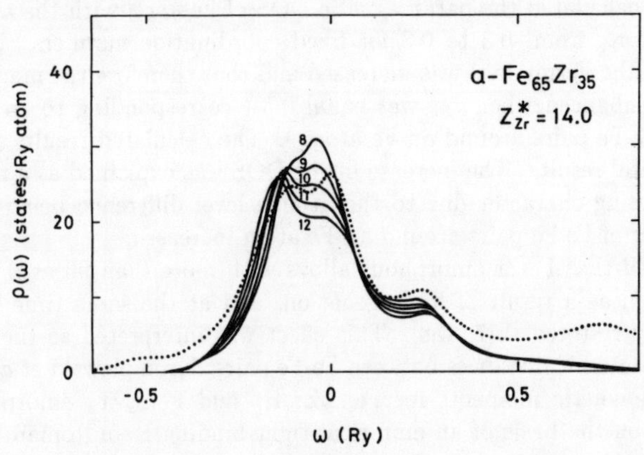

Fig. 1.30: Changes in DOS for amorphous $Fe_{65}Zr_{35}$ alloy when the coordination number around a Fe atom z_{Fe}^{*} is varied from 8 to 12 (solid line). The total DOS calculated by the tight-binding LMTO method combined with the recursion method is also shown (dotted line) [155]

DOS calculated by the tight-binding LMTO method combined with the recursion method is also shown in the figure. In each case, Fe d-states generally form a two-peak structure, with the higher energy peak located near the Fermi level. When z_{Fe}^{*} = 12, the higher energy peak is shoulder-like and lower than the lower energy peak. The development of ferromagnetism is not expected because the Stoner criterion is not satisfied. However, the amplitude of the higher energy peak relative to that of the lower energy peak increases as z_{Fe}^{*} decreases to 8, and the d-band width of Fe becomes narrower. The Stoner criterion is satisfied when z_{Fe}^{*} is less than or equal to 10, and ferromagnetism is therefore expected for these coordination numbers. These coordination numbers are consistent with those expected from the relaxed DRPHS model constructed in this calculation and experimental data for $Fe_{65}La_{35}$ amorphous alloy [59]. These facts indicate that the difference in the coordination number due to the difference in the atomic size is essential for the formation of ferromagnetism in amorphous Fe-Zr alloys. A similar relation between the coordination numbers and the formation of ferromagnetism is also observed in amorphous Co-Y alloys, as will be discussed in Sec. 4.1.4.

Ferromagnetism is also very sensitive to the atomic short-range order. For example, Krebs et $al.$ [156] reported that amorphous Fe_xZr_{1-x} alloys ($0.33 \leq x \leq 0.62$) prepared by sputtering technique were separated into two phases during annealing. In these concentrations, amorphous Fe-Zr alloys becomes more magnetic after the annealing process, although the structural changes are too small to be detected by conventional X-ray diffraction. They et $al.$ suggested that the magnetism was enhanced because the Fe-rich phase was formed during the annealing process. Kakehashi et $al.$ also calculated the paramagnetic DOS of $Fe_{50}Zr_{50}$ with the atomic-short range parameter τ_{Fe} from -0.3 to 0.3 for fixed coordination numbers. They found that the DOS at the Fermi level was increased and that therefore the magnetism was expected to be enhanced when τ_{Fe} was equal 0.3 (corresponding to an increase in the number of Fe-Fe pairs around an Fe atom). The calculated results support the above experimental results. The increase in the DOS was explained as a result of the bonding-antibonding character due to the atomic level difference being suppressed when the number of Fe-Fe pairs around an Fe atom increased.

It was reported that Fe-Zr amorphous alloys with more than 90 at.% Fe became soft in magnetism as a result of hydrogenation, and at the same time the Fe moments are largely restored [157, 158]. This effect was interpreted as the result of a dilatation of interatomic distances between Fe-Fe pairs. Krompiewski et $al.$ [158, 159] calculated the magnetic moments for $Fe_{1-x}Zr_xH_y$ and $Fe_{1-x}Zr_x$ amorphous alloys self-consistently on the basis of an empirical tight-binding Hamiltonian. They concluded that the volume expansion effect was mainly responsible for the enhancement of ferromagnetism due to hydrogenation in amorphous $Fe_{1-x}Zr_x$ ($x < 0.13$).

4.1.3. Fe-Ni amorphous alloys

These materials were recently found by Arai in a range of Fe concentrations $0.64 \leq x \leq 0.72$ [146]. In crystalline Fe-Ni alloys, the saturation magnetization or hyperfine field is well-known to show an abrupt decrease around 65 at.% Fe, where the Invar effects characterized by zero thermal expansion and other anomalies take place. The hyperfine field in amorphous Fe-Ni films was found to show no anomaly around the Invar composition range, in contrast to the crystalline case, and to increase along the so-called Slater-Pauling curve with increasing Fe concentration, as shown in the inset of Fig. 1.31. The strong ferromagnetism in amorphous Fe-Ni alloys was attributed to a band narrowing due to volume expansion caused by the random incorporation of Ar atoms.

Recently, Yu and Kakehashi performed finite temperature calculations in amorphous Fe-Ni alloys on the basis of the single-site theory combined with the geometrical-mean model (see Sec. 2.3.5) [160]. They found, as shown in Fig. 1.31, that the critical concentration of ferromagnetic instability shifts to the Fe side by about 35

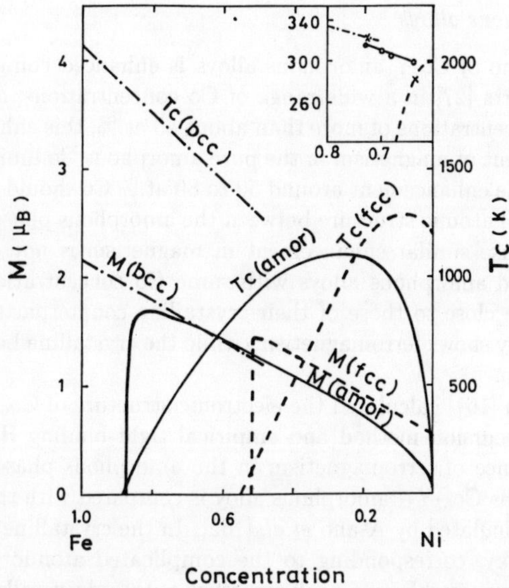

Fig. 1.31: Concentration dependence of the magnetic moment at 150 K and the Curie temperature calculated by a single-site theory for Fe-Ni alloys with amorphous (solid line), fcc (dashed line), and bcc (dashed dotted line) structures [160]. The inset is the corresponding hyperfine field in unit of kOe observed experimentally [146].

at.% Fe, so that strong ferromagnetism is realized around 70 at.% Fe.

According to their analysis, the volume expansion of 1.8% due to structural disorder shifts the critical concentration only by 6 at.%, thus it is not enough to explain the experimental value ($\Delta c^* > 15$ at.%). They found that the shift of the main peak to the lower energy in the nonmagnetic DOS, which was found in amorphous pure metals (See Fig. 1.8), remains even in amorphous alloys, so that amorphous Fe-Ni alloys around 70 at.% Fe have the Fermi level at the main peak, whereas their crystalline fcc counterparts have the Fermi level below the peak. The argument is parallel to that given in Sec. 3.1, and explains the strong ferromagnetism. In particular, it is noteworthy that the calculated $\Delta N = 0.702$ between the average electron numbers of critical concentrations for amorphous and fcc structures in alloys is close to that obtained from the rigid band model ($\Delta N^* = 0.696$, see Sec. 3.1). The shift of the Curie temperature maximum to the higher Fe concentration due to structural disorder was also explained by the same origin for magnetic energy gain.

4.1.4. Co-Y amorphous alloys

The ferromagnetism of Co-Y amorphous alloys is enhanced compared with their crystalline counterparts [27] in a wide range of Co concentrations, as shown in Fig. 1.25. At high Co concentrations of more than about 80 at.%, this enhancement might reflect the enhancement of magnetism in the pure amorphous Co limit as discussed in Sec. 3.3. However, the enhancement around 50 to 80 at.% Co should be attributed to the differences in local atomic structure between the amorphous phase and crystalline phase. This is because similar enhancement in magnetism is not observed in Co-based metal-metalloid amorphous alloys with same Co concentrations, whose local atomic structures are close to those of their crystalline counterparts. In particular, Co_2Y amorphous alloy shows ferromagnetism, while the crystalline Laves-phase Co_2Y shows paramagnetism.

Inoue and Shimizu [161] calculated the electronic structure of $Co_{67}Y_{33}$ amorphous alloy by using the recursion method and empirical tight-binding Hamiltonian, and clarified the appearance of ferromagnetism in the amorphous phase. In Fig. 1.32, the DOS of amorphous $Co_{67}Y_{33}$ amorphous alloy is compared with that of crystalline Laves phase YCo_2 calculated by Asano *et al.* [162]. In the crystalline phase, although the DOS is very spiky, corresponding to the complicated atomic structure of the Laves phase, they have roughly two peak structure with a deep valley. On the other hand, the projected DOS of Co in the amorphous phase has a peak near the Fermi level. It turns out that the DOS at the Fermi level in the amorphous structure is much higher than that of crystalline Co_2Y, so that ferromagnetism is realized in amorphous $Co_{67}Y_{33}$ alloy. The same feature was also found in amorphous Co-Zr alloys [163]. The difference in the DOS results from the difference in the local atomic structure. Inoue *et al.* analyzed the projected DOS's at several Co sites where the coordination number of Y atoms changes from 0 to 8. They found that the amplitude of the DOS at the Fermi level decreased when the coordination number of Y atoms increased. It satisfies the Stoner criterion when the Y coordination number is 4, which is close to the average Y coordination number in the amorphous phase. On the other hand, the DOS at the Fermi level does not satisfy the Stoner criterion when the Y coordination number is 6, which corresponds to that in the crystalline phase.

A more systematic analysis was recently performed by Kakehashi *et al* [155]. They calculated the DOS of amorphous Co_2Y alloy by varying the average coordination numbers around a Co atom z_{Co}^* between 8 and 12 on the basis of Bethe-type approximation which was introduced in Sec. 2.3.5. Fig. 1.33 shows the changes in the total DOS as a function of z_{Co}^*. The DOS at the Fermi level increases with decreasing z_{Co}^*, and satisfies the Stoner criterion between $z_{Co}^* = 10$ and 9. Note that the coordination number around a TM atom is 12 in the crystalline Laves phase, whereas it decreases to ~ 9.5 in the amorphous counterpart [59]. In a binary dense random packing structure, the average coordination number around the smaller atom should be less than 12 that is expected from the close-packed structure, because the larger atom occupies a larger solid angle around an atom than the smaller atom does. This difference in

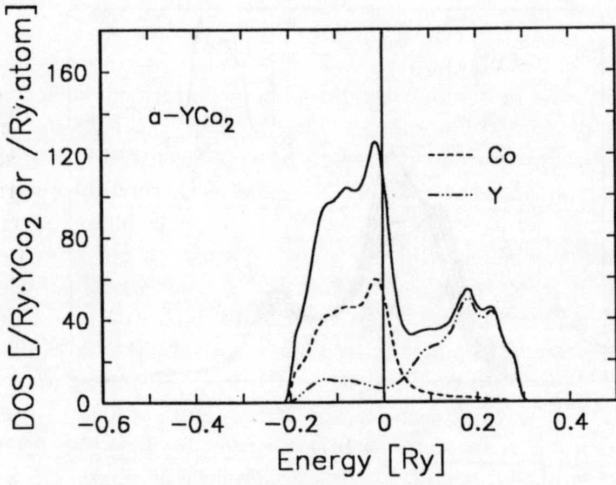

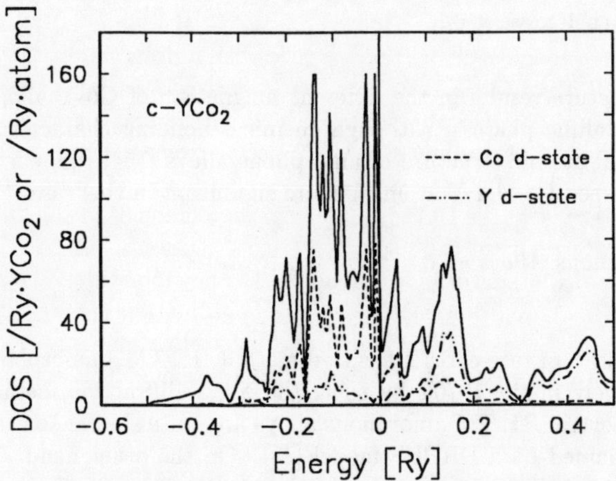

Fig. 1.32: Comparison of the DOS for amorphous $Co_{67}Y_{33}$ alloy [161] and crystalline Laves-phase Co_2Y [162]. The Fermi level is located near the highest peak of the DOS in the amorphous phase, whereas it is away from a peak of the DOS in the crystalline phase.

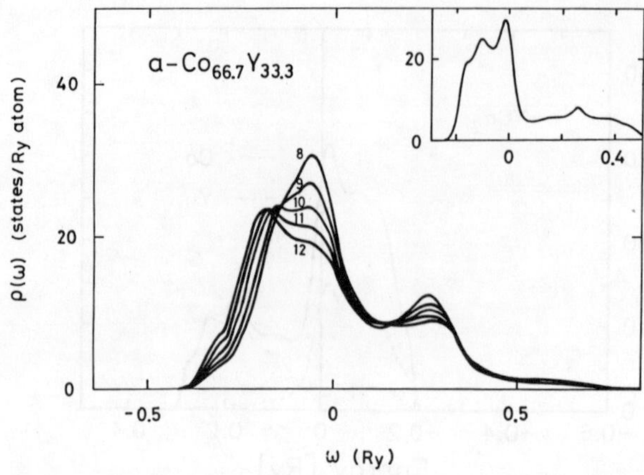

Fig. 1.33: Changes in DOS for amorphous Co$_{67}$Y$_{33}$ alloy varying the coordination number around a Co atom z^*_{Co} from 8 to 12 [155]. The DOS at the Fermi level is developed with an decrease of z^*_{Co}.

the local atomic structure results in the different magnetism of Co$_2$Y alloys in the amorphous and crystalline phases. Although chemical bonding characteristics are important in the local atomic structure of amorphous alloys [154], the effect of the difference in size between Co and Y atoms is more significant in this case.

4.2. RE-TM Amorphous Alloys

4.2.1. General aspects

The atomic structures of rare-earth transition metal (RE-TM) amorphous alloys were studied extensively in the 1970's by using the X-ray diffraction method. The obtained RDF's for several RE-TM amorphous alloys are quite similar to each other, and can be well explained by a DRPHS model [49]. On the other hand, there are many stoichiometric crystalline phases such as RE-TM, RE-TM$_2$, RE-TM$_3$, RE-TM$_4$ and RE-TM$_7$, which are varieties of the Laves-phase RE-TM$_2$ alloy. The local atomic structures of RE-TM amorphous alloys are different from those of their crystalline counterparts. For example, the coordination numbers and nearest neighbor distances in a Gd$_{33}$Fe$_{67}$ amorphous alloy are compared with those in the crystalline Laves phase in Table 1.1 [49]. Although the Fe-Fe and Fe-Gd atomic distances in the amorphous phase are close to those in the crystalline phase, the Gd-Gd distance in the amorphous phase is a little larger than in the crystalline phase. Furthermore, the coordination number of the Fe-Gd pair is much larger in the crystalline phase than

Table 1.1: Comparison of the nearest neighbor distances (r_{ij}) and coordination numbers (N_{ij}) in an amorphous $Gd_{33}Fe_{67}$ alloy and its crystalline counterpart [49]

	amorphous		crystal	
	$r_{ij}(\text{Å})$	N_{ij}	$r_{ij}(\text{Å})$	N_{ij}
Gd-Gd	3.47	6±1	3.20	4
Gd-Fe	3.04	6.5±0.6	3.06	12
Fe-Fe	2.54	6.2±0.5	2.61	6

in the amorphous phase. These differences in the local atomic structure cause the differences in electronic structures and magnetic properties between the amorphous and crystalline phases.

The valence electron configurations and atomic radii of RE atoms are quite similar to those of a Y atom except that RE atoms have open-shell f-electrons. The magnetic properties which can be explained by itinerant ferromagnetism are similar to those of the corresponding amorphous Y-TM alloys (see Sec. 4.1.4). The magnetic moment of an RE atom is mostly built up by f-electrons as $L + 2S$ (μ_B). Here, L and S denote the orbital and spin angular momenta of f-electrons, respectively. The f-electron configuration of a RE atom in a bulk is almost same as that in an isolated atom. It can be well explained by the Hund's rule, because of the localized nature of the f-electrons, except for Ce which shows the valence fluctuation. There exists a negative ($i.e.$ antiferromagnetic) exchange coupling, J_{RT}, between RE and TM spins [164, 165], which originates in the bonding characteristic between RE 5d- and TM 3d-states, as will be discussed in the next section. The negative J_{RT} leads to magnetic couplings between TM and RE magnetic moments in a different way for light and heavy RE-TM alloys [166, 167]. Therefore, in order to discuss the magnetic properties of RE-TM amorphous alloys, it is convenient to classify them into the following three categories according to their f-electron numbers:

(1) Light RE-TM amorphous alloys
Light RE atoms with less than half-filled f-electrons couple ferromagnetically with TM (TM = Fe, Co, and Ni) atoms. This is because positive spin-orbit coupling requires $J = L - S$ for RE atoms, while the orbital moment exceeds the spin moment according to Hund's rule. As a result, the negative J_{RT} tends to align the RE moments parallel to the TM moments. The directions of local RE moments are randomly canted to some extent from the direction parallel to the TM moments, because of random single-ion anisotropy of RE elements. In other words, these materials show non-collinear ferromagnetism.

(2) Heavy RE-TM amorphous alloys

58

The magnetic moments of heavy RE atoms with more than half filled f-electrons couple anti-ferromagnetically with those of TM (TM = Fe, Co, and Ni) atoms, because $J = L + S$ for RE atoms according to the negative spin-orbit coupling, and the negative J_{RT} tends to align RE moments anti-parallel to TM moments. The directions of RE moments are randomly canted from the direction anti-parallel to the TM moments for the same reason as in light RE alloys. Thus, these materials show non-collinear ferrimagnetism.

(3) Gd-TM amorphous alloys

Gd atoms couple anti-ferromagnetically with TM (TM = Fe, Co, and Ni) atoms, and their magnetic moments are anti-parallel to those of TM atoms. In other words, these materials show ferrimagnetism with collinear magnetic moment alignments.

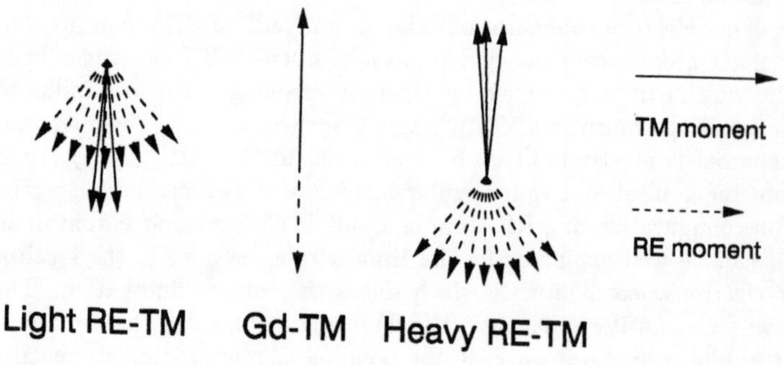

Fig. 1.34: Three types of spin configuration in amorphous RE-TM alloys: light RE-TM alloys (left), Gd-TM alloys (middle), and heavy RE-TM alloys (right)

Fig. 1.34 shows schematic magnetic moment configurations for these materials. Although these three types of RE-TM amorphous alloys exhibit different magnetic moment configurations, Co-base alloys and Fe-base alloys have some common features. Tao et al. [168] found that a Co atom in $Gd_{33}Co_{67}$ amorphous alloy carried a higher average saturation magnetic moment, and that the magnetic ordering temperature was higher in the amorphous alloy than in its crystalline counterpart. This fact was confirmed for a wide range of atomic concentrations [169]. Although it is difficult to measure the magnetic ordering temperature at high Co concentrations, because it exceeds the crystallization temperature, the enhancement of magnetic ordering temperature was also observed in Tb-Co, Dy-Co, and Ho-Co amorphous alloys [10]. The enhancement of magnetism in a Co-base RE-TM amorphous alloy seems to result from the differences in the local atomic structure between the amorphous phase and crystalline phase. The situation is similar to that for the amorphous Co-Y alloys previously discussed.

In Fe-base RE-TM amorphous alloys such as Gd-Fe, Tb-Fe, Ho-Fe, Dy-Fe [10], and Nd-Fe [171], it was observed that the magnetic ordering temperature T_C starts to decrease with increasing Fe concentration at around 60 to 80 at.%, whereas in Co-base RE-TM amorphous alloys T_C keeps increasing with increasing TM concentrations. This suppression of magnetism in Fe-base alloys results from an appearance of anti-ferromagnetic coupling between some Fe-Fe pairs due to the local structural disorder (see Sec. 3.2). Fig. 1.35 shows a summary of the magnetic ordering temperatures of Gd-TM (TM = Fe and Co) amorphous alloys compared with those of their crystalline counterparts [170].

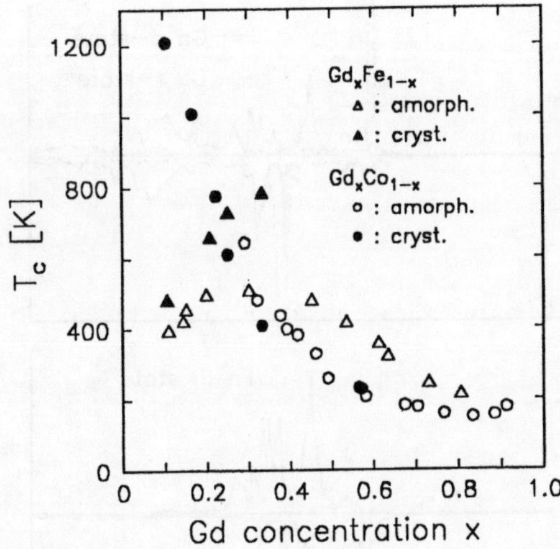

Fig. 1.35: Changes in the magnetic ordering temperatures of amorphous Gd-Fe and Gd-Co alloys as a function of Gd concentration [170]

4.2.2. Gd-Fe alloys

In Gd-TM alloys with TM = Fe, Co, or Ni, the Gd moments couple antiferro-magnetically with the TM moments because of negative exchange coupling. The negative exchange coupling originates in a hybridization between the Gd d-states and TM d-states. The calculated total DOS of the crystalline GdFe$_2$ Laves phase and its projected DOS are shown in Fig. 1.36 [172]. The projected DOS's of Gd$_{33}$Fe$_{67}$ amor-phous alloy are shown in Fig. 1.37 [172]. The DOS's in crystalline and amorphous phases have some common features. The center of Gd d-states is located above the

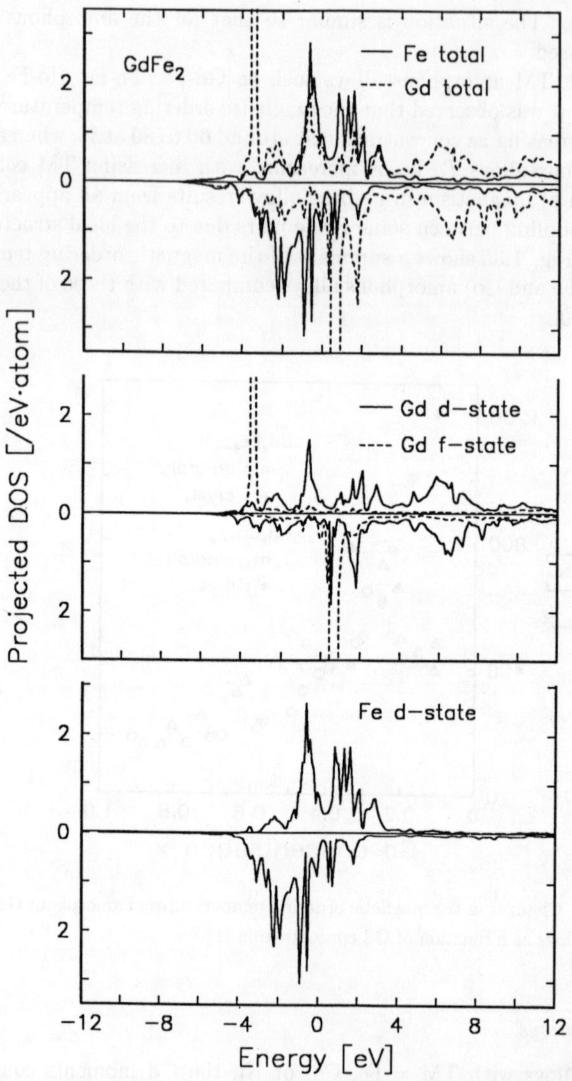

Fig. 1.36: Total and projected DOS's of crystalline Laves phase GdFe$_2$ alloy [172]

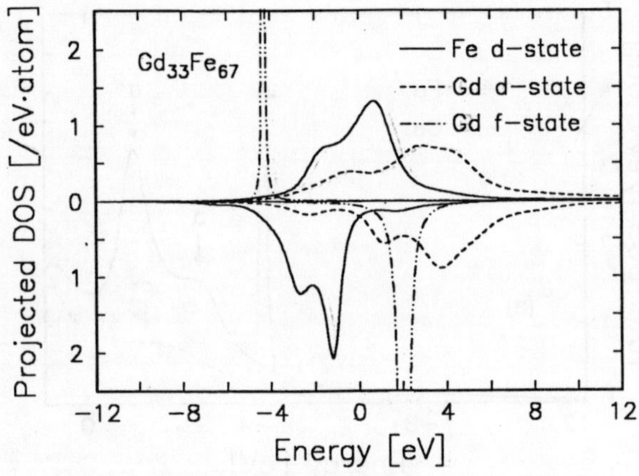

Fig. 1.37: Projected DOS's of amorphous Gd$_{33}$Fe$_{67}$ alloy [172]

Fermi level, while that of Fe d-states is located below the Fermi level. The tails of Gd d-states hybridize strongly with the Fe d-states, and spread below the Fermi level. Since the energy level of the Fe d-states is closer to the Gd d-states in the up-spin band (the minority spin of Fe) than in the down-spin band, the Gd d-states hybridize with the Fe d-states more strongly in the up-spin band than in the down-spin band ($J_{RT} < 0$). Consequently, the DOS of the Gd d-states is accumulated below the Fermi level in the up-spin band more than in the down-spin band. This fact polarizes the Gd d-states, and makes Gd atoms couple with Fe atoms antiferromagnetically *via* the intraatomic d-f exchange interaction. The situation is similar to that for Fe-Zr alloys where Fe atoms also couple anti-ferromagnetically with Zr atoms. Due to the ferrimagnetism in Gd-Fe alloys, the DOS of Fe d-states in the majority spin part is very different from that in the minority spin part. A simple rigid-band splitting picture is no longer applicable to these materials.

The calculated magnetic moments in the amorphous phase are 2.0 μ_B for the Fe site and 7.2 μ_B for the Gd site. The averaged magnetic moment per atom is calculated to be 1.1 μ_B, which agrees well with experimental data [173]. Note that the magnetic moment for the Gd site is larger than 7.0 μ_B, because not only Gd f-states but also Gd d-states are also polarized as discussed above.

The binding energy of occupied f-states obtained from LSDF theory is 3.6 eV in the crystalline phase, and less than 4.3 eV in the amorphous phase. These calculated values are, however, much smaller than the value of 9.4 eV [174] observed in Gd-Fe amorphous alloys by using X-ray photoemission spectroscopy (XPS). This discrepancy seems to result from the many-body correlation effect, especially of f-electrons. Jaswal et al. [175] suggested that the position of the occupied f-states of XPS data

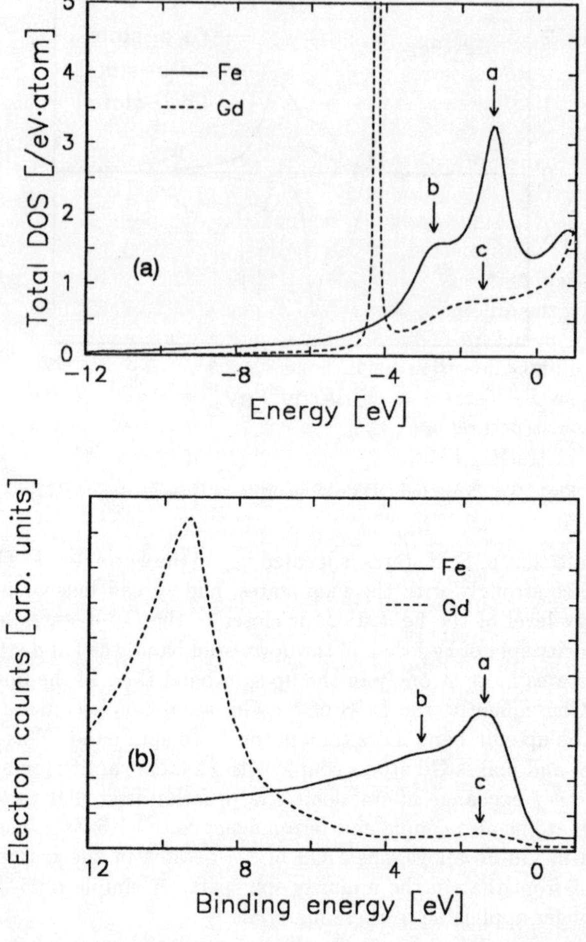

Fig. 1.38: (a) Projected DOS's of a Gd atom (dashed line) and an Fe atom (solid line) obtained from the electronic structure calculation for an amorphous $Gd_{33}Fe_{67}$ alloy [172], and (b) the decomposed XPS spectrum of the Gd part and the Fe part [174]

could be estimated on the basis of Slater transition-state analysis as a single-particle excitation (note that the DOS in the LSDF band calculation does not generally give single-particle excitation spectra).

The projected DOS for Fe d-states in the crystalline Laves phase is quite similar to that of Co d-states in the Co_2Y Laves phase [162] (see Fig. 1.32). Both have a roughly two-peak structure with a deep valley between these two peaks. On the other hand, the projected DOS's for d-states in the amorphous phase have a large peak, which is followed by a small shoulder-like peak on the lower energy side, as shown in Fig. 1.37. Thus the projected DOS of Fe d-states in the amorphous phase is very different from that in its crystalline counterpart. This difference in the projected DOS results from the difference in the local atomic structure, as in the case of Co-Y. The local atomic structure of the $Gd_{33}Fe_{67}$ amorphous alloy is very different from that of its crystalline counterparts, as shown in Table 1.1. The general shape of the projected DOS for Fe d-states in the amorphous phase resembles that in a closed packed crystalline structure, such as fcc or hcp, rather than that in a crystalline bcc or Laves structure. In RE-TM amorphous alloys, it has been suggested that most of the TM atoms tend to form a closely packed tetrahedral structure locally [49].

Güntherodt et $al.$ [174] systematically measured the XPS spectra of $Gd_{1-x}Fe_x$ amorphous alloys, and decomposed each spectrum into a Gd part and an Fe part. The decomposed XPS spectra of a Gd atom and an Fe atom are compared with the projected DOS's of the Gd site and the Fe site obtained from the calculations in Fig. 1.38 (a) and (b), respectively [172]. These two spectra correspond closely to each other except as regards position of occupied f-states. In particular, the projected DOS per Fe atom has a peak around 1 eV (denoted by 'a' in the figure) followed by a shoulder around 3 eV (denoted by 'b' in the figure) below the Fermi level in both the calculated and optically observed DOS's. There is a broad peak corresponding to the tail of Gd d-states (denoted by 'c' in the figure) in both the calculated and optically observed DOS's.

4.2.3. Gd-Co alloys

In Gd-Co amorphous alloys, the magnetic properties such as the magnetic ordering temperature and magnetic moments per Co atom are enhanced in a wide range of Co atomic concentrations than in their crystalline counterparts [168, 169]. It is expected that the enhancement of the magnetism in Gd-Co amorphous alloys stems from the difference in local atomic structure between the amorphous and crystalline phases, as in the case of Co-Y amorphous alloys (see Sec. 4.1.4). This is because the electronic and atomic structures of Gd-Co amorphous alloys are quite similar to those of Co-Y amorphous alloys, except that Gd atoms have f-electrons.

The calculated DOS's of $Gd_{33}Co_{67}$ and $Gd_{18}Co_{82}$ amorphous alloys are shown in Figs. 1.39 (a) and (b), respectively [176]. They have some features in common with the DOS of Gd-Fe amorphous alloy. The tails of Gd d-states hybridize strongly with

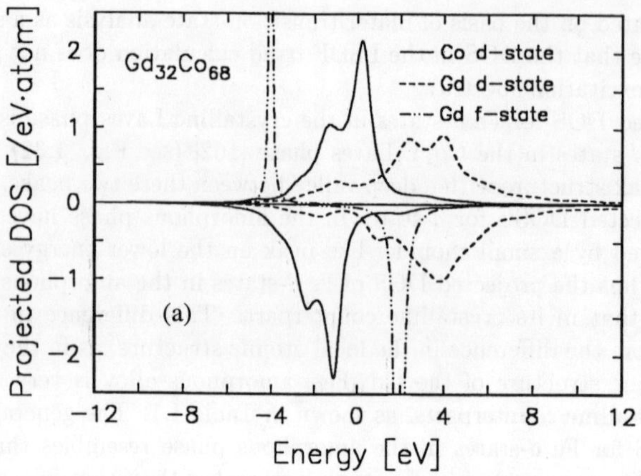

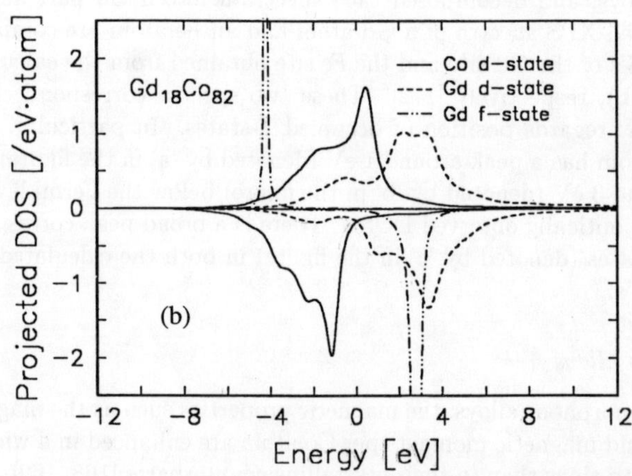

Fig. 1.39: Projected DOS's of amorphous $Gd_{32}Co_{67}$ (a) and $Gd_{18}Co_{82}$ (b) alloys [176]

the Co d-states, and spread below the Fermi level. The Gd atom couples ferrimagnetically with the Co atom because of the difference in the strength of the hybridization in the up-spin and down-spin band, as in the case of the Gd-Fe amorphous alloy. As a result, Gd d-states as well as f-states are polarized. The binding energy of occupied f-states is about 4 eV, which is much lower than the experimental value of 8.0 eV [175] obtained by using XPS measurement. Except for the position of f-states, our calculated DOS agrees well with the optically observed DOS by using ultraviolet photoemission spectroscopy (UPS) [175]. In particular, Co d-states have a sharp peak at 1 eV below the Fermi level in both the calculated DOS and UPS data.

Let us focus on the changes in the projected DOS's of Co d-states with increasing Gd concentrations. Fig. 1.40 shows the projected DOS's of Co d-states in amorphous $Gd_{32}Co_{68}$ (top), in $Gd_{18}Co_{82}$ (middle), and in amorphous pure Co (bottom) [176]. The band widths of Co d-states decrease with increasing Gd concentrations, because the coordination numbers of Co-Co pair decrease. The exchange splitting of Co d-states also decreases with increasing Gd concentration. Corresponding to this change, the highest peak of the minority spin Co d-state shifts toward the low energy side with an increase in the Gd concentration, while that of the majority spin stays about 1 eV below the Fermi level. This means that the highest peak of Co in the *paramagnetic* DOS shifts toward the low energy side and falls below the Fermi level with increasing Gd concentration. Therefore, the decrease in the *paramagnetic* DOS at the Fermi level due to the peak shift seems to result in the decrease of exchange splitting of Co d-states.

When the TM moment of Gd-TM alloys is evaluated from the experimental value for the saturation magnetization, the Gd moment is frequently assumed to be 7.0 μ_B. This assumption neglects any polarization of conduction electrons. However, the calculated magnetic moments per Gd atom are 7.16 μ_B for $Gd_{18}Co_{82}$ and 7.14 μ_B for $Gd_{32}Co_{68}$ amorphous alloys. These values are larger than 7.0 μ_B, because of mainly the polarization of Gd d-states. Fukamichi *et al.* [177] estimated the magnetic moment of Gd in Gd-rich Gd-Co amorphous alloys on the basis of the bulk magnetization and pressure derivative of the Curie temperature data. They found that the Gd moment was 7.49 μ_B – higher than 7.0 μ_B – and that not only f-electrons but other electrons were also polarized.

In Fig. 1.41, the calculated magnetic moments per Co atom for pure Co, $Gd_{18}Co_{82}$, and $Gd_{32}Co_{68}$ amorphous alloys are compared with experimental data. The calculated Co moments are a little higher than experimental ones at high Gd concentrations, because the Gd moment is assumed to be 7.0 μ_B in the experimental data. The magnetic moment per Co atom decreases with an increase in the Gd concentration. This reduction in the magnetic moment was been explained as a result of charge transfer from Gd atoms to minority Co d-states, and its band filling effect [168]. However, the numbers of charge transfers from a Gd atom to a Co atom are 0.11 electrons in $Gd_{18}Co_{82}$ and 0.12 electrons in $Gd_{32}Co_{68}$, respectively. These numbers are too small to explain the reduction in the magnetic moment per Co atom. Our

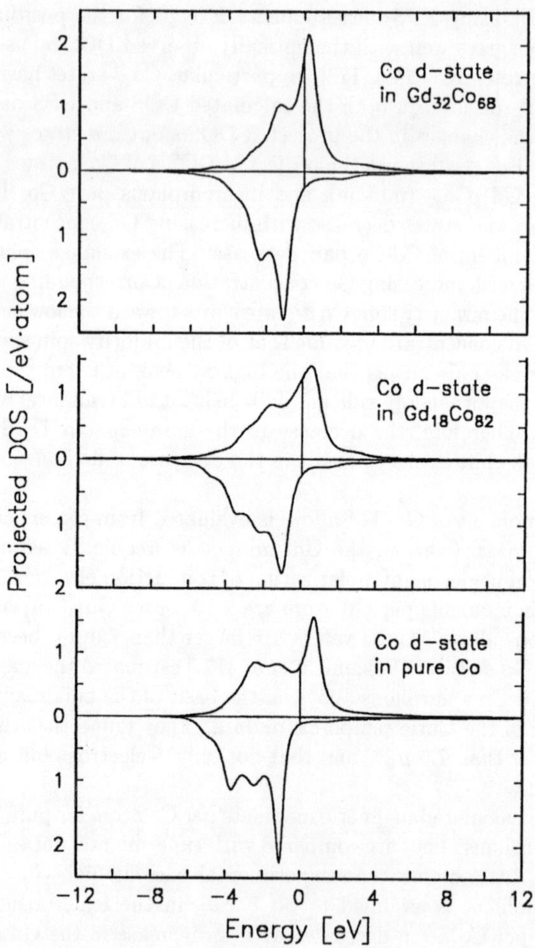

Fig. 1.40: Changes in projected DOS's of Co *d*-states for amorphous Gd$_{33}$Co$_{67}$ (top) and Gd$_{18}$Co$_{82}$ (middle) alloys [176], and for amorphous pure Co (bottom) [132]

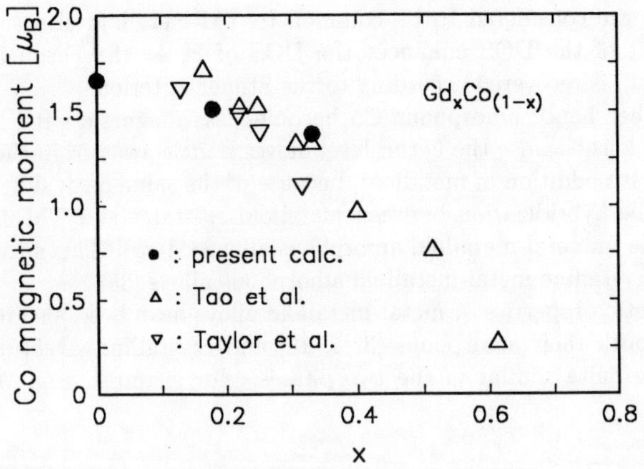

Fig. 1.41: Comparison of the value of the magnetic moment per Co atom obtained by electronic structure calculations (filled circles) [176] and by experiments (open triangles) [168, 169]

results suggest that the this reduction in the magnetic moment is caused by the reduction in the exchange splitting of the Co d-states, as mentioned above, rather than by the charge transfer.

4.3. TM-Metalloid Alloys

4.3.1. General aspects

A pure amorphous Fe is expected not to be ferromagnetic, as mentioned before, because the Fermi level falls on low energy side of the main peak in the DOS. The ferromagnetism is however rapidly restored by adding only a few (less than 17 at.%) metalloid atoms as in the case of early transition metal-transition metal amorphous alloys. Fujiwara [34] calculated the electronic structure of Fe$_{80}$B$_{20}$ amorphous alloy in the *paramagnetic* state, and explained this recovery of ferromagnetism as an effect of the addition of B atoms to the electronic structures. His explanation can be summarized as follows:

(1) B sp-states hybridize strongly with Fe d-states. This hybridization splits B sp-states into bonding and anti-bonding states, and shifts the main peak of DOS for Fe d-states toward the low energy side. The Fermi level becomes close to the main peak position of the DOS.

(2) Fe-Fe distances increase, and the coordination numbers of Fe-Fe pairs decrease with addition of B atoms. As a result, the d-bandwidth of Fe becomes narrower.

These effects are considered to be common to TM-metalloid alloys. In particular, the peak shift of the DOS enhances the DOS of Fe at the Fermi level, and thus ferromagnetism is recovered according to the Stoner criterion.

On the other hand, amorphous Co becomes less magnetic with the addition of metalloid. This is because the Fermi level moves a little away from the peak position of the DOS with addition of metalloid, because of the same peak shift as in the Fe-B case. Thus the hybridization between metalloid sp-states and TM d-states plays a significant role in metal-metalloid amorphous alloy systems. This situation is almost same as for crystalline metal-metalloid amorphous alloys [35].

The magnetic properties of metal-metalloid alloys have been also investigated extensively in both their amorphous [3, 5, 10] and crystalline structures [178]. The properties are quite similar in the two phases. For example, Fig. 1.42 shows the

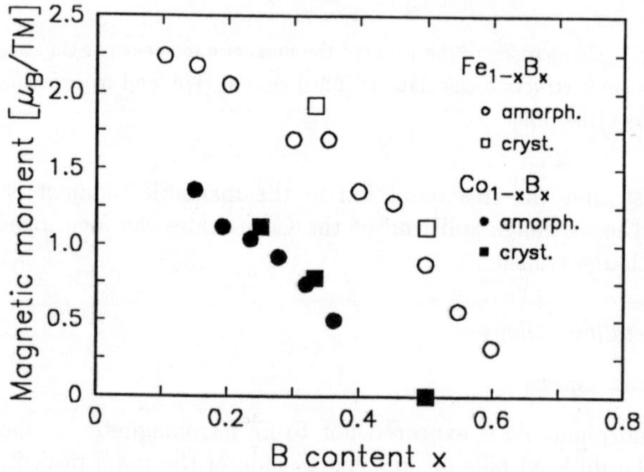

Fig. 1.42: Changes in the saturation magnetization per TM atom in amorphous and crystalline Fe-B and Co-B alloys as a function of metalloid concentration [10, 180]

changes in the magnetic moments per TM atom as a function of metalloid concentration in Fe-B and Co-B alloys [10, 147]. The data for the crystalline phases lie almost on the curve for the amorphous phases.

The similarity of the magnetism in the amorphous and crystalline counterparts is considered to originate in the similarity of the local atomic structure. In fact, it has been found that there is no direct contact between P atoms in a Co-P amorphous alloy [181]. The same results have also been obtained for other metal-metalloid amorphous alloys [182, 183]. It has been concluded that direct contact between metalloid atoms is prohibited from 0 at.% to about 20 at.% metalloids in metal-metalloid amorphous alloys [184]. Fujiwara *et al.* analyzed the local atomic structure of these materials by using a relaxed DRPHS model [185]. They showed that metalloid atoms were captured in the Bernal holes (see Sec. 2.1.1) formed by metal atoms. This local atomic structure is quite similar to that of trigonal-prismatic crystalline Fe_3P or Fe_2P.

Mizoguchi et al. [11, 12] systematically measured the magnetic properties of pseudo-binary transition metal amorphous alloys in the form $(TM_x^1 TM_{1-x}^2)_{80}B_{10}P_{10}$. They found that there was a relation similar to the Slater-Pauling curve between the average magnetic moment for a TM atom and the average d-electron number even, in pseudo-binary amorphous TM alloys. Fig. 1.43 shows saturation magnetic moments

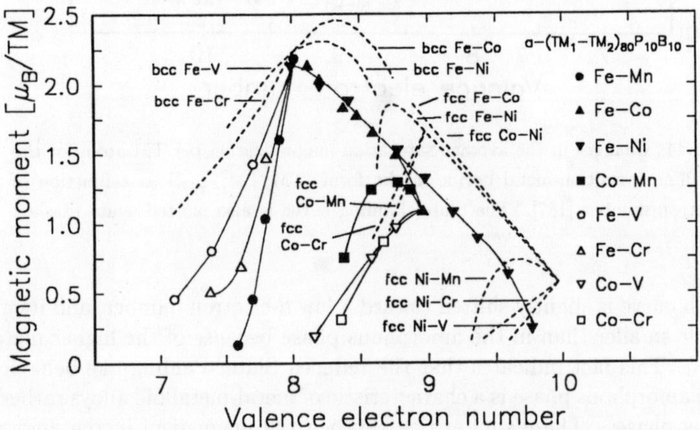

Fig. 1.43: Changes in the average saturation magnetization per TM atom for the pseudo-binary transition metal amorphous alloys as a function of d-electron number [12]. The well-known Slater-Pauling curve is also plotted (dashed line).

as a function of d-electron numbers. The well-known Slater-Pauling curves are also plotted in the figure. It is clearly shown that each curve for amorphous alloys can be almost explained simply by shifting the corresponding curve for crystalline alloys by 0.5 electrons to lower d-electron numbers so that the saturation magnetizations are reduced. This shift of 0.5 electrons was first explained as a result of a charge transfer from metalloid atoms to transition metal d-states, and its band filling effect [186]. Recent band calculation, however, has revealed that the effect of adding metalloid atoms is more complicated as we explained above [34]. We will argue this issue in detail in Sec. 4.3.3.

The Slater-Pauling like behavior of the magnetic moment has been also observed in crystalline $(TM_x^1 TM_{1-x}^2)_2B$ [187] alloys. Fig. 1.44 shows the changes in the magnetic moments of $(TM_x^1 TM_{1-x}^2)_2B$ alloys as a function of averaged d-electron numbers [187]. The Slater-Pauling curve is plotted again. The figure is quite similar to Fig. 1.43,

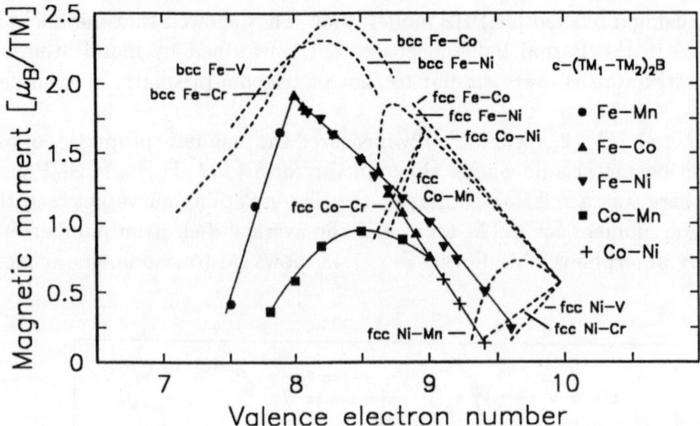

Fig. 1.44: Changes in the average saturation magnetization per TM atom for the crystalline transition-metal boride in the form $(\mathrm{TM}_x^1 \mathrm{TM}_{1-x}^2)_2\mathrm{B}$ as a function of d-electron number [187]. The Slater-Pauling curve is also plotted again (dashed line).

although each curve is slightly shifted toward a low d-electron number, and its amplitude is a little smaller than in the amorphous phase because of the higher metalloid concentration. This fact indicates that the reduced Slater-Pauling like behavior observed in the amorphous phase is a characteristic of metal-metalloid alloys rather than of amorphous phases. These similarities between the magnetism in the amorphous phase and that in the crystalline phase stem from the similarity in the local atomic structure of each phase.

4.3.2. Fe-B alloys

There have been several electronic structure calculations in the ground state for amorphous Fe-B alloys [34, 188, 189, 190, 35]. In Fig. 1.45, the calculated DOS of amorphous $\mathrm{Fe}_{80}\mathrm{B}_{20}$ alloy (a) is compared with that of crystalline $\mathrm{Fe}_2\mathrm{B}$ (b) [5]. The general features of their DOS's are almost same, except for the fine structures observed in the crystalline phase. This is because the local atomic structure of amorphous $\mathrm{Fe}_{80}\mathrm{B}_{20}$ is quite similar to that of crystalline $\mathrm{Fe}_2\mathrm{B}$ as discussed in the previous section. The calculated average magnetic moment per Fe atom is 2.1 μ_B in the amorphous phase, and agrees well with the experimental values [15, 179].

Some information about the electronic structure is provided by the low temperature specific heat, of which temperature dependency can be described by

$$C = \gamma_e T + \zeta T^{3/2} + \gamma_l T^3. \tag{1.112}$$

Here the second term is due to spin-wave excitation, and appears for **ferromagnetic** materials. The third term is the well-known lattice contribution. The first term is related to the DOS at the Fermi level through the following equation:

$$\gamma_e = \frac{1}{3}\left[1 + \lambda + \left(\frac{m*}{m} - 1\right)\right]\pi^2 k_{\rm B}^2 \rho(E_f), \qquad (1.113)$$

where λ, m^*/m, and $\rho(E_f)$ denote electron-phonon coupling, the mass enhancement factor due to electron correlations, and the DOS at the Fermi level, respectively.

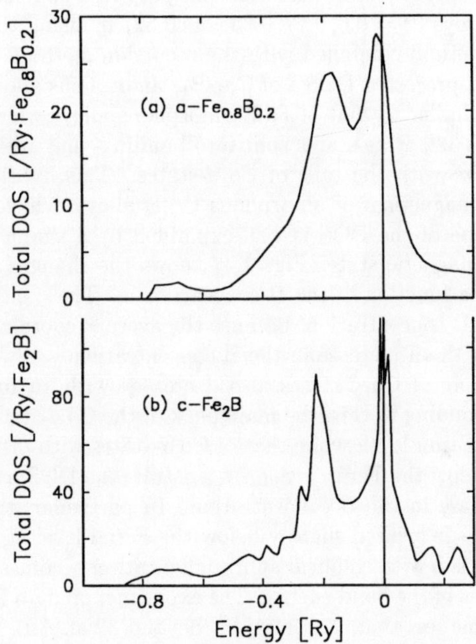

Fig. 1.45: Comparison of the total DOS for amorphous Fe$_{80}$B$_{20}$ alloy with that for crystalline Fe$_2$B alloy [35]

The T-linear specific heat of amorphous Fe-B alloys decreases with increasing B concentration [191]. Their calculated DOS's at the Fermi level also decrease with an increase in the B concentration [188], and are qualitatively consistent with experimental data. Fujiwara [188] pointed out, however, that the rate of decrease was much more rapid in the experimental data, and attributed the high specific heat at low B concentrations to magnetic scattering, that is, the mass enhancement due to spin

72

fluctuations. The role of structural disorder in the mass enhancement, however, has not yet been clarified.

4.3.3. Co-B alloys

Amorphous pure Co is expected to show the strong ferromagnetism with an enhanced Curie temperature of 1850 K, as explained in Sec. 3.3. When B is added, experimental data indicate that $Co_{1-x}B_x$ amorphous alloys show ferromagnetism at least in the range $0 \leq x \leq 0.3$ [192]. The system is, therefore, a good example for the study of ferromagnetism in amorphous alloys.

Recently, Tanaka *et al.* [193] calculated the spin-polarized electronic structures of amorphous $Co_{1-x}B_x$ alloys ($x = 0.0, 0.17, 0.23,$ and 0.32) by using the self-consistent tight-binding LMTO method combined with the recursion method. Fig. 1.46 shows the calculated total and projected DOS's of $Co_{83}B_{17}$ amorphous alloy. On the whole, the bonding nature is similar to that of Fe-B amorphous alloys which was discussed in Sec. 4.3.1 and Sec. 4.3.2. B *sp*-states split into bonding- and anti-bonding states, and B *p*-states hybridize with the tails of Co *d*-states. This hybridization plays a significant role in the magnetism of amorphous Co-B alloys, which we will discuss later. The general shape of the DOS is well explained by a simple splitting of the rigid band in the paramagnetic state. Fig. 1.47 shows the changes in the projected DOS of Co *d*-states as a function of the B concentration. The bandwidth decreases with an increase in the B concentration, because the average coordination number of Co-Co pair decreases with an increase in the B concentration.

The exchange splitting of Co *d*-states also decreases with an increase in the B concentration. Corresponding to this, the main peak of the Co *d*-states with minority spin shifts close to the Fermi level, while that of Co *d*-states with majority spin stays almost about -1 eV below the Fermi level. As a result, the DOS at the Fermi level increases with an increase in the B concentration. In particular, the main peak of the minority Co *d*-states is located slightly below the Fermi level at 32 at.% B. This means that ferromagnetism with collinear spin configuration becomes unstable at this B concentration in terms of the band energy. The experimental data for the spin-wave stiffness constant become less than 200 [meVÅ^2] beyond 32 at.% B, suggesting weak ferromagnetism [180].

In Fig. 1.48, the calculated values of magnetic moments per Co atom in several atomic concentrations are compared with experimental values [193]. The calculated values decrease with an increase in the B concentration, and agree well with the experimental ones. The theoretical values for high B concentrations are a little higher than the experimental ones. This seems to be related to the weakening of the ferromagnetism with the addition of B atoms.

The decrease in the magnetic moment was explained as resulting from the interatomic charge transfer from metalloid atoms to the minority TM *d*-state, and its band-filling effect [186]. On the other hand, Alben, Budnick, and Cargill suggested

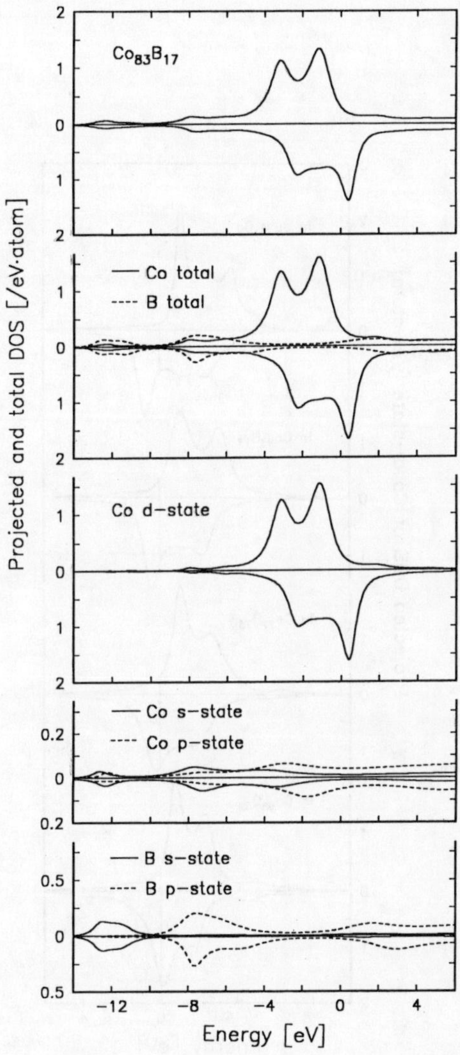

Fig. 1.46: Calculated total and projected DOS's of amorphous $Co_{83}B_{17}$ alloy [193]

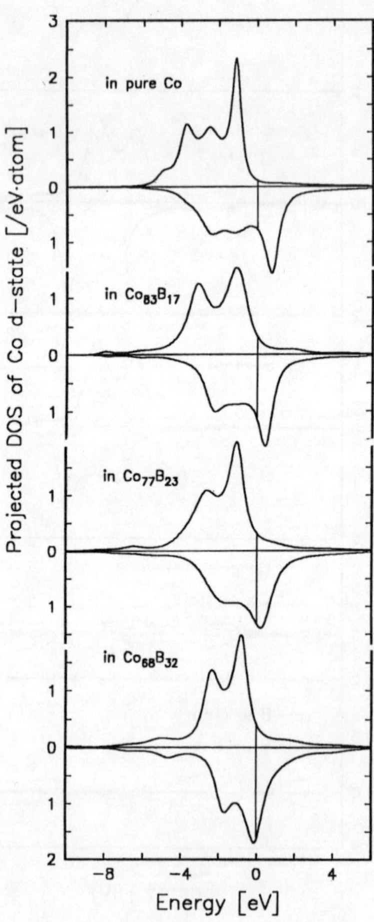

Fig. 1.47: Changes in the projected DOS of Co d-states for amorphous $Co_{1-x}B_x$ alloys with $x = 0.0$, 0.17, 0.23, and 0.32 [193]

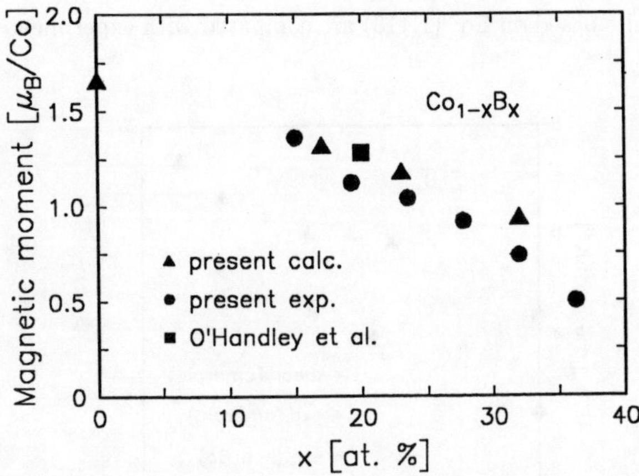

Fig. 1.48: Magnetic moment per Co atom for amorphous Co-B alloys obtained from electronic structure calculations (filled triangles) are compared with experimental ones (filled circles and square) [193].

that the decrease in the magnetic moment originates in the loss of d-character because of the hybridization with metalloid p-states [194]. The results from the above electronic structure calculations supports the latter explanation. In fact, there is an interatomic charge transfer from B atoms to a Co atom, which increases with increasing B concentration. However, the internal charge transfer from the majority Co d-band to the minority d-band decreases the magnetic moment per Co atom more significantly than the interatomic charge transfer. The internal charge transfer originates in the decrease of exchange splitting of Co d-states. The magnetic moment per Co atom is almost proportional to the exchange splitting of Co d-states [193], and is well explained by the generalized Stoner model [36].

Tanaka *et al.* calculated *paramagnetic* DOS's of amorphous Co$_{1-x}$B$_x$ ($x = 0$, 0.17, 0.23, and 0.32) alloys to clarify the effect of adding B atoms on the magnetism. The DOS at the Fermi level decreases with increasing B concentration for $x = 0.17$ and 0.23, because the hybridization between the Co d-states and B p-states suppresses the amplitude of the highest peak around the Fermi level. On the other hand, for $x = 0.32$, the DOS at the Fermi level decreases, because the highest peak shifts toward the lower energy side and away from the Fermi level on account of the hybridization. Both effect decrease the magnetization, as expected from the Stoner model. Thus the hybridization between the Co d-states and B p-states plays a significant role in the above-mentioned decrease in magnetization and exchange splitting.

The T-linear specific heat γ_e of amorphous Co-B alloys increases with an increase in the B concentration in contrast to the case for amorphous Fe-B alloys. In Fig. 1.49, theoretical results based on Eq. (1.113) are compared with experimental data. In the

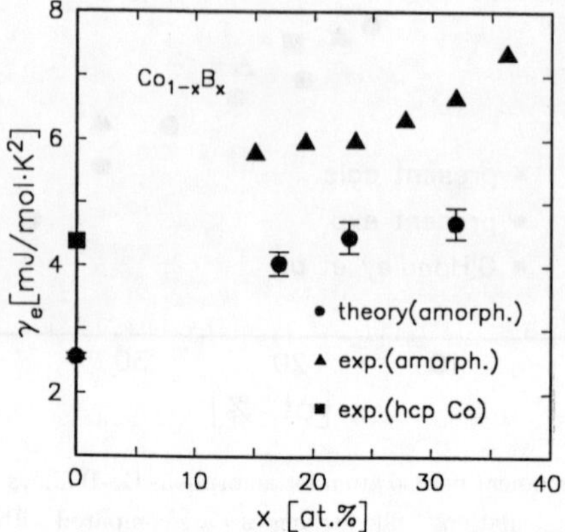

Fig. 1.49: Comparison of temperature linearly dependent specific heat coefficient γ_e obtained from the electronic structure calculations and experimental ones [193]. In the calculation, the electron-phonon enhancement is taken into account. The experimental value for hcp Co is also included for comparison.

calculations, electron-phonon coupling parameters λ were estimated by employing a rigid muffin-tin approximation (RMTA), which had been developed by Gaspari and Gyorffy [195] and parameterized by Skriver [196] within the atomic sphere approximation (ASA). Error bars in the calculated γ_e correspond to uncertainty of the Debye temperatures obtained experimentally. Both the measured and calculated γ_e increase similarly with increasing B concentration. The increase in the γ_e is due to the increase in the *spin-polarized* DOS at the Fermi level. The DOS at the Fermi level increases with an increase in the B concentration because of the band narrowing and the decrease in exchange splitting, as mentioned above. The experimental γ_e however keeps increasing even beyond 30 at.%, whereas the calculated γ_e seems to reach saturation around this B concentrations. It is expected that the enhancement of the experimental γ_e beyond 30 at.% B originates in the mass enhancement due to spin fluctuations. This is because the ferromagnetism is expected to be weakened in these B concentrations as discussed above.

References

[1] H.O. Hooper and A.M. de Graaf, *Amorphous Magnetism* (Plenum, New York, 1973).

[2] R.A. Levy and R. Hasegawa (ed.), *Amorphous Magnetism II* (Plenum, New York, 1977).

[3] Luborsky, *Ferromagnetic Materials* Vol. 1, ed. by K.H.J. Bushow and E.P. Wohlfarth (North-Holland, Amsterdam, 1980) p. 451.

[4] J. Chappert, in *Magnetism of Metals and Alloys*, ed. by M. Cyrot (North-Holland, Amsterdam, 1982) p. 487.

[5] K. Moorjani and J.M.D. Coey, *Magnetic Glasses* (Elsevier, Amsterdam, 1984).

[6] T. Kaneyoshi, *Amorphous Magnetism* (CRC Press, Boca Raton, 1984).

[7] D. Adler, H. Fritzsche, and S.R. Ovshinsky, *Physics of Disordered Materials* (Plenum, New York, 1985), Part V.

[8] H. Matyja and P.G. Zielenski (ed.), *Amorphous Metals* (World Scientific, Singapore, 1985).

[9] S.R. Elliot, *Physics of Amorphous Materials* (Longman Scientific & Technical, Essex, 1990).

[10] P. Hansen, in *Handbook of Magnetic Materials*, Vol. 6, ed. by K.H.J. Bushow (North-Holland, Amsterdam, 1991) p. 289.

[11] T. Mizoguchi, K. Yamaguchi, and H. Miyajima, in *Amorphous Magnetism*, ed. by H.O. Hooper and A.M. de Graaf (Plenum, New York, 1973).

[12] T. Mizoguchi, AIP Conf. Proc. **30** (1976) 286.

[13] R. Hasegawa, R.C. O'Handley, L. Tanner, R. Ray, and S. Kavesh, Appl. Phys. Lett. **29** (1976) 219.

[14] R.C. O'Handley, R. Hasegawa, R. Ray, and C.-P. Chou, Appl. Phys. Lett. **29** (1976) 330.

[15] R.C. O'Handley, R. Hasegawa, R. Ray, and C.-P. Chou, J. Appl. Phys. **48** (1976) 2095.

[16] J. Friedel, Nuovo Cimento **10**, Suppl. **2** (1958) 287.

[17] A.R. Williams, V.L. Morruzi, A.P. Malozemoff, and K. Terakura, IEEE Trans. Magn. **Mag-19** (1983) 1983.

[18] A.P. Malozemoff, A.R. Williams, and V.L. Moruzzi, Phys. Rev. B **29** (1984) 1620.

[19] M. Sostarich, J. Appl. Phys. **67** (1990) 5793.

[20] T. Masumoto, S. Ohnuma, K. Shirakawa, M. Nose, and K. Kobayashi, J. de Phys. C8 (1980) 686.

[21] K. Fukamichi, T. Goto, H. Komatsu, and H. Wakabayashi, Proc. 4 th Int. Conf. on Phys. Magn. Mater. (Poland), ed. W. Gorkowski, H.K. Lachowics, and H. Szymczak (World Scientific Pub., Singapore, 1989) p. 354.

[22] H. Hiroyoshi and K. Fukamichi, Phys. Letters, **85A** (1981) 242; J. Appl. Phys. **53** (1982) 2226.

[23] N. Saito, H. Hiroyoshi, K. Fukamichi, and Y. Nakagawa, J. Phys. F**16** (1986) 911.

78

[24] J.M.D. Coey, D.H. Ryan, and R. Buder, Phys. Rev. Lett. **26** (1987) 385.

[25] D. H. Ryan, J.M.D. Coey, E. Batalla, Z. Altounian, and J.O. Ström-Olsen, Phys. Rev. B **35** (1987) 8630.

[26] H. Wakabayashi, K. Fukamichi, H. Komatsu, T. Goto, and K. Kuroda, Proc. Int. Symposium on Physics of Magnetic Materials (World Scientific, Singapore, 1987), p. 342.

[27] K. Fukamichi, T. Goto, and U. Mizutani, IEEE Transactions on Magnetism, **MAG-23** (1987) 3590.

[28] N. Mott, *Metal-Insulator Transition* (Taylor & Francis, London, 1990).

[29] P. Fulde, *Electron Correlations in Molecules and Solids*, Solid-State Sciences Vol. 100 (Springer-Verlag, Berlin, 1991).

[30] H. Capellmann (ed.), *Metallic Magnetism*, Topics in Current Physics Vol. 42 (Springer-Verlag, Berlin, 1987).

[31] J.M. Ziman, *Models of Disorder* (Cambridge Univ. Press., Cambridge, 1979).

[32] M. Cyrot, *Magnetism of Metals and Alloys* (North Holland, Amsterdam, 1982).

[33] Y. Kakehashi, Prog. Theor. Phys. Suppl. **101** (1990) 105.

[34] T. Fujiwara, J. Non-Cryst. Solids, **61&62** (1984) 1039.

[35] H.J. Nowak, O.K. Andersen, T. Fujiwara, O. Jepsen, and P. Vargas, Phys. Rev. **B44** (1991) 3577.

[36] O.K. Andersen, O. Jepsen, and D. Glötzel, in *Highlights of Condensed Matter Theory*, ed. by F. Bassani, F. Fumi, and M.P. Tosi (North-Holland, New York, 1985).

[37] H.L. Skriver, *The LMTO Method* (Springer-Verlag, Berlin, 1984).

[38] R. Haydock, V. Heine, and M.J. Kelly, J. Phys. **C8** (1975) 2591.

[39] V. Heine, Solid State Phys. **35** (1980) p. 1.

[40] Y. Kakehashi, Phys. Rev. **B40** (1989) 11059.

[41] Y. Kakehashi, Phys. Rev. **B41** (1990) 9207.

[42] Y. Kakehashi, Phys. Rev. **B42** (1991) 10820.

[43] M. Cyrot, J. de Phys. (Paris) **33** (1972) 25.

[44] J. Hubbard, Phys. Rev. **B19** (1979) 2626; **20** (1979) 4584; **23** (1981) 597.

[45] H. Hasegawa, J. Phys. Soc. Jpn. **46** (1979) 1504; **49** (1980) 178.

[46] F. Matsubara, Prog. Theor. Phys. **52** (1974) 1124.

[47] S. Katsura, S. Fujiki, and S. Inawashiro, J. Phys. **C12** (1979) 2839.

[48] J.C. Slater, *The Selfconsistent Field for Molecules and Solids*, in Quantum Theory of Molecules and Solids, Vol. 4 (McGraw-Hill, New York, 1974).

[49] G.S. Cargill, Solid State Phys. **30** (1975) p. 227.

[50] Y. Waseda, *The Structure of Non-Crystalline Materials* (McGraw-Hill, New York, 1980).

[51] J.D. Bernal, *Nature*, **183** (1959) 141; **185** (1960) 68.

[52] J.D. Bernal, Proc. Roy. Soc. **A 280** (1964) 299.

[53] J.L. Finney, Proc. Roy. Soc. Ser. **A 319** (1970) 479; ibid. 495.

[54] C.H. Bennett, J. Appl. Phys. **43** (1972) 272.

[55] T. Ichikawa, Phys. Stat. Sol. (a) **29** (1975) 293.

[56] L.v. Heimendahl, J. Phys. **F5** (1975) L141.

[57] G. A. N. Connel, Solid State Commun. **16** (1975) 109.

[58] R. Yamamoto and M. Doyama, J. Phys. **F9** (1979) 617.

[59] M. Matsuura, H. Wakabayashi, T. Goto, H. Komatsu, and K. Fukamichi, J. Phys. Cond. Matter **1** (1989) 2077.

[60] See for example, D.W. Heermann, *Computer Simulation Methods in Theoretical Physics* (Springer-Verlag, Heidelberg, 1990).

[61] R. Car and M. Parrinello, Phys. Rev. Lett., **25** (1985) 2471.

[62] R. Car and M. Parrinello, Phys. Rev. Lett., **60** (1988) 204.

[63] D. Hohl and R.O. Jones, J. Non-Cryst. Solids **117** & **118** (1990) 922.

[64] A. Selloni, P. Carnevali, R. Car, and M. Parrinello, Phys. Rev. Lett. **59** (1987) 823.

[65] L.M. Roth, Phys. Rev. B**9** (1974) 2476.

[66] S. Asano and F. Yonezawa, J. Phys. **F10** (1980) 75.

[67] P. Soven, Phys. Rev. **156** (1967) 809.

[68] H. Eherenreich and L. Schwartz, Solid State Phys. **31** (1976) 146.

[69] P. Hohenberg and W. Kohn, Phys. Rev. **136** (1964) B 864.

[70] L. Hedin and B.I. Lundqvist, J. Phys. **C4** (1971) 2064.

[71] O. Gunnarson and B.I. Lundqvist, Phys. Rev. **B13** (1976) 4274.

[72] S.K. Bose, S.S. Jaswal, O.K. Andersen, and J. Hafner, Phys. Rev. B**37** (1988) 9955.

[73] S. Ferreira, J. Duarte, and S. Frota-Pessôa, Phys. Rev. B**41** (1990) 5627; P.R. Peduto, S. Frota-Pessôa, and M.S. Methfessel, to be published in Phys. Rev. B.

[74] I. Turek, J. Phys., Condens. Matter, **2** (1990) 10559.

[75] W.Y. Ching, L.W. Song, and S.S. Jaswal, Phys. Rev. B**30** (1984) 544.

[76] W.Y. Ching, G.L. Zhao, and Y. He, Phys. Rev. B**42** (1990) 10878.

[77] S.F. Edwards and P.W. Anderson, J. Phys. **F5** (1975) 965; see also, K. H. Fisher and J. A. Hertz, *Spin Glasses* (Cambridge Univ. Press, Cambridge, 1991).

[78] D. Sherrington and S. Kirkpatrick, Phys. Rev. Lett. **35** (1975) 1972.

[79] S. Kirkpatrick, C.D. Gelatt, and M.P. Vecchi, Science **13** (1985) 671.

[80] N. Mott, Adv. Phys. **13** (1964) 325.

[81] W. Kohn and L.J. Sham, Phys. Rev. **140** (1965) A 1135.

[82] O.K. Andersen, Phys. Rev. B**12** (1975) 3060.

[83] J.C. Slater, Phys. Rev. **51** (1937) 846.

[84] J. Korringa, Physica **13** (1947) 392.; W. Kohn and J. Rostoker, Phys. Rev. **94** (1954) 1111.

[85] N. Beer and D. G. Pettifor, in *The Electronic Structure of Complex systems*, eds. P. Phariseau W. M. Temmerman (Plenum, New York, 1984).

[86] A. Bieber and F. Gautier, Solid State Commun. **39** (1981) 149.

[87] T. Oguchi, K. Terakura, and N. Hamada, J. Phys. **F13** (1983) 145.

[88] Y. Kakehashi, Phys. Rev. **B34** (1986) 3243.

[89] E.C. Stoner, Proc. R. Soc. **A165** (1938) 372.

[90] E.P. Wohlfarth, Rev. Mod. Phys. **25** (1953) 211.

[91] O. Gunnarson, J. Phys. **F6** (1976) 587.

[92] J.H. Samson, J. Phys. (Paris) **45** (1984) 1675.

[93] M.C. Gutzwiller, Phys. Rev. Lett. **10** (1963) 159; Phys. Rev. **134** (1964) A293; **137** (1965) A1726.

[94] J. Hubbard, Proc. R. Soc. London, Ser. A **276** (1963) 238; **277** (1964) 237.

[95] Y. Kakehashi and P. Fulde, Phys. Rev. **B32** (1985) 1595.

[96] Y. Kakehashi, Phys. Rev. **B38** (1988) 6928.

[97] H. Hasegawa, J. Phys. Condensed Matter **1** (1990) 9325.

[98] Y. Kakehashi, J. Phys. Soc. Jpn. **50** (1981) 1505.

[99] Y. Kakehashi, J. Magn. Magn. Mater. **37** (1983) 189.

[100] Y. Kakehashi, unpublished.

[101] P.W. Anderson, Phys. Rev. **115** (1959) 2.

[102] H. Miwa, Prog. Theor. Phys. **52** (1974) 1.

[103] F. Brouers, F. Ducastelle, F. Gautier, and J. Van der Rest, J. Phys. **F3** (1973) 2120.

[104] V. Heine, Phys. Rev. **153** (1967) 673.

[105] U.K. Poulsen, J. Kollár, and O.K. Andersen, J. Phys. **F6** (1976) L241.

[106] Y. Kakehashi and H. Tanaka, J. Magn. Magn. Mater. **104-107** (1992) 91.

[107] H. Shiba, Prog. Theor. Phys. **46** (1971) 77.

[108] P. Oelhafen, E. Hansen, H.-J. Güntherodt, and K.H. Bennemann, Phys. Rev. Lett. **43** (1979) 1134.

[109] Y. Kakehashi, Phys. Rev. **B35** (1987) 4973.

[110] Y. Kakehashi, J. Phys. Soc. Jpn. **50** (1981) 3177.

[111] T. Fujiwara, Nippon Butsuri Gakkaishi **40** (1985) 209.

[112] V.L. Moruzzi, J.F. Janak, and A.R. Williams, *Calculated Electronic Properties of Metals* (Pergamon, New York, 1978).

[113] J.F. Janak, Phys. Rev. **B16** (1977) 255.

[114] Y. Kakehashi, J. Magn. Magn. Mater. **66** (1987) L163.

[115] Y. Kakehashi, Phys. Rev. **B38** (1988) 474.

[116] E.F. Wassermann, in *Handbook of Magnetic Materials*, Vol. 5, ed. by K.H.J. Bushow and E.P. Wohlfarth (North Holland, Amsterdam, 1990) p. 237.

[117] This relation is obtained by integrating the thermodynamical relation of specific heat $T\partial S/\partial T = \partial E/\partial T$ over temperature from zero to a paramagnetic state, and replacing the temperature factor T at the left-hand side with a characteristic temperature T_C where the specific heat shows a maximum.

[118] Y. Kakehashi, Phys. Rev. B47 (1993) 3185 .

[119] Y. Nakagawa, J. Phys. Soc. Jpn. 11 (1956) 855; 12 (1975) 700.

[120] P.R. Rhodes and E.P. Wohlfarth, Proc. R. Soc. London 273 (1963) 247.

[121] U. Krey, U. Krauss, and S. Krompiewski, J. Magn. Magn. Mater. 103 (1992) 37.

[122] J.A. Fernandez-Baca, J.J. Rhyne, G.E. Fish, M. Hennion, and B. Hennion, J. Appl. Phys. 67 (1990) 5223; S. N. Kaul, J. Phys.: Condens. Matter 3 (1991) 4027.

[123] D.H. Ryan, J.M.D. Coey, and J.O. Ström-Olsen, J. Magn. Magn. Mater. 67 (1987) 148.

[124] T. Goto, C. Murayama, N. Mori, H. Wakabayashi, K. Fukamichi, and H. Komatsu, J. de Phys. (Paris) C8 (1988) 1143.

[125] H. Tange, Y. Tanaka, M. Goto, and K. Fukamichi, J. Magn. Magn. Mater. 81 (1989) L243.

[126] Y. Kakehashi, J. Magn. Magn. Mater. 103 (1992) 78.

[127] J.P. Reboullat, IEEE Trans. on Magnetics, MAG-8 (1972) 630.

[128] Y. Kakehashi, J. Phys. Soc. Jpn. 50 (1981) 1925.

[129] Y. Kakehashi, J. Phys. Soc. Jpn. 50 (1981) 2236.

[130] Y. Kakehashi, Physica B161 (1989) 143.

[131] Y. Kakehashi, J. Phys. Soc. Jpn. 50 (1981) 792.

[132] H. Tanaka and S. Takayama, J. Phys.:Condens. Matter 4 (1992) 8203.

[133] T. Jarborg and M. Peter, J. Magn. Magn. Mater. 42 (1984) .

[134] V.L. Moruzzi, P.M. Marcus, K. Schwarz, and P. Mohn, J. Magn. Magn. Mater. 54-57 (1986) 955.

[135] M. Podgórny and J. Goniakowski, Phys. Rev. B42 (1990) 6683.

[136] R.A. Reck and D.L. Fry, Phys Rev. 184 (1969) 492.

[137] Y. Kakehashi, H. Tanaka, and M. Yu, unpublished.

[138] K. Fukamichi, private communication.

[139] P.R. Peduto and S. Frota-Pessôa, Phys. Rev. B (1992), to be published.

[140] T. Ichikawa, phys. stat. sol. (a)19 (1973) 707.

[141] A. Liénard and J.P. Rebouillat, J. Appl. Phys. 49 (1978) 1680.

[142] K. Fukamichi, H. Hiroyoshi, K. Shirakawa, T. Masumoto, and T. Kaneko, IEEE Trans. Magn. MAG-22 (1986) 424.

[143] K. Fukamichi, H. Hiroyoshi, T. Kaneko, T. Masumoto, and K. Shirakawa, J. Appl. Phys. **53** (1982) 8107.

[144] S. H. Liou and C. L. Chien, J. Appl. Phys. **55** (1984) 1820.

[145] N. Kataoka, K.Sumiyama, and Y. Nakamura, Trans. of Japan Inst. of Metals vol. **27** (1986) 823; B. Fultz, G. Le Caër, and P. Matteazzi, J. Mater. Res. **4** (1989) 1450.

[146] A. Arai, J. Appl. Phys. **64** (1988) 3143.

[147] T. Nakamichi, J. Phys. Soc. Jpn. **25** (1967) 1189.

[148] K. Sumiyama, Y. Hashimoto, and Y. Nakamura, Trans. JIM, **24** (1983) 61.

[149] H. Inaba, S. Murayama, K. Hoshi, and Y. Obi, J. Magn. Magn. Mater. **90&91** (1990) 340.

[150] N. Heiman and N. Kazama, Phys. Rev. **B19** (1979) 1623.

[151] K. H. J. Buschow, Proc. of MRS Conf. ed. M. von Allen, Strasbourg (Les Edition des Physique, Les Ulis, 1984) p.313.

[152] L. Krusin-Elbaum, A. P. Malozemoff, and R. C. Taylor, Phys. Rev. **B27** (1983) 562.

[153] T. M. Hayes, J. W. Aller, and J. B. Boyce, Phys. Rev. **B23** (1981) 4691.

[154] I. Turek, J. Hafner, and Ch. Hausleitner, J. Magn. Magn. Mater. **109** (1992) L145; J. Hafner, Ch. Hausleintner, W. Jank, and I. Turek, to be published in J. Non-Cryst. Solids (1992).

[155] Y. Kakehashi, H. Tanaka, and M. Yu, Phys. Rev. **B47** (1993) 7736.

[156] H. U. Krebs, W. Biegel, A. Bienenstock, D. J. Webb, and T. H. Geballe, Mater. Sci. Eng. **97** (1988) 163.

[157] H. Fujimori, K. Nakanishi, H. Hiroyoshi, and N. S. Kazama, J. Appl. Phys. **53** (1982) 7792.

[158] Yu Boliang, D. H. Ryan, J. M. D. Coey, Z. Altounian, J. O. Ström-Olsen, and F. Razavi, J. Phys. F: Met. Phys. **13** (1983) L 217.

[159] U. Krauss and U. Krey, Phys. Rev. **B39** (1989) 2819.

[160] Y. Kakehashi and M. Yu, J. Appl. Phys. **73** (1993) 3426.

[161] J. Inoue and M. Shimizu, J. Phys. F: Met. Phys. **15** (1985) 1525.

[162] S. Asano and S. Ishida, J. Phys. F: Met. Phys. **18** (1988) 501.

[163] J. P. Xanthakis, R. L. Jacobs, and E. Babić, J. Phys. F **16** (1986) 323.

[164] I. A. Campbell, J. Phys.:F **2** (1972) L47.

[165] K. H. J. Bushow, Rep. Prog. Phys. **40** (1977) 1179.

[166] R. C. Taylor, T. R. McGuire, J. M. D. Coey, and A. Gangulee, J. Appl. Phys. **49** (1978) 2885.

[167] J. J. Rhyne, in *Handbook on the Physics and Chemistry of Rare Earths*, p. 270, eds. K. A. Gschneidner Jr. and L. Ering, North-Holland, (1979).

[168] L. J. Tao, S. Kirkpatrick, R. J. Gambino, and J. J. Cuomo, Solid State Commun. **13** (1973) 1491.

[169] R. Taylor and A. Gangulee, J. Appl. Phys. **47** (1976) 4666.

[170] P. Hansen and H. Heitmann, IEEE Trans. Magn. **MAG-25** (1989) 4390.

[171] K. Nagayama, H. Ino, N. Saito, Y. Nakagawa, E. Kita, and K. Shiratori, J. Phys. Soc. Jpn. **59** (1990) 2483.

[172] H. Tanaka, S. Takayama, and T. Fujiwara, Phys. Rev. **B46** (1992) 7390.

[173] K. H. J. Buschow, M. Brouha, J. M. N. Biesterbos, and A. G. Dirks, Physica **91B** (1977) 261.

[174] G. Güntherodt and N. J. Shevchik, AIP Conf. Proc. No.29, (1976) 174.

[175] S. S. Jaswal, D. J. Sellmyer, M. Engelhardt, and Z. Zhao, Phys. Rev. **B35** (1991) 996.

[176] H. Tanaka and S. Takayama, J. Appl. Phys. **70** (1991) 6577.

[177] K. Fukamichi, K. Shirakawa, Y. Satoh, T. Masumoto, and T. Kaneko, J. Magn. Magn. Mater. **54-57** (1986) 231.

[178] O. Beckmann and L. Lundgren, in em Handbook of Magnetic Materials Vol. 6, ed. K. H. J. Buschow (North-Holland, Amsterdam 1991).

[179] C. L. Chien and K. M. Unruh, Nucl. Instrum. & Methods **199** (1982) 193.

[180] U. Mizutani, M. Hasegawa, K. Fukamichi, Y. Hattori, Y. Yamada, H. Tanaka, and S. Takayama, Phys. Rev. **B47** (1993) 2678.

[181] J. F. Sadoc and J. Dixmier, Mater. Sci. Eng. **23** (1976) 187.

[182] Y. Waseda, H. Okazaki, and T. Masumoto, Sci. Rep. Inst. Tohoku Univ. **26A** (1976) 102.

[183] Y. Waseda, H. Okazaki, and T. Masumoto, Sci. Rep. Inst. Tohoku Univ. **26A** (1977) 202.

[184] Y. Waseda, Prog. Mater. Sci. **26** (1981) 1.

[185] T. Fujiwara and Y. Ishii, J. Phys. F: Met. Phys. **10** (1980) 1901.

[186] K. Yamaguchi and T. Mizoguchi, J. Phys. Soc. Japan **39** (1975) 541.

[187] M. C. Cadeville and A. J. P. Meyer, Compt. Rend. **255** (1962) 3391.

[188] T. Fujiwara, J. Phys. F: Met. Phys. bf 12 (1982) 661.

[189] S. Krompiewski, U. Krey, U. Krauss, and H. Ostermeier, J. Magn. Magn. Matter. 73 (1988) 5.

[190] W. Y. Ching and Y.-N. Xu, J. Appl. Phys. **70** (1991) 6305.

[191] M. Matsuura and U. Mizutani, J. Magn. Magn. Mater. **31-34** (1983) 1481.

[192] R. Hasegawa and R. Ray, J. Appl. Phys. **49** (1978) 4174.

[193] H. Tanaka, S. Takayama, M. Hasegawa, T. Fukunaga, U. Mizutani, M. Fujita, and K. Fukamichi, Phys. Rev. **B47** (1993) 2671.

[194] R. A. Alben, J. I. Budnick, and G. S. Cargill, in *Metallic Glasses*, eds. J.J. Gilman and H. J. Leamy (American Society of Metals, Metals Park, Ohio, 1978).

[195] G. P. Gaspari and B. L. Gyorffy, Phys. Rev. Lett. **28** (1972) 801.

[196] H. L. Skriver and I. Mertig, Phys. Rev. **B41** (1990) 6553.

ELECTRONIC STRUCTURE CALCULATIONS IN MAGNETIC METALLIC GLASSES

W.Y. CHING

Department of Physics
University of Missouri-Kansas City
Kansas City, Missouri 64110, USA

1. Introduction

In the last quarter of century, the study of disordered magnetic materials becomes a major field of study in condensed matter physics [1-5]. Quantum theory of magnetism is treated either in the localized model using the Heisenberg Hamiltonian, or in the itinerant band picture [3,4]. In the early days, theoretical treatment using the band model was difficult because of the lose of Bloch theorem based on which the foundation of entire solid state physics has been built on. Furthermore, practical calculations of electronic states inadvertently involved many justified and unjustified approximations which led to poor results and were unfairly attributed to the failure of the itinerant model itself. It was quite impossible to have any microscopic understanding on the electronic states in a real disordered system. Rapid progress was made in late seventies and early eighties in both the conceptual understanding and in the actual calculation of electronic properties of specific systems. Among them, the disordered magnetic alloys and amorphous metallic glasses (MG) played a very important role in understanding the physics of noncrystalline solids [2-5]. The historical development has been described in Chapter 1 by Kakehashi and Tanaka [6]. In the same chapter, the properties and the theory of many amorphous magnetic systems have been discussed. In this chapter, we take a rather different approach. Since the electronic structure is the foundation for studying physical properties of any material, we outline a very specific computational method for electronic structure calculation and provide the results obtained by using such a method. It is not our intention to review the properties of various amorphous glasses or alloys, or assess the merits of different methods and their calculated results. Experimental results are mentioned only in connection with calculated quantities that serve as illustrative examples. Thus the scope of this chapter is rather narrow but highly focused.

In recent years, the direct space self-consistent orthogonalized linear combination of atomic orbitals method (OLCAO) has developed into a highly competitive modern method of band structure calculation [7]. The method is based on the local density approximation (LDA) of the density functional theory [8-9] whose success in predicting ground state properties of a variety of crystals have been quite phenomenal. The OLCAO-LDA method has been adapted to study the electronic structures of non-crystalline solids using large periodic model structures since early seventies. Over the years, the method has been systematically refined and applied to many diversified amorphous solids including amorphous Si [10-13], hydrogenated a-Si [14], impurities in a-Si [15], insulating glasses [16-21], metallic glasses [22-27], and more recently, novel carbon structures [28-29]. The effective potential used in amorphous calculation is in the form of superposition of atomic-like potentials which are usually obtained from self-consistent calculations on crystalline phases, or on smaller model structures. Since all interaction integrals, regardless of interatomic distances and the nature of local bonding patterns, are evaluated exactly without any empirical parameters, the method is essentially first-principles in nature. This is particularly important for MG systems because both the short range order and the intermediate range order must be reflected in the calculation. This could also imply that the resulting electronic wave functions contain all the information about the quantum interference and multiple scattering effect which are of paramount importance in the transport properties of MG The use of atomic orbital basis also facilitates the interpretation of the calculated results. From a pure computational standing point, it is far more economic to use atomic orbitals than say, plane waves for basis expansion. Thus, for a given amount of computer resource, much larger models of MG can be studied using the OLCAO method. For a model structure with periodic boundary conditions, the technical aspect of the electronic structure calculation is not different from that of a very complex crystal with a large number of atoms in the unit cell.

The organization of this chapter is as follows. We first describe the OLCAO method in the LDA formulation in section 2. Section 3 describes the results obtained for different classes of metallic glasses and section 4 for results on magnetic alloys and glasses. These results demonstrate the capabilities and advantages of the method. Section 5 is for some conclusions and a brief discussion on future directions using this very promising method for the study of amorphous magnetic materials.

2. The Orthogonalized Linear Combination of Atomic Orbitals Method

2.1. The Atomic Basis Functions

We shall describe the OLCAO method strictly as a band structure method and therefore retain the wave vector $\mathbf{k}$ description of the electronic state. In applying to an amorphous solid whose structure is represented by a large model with periodic boundary conditions, only the solution at $\mathbf{k}=0$ is needed. The use of the Bloch theorem circumvents the problem of free surface encountered with finite cluster models and makes it possible to describe the system as an infinitely extended system. As long as the periodic model is large enough, say much larger than the mean free path of the electron in the amorphous solid, this is a very effective way of studying the electronic structure of a disordered solid.

The solid state wave function of band index n and wave vector $\mathbf{k}$ is expanded in terms of Bloch functions $b_{i\gamma}(\mathbf{k},\mathbf{r})$:

$$\Psi_{nk}(\mathbf{r}) = \sum_{i,\gamma} C_{i\gamma}^{n}(\mathbf{k})b_{i\gamma}(\mathbf{k},\mathbf{r}), \tag{2.1}$$

where γ labels the atoms in the cell and i represents the orbital quantum number (ℓ,m). γ represents different types of atoms as well as different atoms of the same type. The Bloch function is constructed from the linear combination of atomic orbitals u_i centered at each atomic site:

$$b_{i\gamma}(\mathbf{k},\mathbf{r}) = (1/\sqrt{N})\sum_{v} \exp(i\,\mathbf{k}\cdot\mathbf{R}_v)u_i(\mathbf{r}-\mathbf{R}_v-\mathbf{t}_\gamma). \tag{2.2}$$

Here $\mathbf{R}_v$ represents the lattice of the supercell and $\mathbf{t}_\gamma$ is the position of the γth atom in the cell. For amorphous calculation, only the $\mathbf{k}=0$ case is important and (2.2) is a real function. The atomic orbital u_i consists of the radial part and the angular part. The radial part is represented by the linear combination of a suitable number of Gaussian-type of orbitals (GTO) and the angular part is represented by spherical harmonics.

$$u_i(\mathbf{r}) = [\sum_{j}^{N} A_j r^{n-1} \exp(-\alpha_j r^2)]\cdot Y_{\ell m}(\theta,\phi). \tag{2.3}$$

Each of the GTO in (2.3) is characterized by a decaying exponent α_j. The way the α_j are chosen for preparing u_i deserves some comments. The simple way is to choose a set of predetermined exponentials $\{\alpha_j\}$ ranging from a minimum α_{min} to a maximum α_{max}, distributed in a geometric seires. The number of exponents N used for each atomic orbital and the chosen values of α_{min} and α_{max} depend on the type of atom and the material under study. They are usually guided by past experience. Typical values for N are from 16 to 20, and for α_{min} and α_{max} are 0.15 and 10^6 respectively. Once the exponential set $\{\alpha_j\}$ has been fixed, the expansion coefficients A_j can be obtained in several ways. It can be obtained as eigenvalues in solving the single-atom problem self-consistently within the LDA, or by linearly fitting to the atomic wave functions calculated by the self-consistent Hartree-Fock method [30]. For amorphous calculations, it is desirable to use the same $\{\alpha_i\}$ set for all the atoms and for all the orbitals. This greatly reduces the total number of analytical integrals that need to be evaluated which is enormous for a model structure containing several hundreds of atoms.

The set of atomic orbitals u_i in Eq. (2.2) may include the core orbitals, the occupied valence orbitals, and additional empty orbitals. For calculation with amorphous systems, a minimal basis is usually sufficient which includes all core orbitals and the orbitals in the valence shell of the atom, occupied or unoccupied. For example, a minimal basis set for Fe consists of 1s, 2s, $2p_x$, $2p_y$, $2p_z$, 3s, $3p_x$, $3p_y$, $3p_z$ (core orbitals) and 4s, $4p_x$, $4p_y$, $4p_z$, $3d_{xy}$, $3d_{yz}$, $3d_{zx}$, $3d_{x^2-y^2}$, $3d_{3z^2-r^2}$ (valence orbitals). If additional empty orbitals of the next shell are added, the basis set is refereed to as a full basis set. For most purposes, a minimal basis is quite adequate to give accurate electronic structure results and it retains the full advantage of the Mulliken scheme [31] for charge analysis. A full basis is usually employed if high lying unoccupied states are of importance such as in the case of optical transition calculations. For special purpose high precision calculations, u_i can be further optimized as practiced by many quantum chemists in molecular calculations [32]. Additional single GTO may be added to the full basis to increase the variational freedom of the basis set. However, this may lead to numerical instability due to the near linear dependence of the basis set.

2.2. The Site-decomposed Potential Functions

In the OLCAO method, the potential is constructed according to the LDA of the density functional theory [8-10, 33-36]. The one-electron Kohn-Sham equation of the system is given by, in atomic units:

$$\left\{-\vec{\nabla}^2 + V_{e-n}(\mathbf{r}) + V_{e-e}(\mathbf{r}) + V_{xc}[\rho(\mathbf{r})]\right\}\Psi_{nk}(\mathbf{r}) = E_{nk}\Psi_{nk}(\mathbf{r}). \tag{2.4}$$

where $V_{e-n}(\mathbf{r})$, $V_{e-e}(\mathbf{r})$, $V_{xc}[\rho(\mathbf{r})]$ in (2.4) are the electron-nuclear, electron Coulomb, and exchange-correlation parts of the potential respectively. They depend on the electron density $\rho(\mathbf{r})$. The electron density of the solid is obtained from $\rho(\mathbf{r}) = \sum_{occ} |\Psi_{nk}(\mathbf{r})|^2$ so that equation (2.4) must be solved self-consistently.

The LDA assumes that the exchange and correlation part of the potential $V_{xc}(\mathbf{r})$, which effectively includes all the many body interactions, is derivable from an exchange-correlation energy functional ε_{xc} for the exchange-correlation energy $E_{xc}(\mathbf{r})$

$$E_{xc}(\mathbf{r}) = \int \rho(\mathbf{r})\varepsilon_{xc}[\rho(\mathbf{r})]\, d\mathbf{r}. \tag{2.5}$$

In the local approximation and by making use of the result for a system of uniform electron gas, $V_{xc}(\mathbf{r})$ takes the form:

$$V_{xc}(\mathbf{r}) = d(\rho\varepsilon_{xc}[\rho])/d\rho = -\frac{3}{2}\alpha[\frac{3}{\pi}\rho(\mathbf{r})]^{1/3}. \tag{2.6}$$

In the usual Kohn-Sham approximation for V_{xc} [9], α in (2.6) equals 2/3. There exist several other forms of V_{xc} which were obtained with different treatments aimed at improving the exchange and correlation energy term [33-38]. We adopt the Kohn-Sham potential and use the Wigner interpolation formula to account for additional correlation effect [39] such that:

$$V_{xc}(\mathbf{r}) = \rho(\mathbf{r})^{1/3}\left[-0.984 - \frac{0.944 + 8.90\rho(\mathbf{r})}{1 - 12.57\rho(\mathbf{r})^{1/3}}\right]^{1/3}. \tag{2.7}$$

The total energy of the system in the LDA approximation can be evaluated from the following expression:

$$E_T = \sum_{n,k}^{occ.} E_n(\mathbf{k}) + \int \rho(\mathbf{r})(\varepsilon_{xc} - V_{xc} - V_{e-e}/2)d\mathbf{r} + 1/2\sum_{\gamma,\delta} Z_\gamma Z_\delta/(\mathbf{R}_\gamma - \mathbf{R}_\delta), \tag{2.8}$$

where the first term is the sum over one-electron states and the last term is a sum over the lattice. While the total energy calculation for simple crystals becomes quite routine, application of equation (2.8) directly to amorphous system is rarely done. A more practical approach is to extract meaningful interatomic pair potentials for amorphous systems based on accurate total energy calculations on simpler systems.

An important feature of the OLCAO method is the real space description of the charge density $\rho_{cry}(\mathbf{r})$ and the one-electron potential $V_{cry}(\mathbf{r})$ as sums of atom-centered functions consisting of Gaussians [40]. When carefully constructed, the site-specific atom-centered potential functions are transferable. Thus, it is possible to obtain the self-consistent potential from the calculation for a simpler system and use it in the calculation for a larger and more complicated system. We may write:

$$\rho_{cry}(\mathbf{r}) = \sum_A \rho_A(\mathbf{r} - t_A), \quad \rho_A(\mathbf{r}) = \sum_{j=1}^N B_j \exp(-\beta_j r^2). \tag{2.9}$$

$$V_{coul}(\mathbf{r}) = \sum_A V_C(\mathbf{r} - t_A), \quad V_C(\mathbf{r}) = -(Z_A / r)\exp(-\zeta r^2) - \sum_{j=1}^N D_j \exp(-\beta_j r^2). \tag{2.10}$$

$$V_{xc}(\mathbf{r}) = \sum_A V_x(\mathbf{r} - t_A), \quad V_x(\mathbf{r}) = \sum_{j=1}^N F_j \exp(-\beta_j r^2). \tag{2.11}$$

$$V_{cry}(\mathbf{r}) = \sum_A V_A(\mathbf{r} - t_A), \quad V_A(\mathbf{r}) = V_C(\mathbf{r}) + V_x(\mathbf{r}). \tag{2.12}$$

Z_A in (2.10) is the mass number of the atom at the site and the first term in $V_C(\mathbf{r})$ describes the potential near the nuclei. The exponential set $\{\beta_j\}$ in (2.9) to (2.11) is predetermined for each atom while the coefficients B_j, D_j, F_j etc. are determined by the least square fitting. Since the accuracy of the calculation depends largely on how accurate the true charge density can be represented by (2.9), the optimal choice of the fitting set $\{\beta_j\}$ is very important. One way to determine $\{\beta_j\}$ for each atom is to fit non-linearly to the atomic charge density calculated from the accurately determined Hartree wave functions [30] and then add an extra small exponential of the order of 0.1 to 0.02 for a better description of the potential function. The actual value chosen for the smallest β_j depends on the system under study, and is guided by the desire to minimize the total fitting error. The accuracy of the fit is carefully monitored by comparing the integrated charge from the fitted functions with the total number of electrons in the unit cell. For electronic structure studies, accuracy with fitting error of 0.001 electron per valence electron is more than adequate. For high precision total energy calculations, a still higher fitting accuracy may be necessary [41,42].

It should be noted that although the transferable atom-centered functions $V_A(\mathbf{r})$ and $\rho_A(\mathbf{r})$ in (2.9) and (2.12) consist of spherical Gaussian functions around each atom, the actual superposition of them is non-spherical and can accurately represent different types of bonding in different types of structural configurations without any shape approximation. The site-decomposed charge density $\rho_A(\mathbf{r})$ and potential $V_A(\mathbf{r})$ in real space are very useful for calculating many physical observables. They can also be utilized in creative simulational studies. More importantly, the expansion of ρ_A and V_A in terms of Gaussians facilitates the analytic evaluation of multi-center integrals in the Hamiltonian matrix elements via the technique of Gaussian transformation [43,44]. This will be discussed below.

In solving (2.4) iteratively, the new charge density is calculated from the occupied state wave functions. It is a common practice to blend the old and the new charge densities for the next iteration to improve numerical stability. The coefficients B_j, D_j, F_j in (2.9)-(2.11) are constantly up-graded until convergence is obtained when the energy eigenvalues change by less than a predetermined value, typically 0.0001 eV. Various acceleration schemes for fast convergence can be utilized. In general, insulating systems converge much faster because the existence of band gap unambiguously separates the occupied and unoccupied states. In metallic systems, the presence of a Fermi level may complicates the matter slightly, and a sufficiently large number of $\mathbf{k}$-space sampling points may be necessary to determine the Fermi level accurately.

2.3. The Technique of Gaussian Transformation

The band structure of the crystal is obtained by solving the Kohn-Sham equation (2.4) at various $\mathbf{k}$ points in the Brillouin zone, or equivalently, by solving the secular equation:

$$\left| H_{i\gamma,j\delta}(\mathbf{k}) - S_{i\gamma,j\delta}(\mathbf{k})E(\mathbf{k}) \right| = 0, \tag{2.13}$$

where $S_{i\gamma,j\delta}(\mathbf{k})$ and $H_{i\gamma,j\delta}(\mathbf{k})$ are the overlap and Hamiltonian matrix respectively:

92

$$S_{i\gamma,j\delta}(\mathbf{k}) = \langle b_{i\gamma}(\mathbf{k},\mathbf{r}) | b_{j\delta}(\mathbf{k},\mathbf{r}) \rangle$$

$$= \sum_{\mu} \exp(-i\mathbf{k}\cdot\mathbf{R}_{\mu}) \int u_i(\mathbf{r}-\mathbf{t}_{\gamma}) u_j(\mathbf{r}-\mathbf{R}_{\mu}-\mathbf{t}_{\delta}) d\mathbf{r}. \tag{2.14}$$

$$H_{i\gamma,j\delta}(\mathbf{k}) = \langle b_{i\gamma}(\mathbf{k},\mathbf{r}) | H | b_{j\delta}(\mathbf{k},\mathbf{r}) \rangle.$$

$$= \sum_{\mu} \exp(-i\mathbf{k}\cdot\mathbf{R}_{\mu}) \int u_i(\mathbf{r}-\mathbf{t}_{\gamma})[-\nabla^2 + V_{coul}(\mathbf{r}) + V_{ex}(\mathbf{r})] u_j(\mathbf{r}-\mathbf{R}_{\mu}-\mathbf{t}_{\delta}) d\mathbf{r}.$$

$$\tag{2.15}$$

For calculations with amorphous solids represented by large periodic models, the lattice sums in (2.14) and (2.15) converge very rapidly and entail the summation over the neighboring cells only. Thus, in spite of a large number of atoms in the model, the total number of interaction integrals need to be evaluated is still kept at a manageable level.

With the one-electron LDA potential expressed as site-decomposed atom-centered Gaussian functionals as in (2.9) to (2.12), let us consider for the time being u_i to be an s-type function consisting of the simplest Gaussian orbital with exponential α_1 centered at the atomic site $\mathbf{A}$,

$$|s_A\rangle = \exp(-\alpha_1 r_A^2); \qquad \mathbf{r}_A = \mathbf{r} - \mathbf{A}. \tag{2.16}$$

The merit of using GTO in the basis expansion and in the potential function representation lies in the fact that all the interaction integrals can be expressed in analytical forms. It can be shown that the integrals involved in (2.14) and (2.15) are of the following four types:

Overlap Integral:

$$I1 = \langle s_A | s_B \rangle = \int \exp(-\alpha_1 r_A^2) \exp(-\alpha_2 r_B^2) d\mathbf{r}. \tag{2.17}$$

Kinetic Energy Integral:

$$I2 = \left\langle s_A \left| -\nabla^2 \right| s_B \right\rangle = \int \exp(-\alpha_1 r_A^2)(-\nabla^2) \exp(-\alpha_2 r_B^2) d\mathbf{r}. \tag{2.18}$$

Three-Center Integral:

$$I3 = \left\langle s_A \left| e^{-\alpha r_C^2} \right| s_B \right\rangle = \int \exp(-\alpha_1 r_A^2) \exp(-\alpha_3 r_C^2) \exp(-\alpha_2 r_B^2) d\mathbf{r}. \tag{2.19}$$

Three-center over r integral:

$$I4 = \left\langle s_A \left| \frac{1}{r_C} e^{-\alpha r_C^2} \right| s_B \right\rangle = \int \exp(-\alpha_1 r_A^2) \frac{1}{r_C} \exp(-\alpha_3 r_C^2) \exp(-\alpha_2 r_B^2) d\mathbf{r}. \tag{2.20}$$

In addition, the matrix elements of the momentum operator can be expressed in the similar fashion:

Momentum Integral:

$$I5 = \left\langle s_A \left| \mathbf{P} \right| s_B \right\rangle = -i\hbar \int \exp(-\alpha_1 r_A^2)(\frac{\partial}{\partial x_B}, \frac{\partial}{\partial y_B}, \frac{\partial}{\partial z_B}) \exp(-\alpha_2 r_B^2) d\mathbf{r}. \tag{2.21}$$

I1, I2 and I5 are two center integrals involving a Gaussian at site **A** and another at site **B**. I3 and I4 are three-center integrals involving another Gaussian from the potential function centered at site **C**.

The two-center integral I1 can be considered as a special case of I3 with $\alpha_3 = 0$. By applying the technique of Gaussian transformation as illustrated in Fig. 2.1, it can be shown [11] that I3 can be reduced to a closed form:

$$I3 = \left\langle s_A \left| e^{-\alpha_3 r_C^2} \right| s_B \right\rangle = \left[\frac{\pi}{\alpha_T} \right]^{3/2} \exp[\alpha_T \mathbf{E}^2 - \alpha_1^2 - \alpha_2^2 - \alpha_3^2], \tag{2.22}$$

where $\quad \alpha_T = \alpha_1 + \alpha_2 + \alpha_3; \quad \mathbf{E} = (\alpha_1 \mathbf{A} + \alpha_2 \mathbf{B} + \alpha_3 \mathbf{C}) / (\alpha_1 + \alpha_2 + \alpha_3).$

The kinetic energy integral I2 and the momentum integral I5 can be obtained from the overlap integral I1 by direct differentiation:

$$\left\langle s_A \left| -\nabla^2 \right| s_B \right\rangle = \left[6\alpha_2 + 4\alpha_2^2 \frac{\partial}{\partial \alpha_2} \right] \left\langle s_A | s_B \right\rangle \tag{2.23}$$

$$\left\langle s_A \left| p_X \right| s_B \right\rangle = \left\langle s_A \left| -i\hbar \frac{\partial}{\partial x_B} \right| s_B \right\rangle = -i\hbar \frac{\partial}{\partial x_B} \left\langle s_A | s_B \right\rangle \tag{2.24}$$

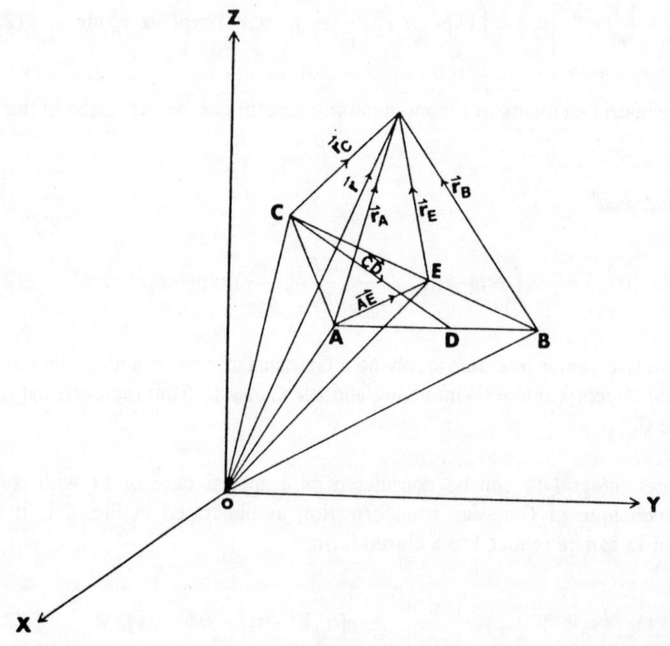

FIG. 2.1.   Vector relations between various points for the reduction of multicenter integrals.

I4 can not be expressed in closed form, but can be expressed in terms of standard error functions [45].

$$I4 = \left\langle s_A \left| \frac{1}{r_C} e^{-\alpha_3 r_C^2} \right| s_B \right\rangle$$

$$= \left[\frac{2\pi}{\alpha_T t} \right] \exp[-\alpha_1 |A - C|^2 - \alpha_2 |B - C|^2 + (\alpha_1 + \alpha_2)^2 |D - C|^2 / \alpha_T] \int_0^t e^{-z^2/2} dz.$$

$$(2.25)$$

where $D = (\alpha_1 A + \alpha_2 B) / (\alpha_1 + \alpha_2)$, and $t = \sqrt{2}(\alpha_1 + \alpha_2)|D - C|^2 / \sqrt{\alpha_T}$.

The analytic expressions in (2.17) to (2.21) pertain to four types of integrals between s-type of GTO. The corresponding expressions for integrals involving p-type, d-type and f-type of GTOs can be obtained from those of lower ℓ quantum number by successive differentiations with respect to the components of lattice coordinates. For example,

$$\left\langle p_A^X \middle| s_B \right\rangle = \int x_A \exp(-\alpha_1 r_A^2) \exp(-\alpha_2 r_B^2) dr$$

$$= \int \frac{1}{2\alpha_1} \frac{\partial}{\partial A_x} \exp(-\alpha_1 r_A^2) \exp(-\alpha_2 r_B^2) dr = \frac{1}{2\alpha_1} \frac{\partial}{\partial A_x} \left\langle s_A \middle| s_B \right\rangle,$$

$$(2.26)$$

$$\left\langle p_A^X \middle| p_B^Y \right\rangle = \frac{1}{2\alpha_2} \frac{\partial}{\partial B_y} \left\langle p_A^X \middle| s_B \right\rangle, \qquad \text{etc.} \qquad (2.27)$$

Analytic formulae for integrals of the type I1 to I4 up to $\ell=3$ (f orbitals) [38] have been derived. Repeated differentiations quickly result in complicated messy formulae. It is desirable to use special computer software to derive such expressions and check for accuracy.

It is possible to express the cartesian GTO in the most general form

$$x^n y^l z^m \exp(-\alpha r^2), \qquad (2.28)$$

and obtain general expressions for the integrals discussed above. For example, the generalized overlap integrals can be written as [47]:

$$S_{nlm,n'l'm'}(\alpha_1,\alpha_2,\mathbf{A},\mathbf{B}) = \int x_A^n y_A^l z_A^m \exp(-\alpha_1 r_A^2) x_B^{n'} y_B^{l'} z_B^{m'} \exp(-\alpha_2 r_B^2) d\tau$$

$$= S_{n,n'}(\alpha_1,\alpha_2,A_x,B_x) S_{l,l'}(\alpha_1,\alpha_2,A_y,B_y) S_{m,m'}(\alpha_1,\alpha_2,A_z,B_z),$$

$$(2.29)$$

where:

$$S_{n,n'}(\alpha_1,\alpha_2,A_x,B_x) = \exp(-h\overline{AB}x) \sum_s \binom{n}{s} \overline{AD}_x^{n-s} \sum_{s'} \binom{n'}{s'} \overline{BD}_x^{n'-s'} F_{s+s'}(\beta)$$

where:

$$\beta = \alpha_1 + \alpha_2, \quad h = \alpha_1\alpha_2 / \beta, \quad D = (\alpha_1 A + \alpha_2 B)/\beta, \quad AB_x = B_x - A_x, \quad \text{and}$$

$$F_n(\beta) = N_n \beta^{-(n+1)/2}, \quad N_n = \pi^{1/2} \begin{cases} 1 & n = 0, \\ 0 & \text{if} \quad n = \text{odd}, \\ (n-1)!!/ 2^{n/2} & n = \text{even}. \end{cases}$$

$$(2.30)$$

For cubic or rectangular cells used for amorphous solids, Lafon claimed [47] that all the integrals can be factorized as above such that the number of multicenter integrals that need to be evaluated goes with 3N, not N^3 where N is the total number of orbitals. This approach has not been sufficiently tested but could potentially further facilitate the large scale computation associated with complex systems using GTO.

2.4. The Technique of core-orthogonalization

For amorphous systems, the matrix equation (2.13) is large and its solution requires large amount of computer time. However, in most cases, core states of the solid are of little physical interest. By using an orthogonalization to the core procedure [48], the dimension of equation can be reduced to those involving only the valence (or non-core) orbitals. For the purpose of orthogonalization, we arbitrary define a core state as one whose orbital energy is less than the oxygen-2s atomic level. We briefly describe the orthogonalization process below.

The overlap and the Hamiltonian matrices in (2.13) are rearranged by interchanging the lows and columns such that all the core states are in the upper-left side and all the non-core states are on the lower right side as illustrated in Fig. 2.2. The matrix elements between Bloch sums in (2.13) can be divided into three groups: (1) core-core, (2) core-valence and valence-core, and (3) valence-valence. By assuming that the matrix elements between core Bloch sums on different sites to be effectively zero, and by imposing the orthogonalization condition, the dimension of the matrix equation can be effectively reduced. The orthogonalized matrix elements in the valence-valence block can be expressed as the original non-orthogonalized elements plus correction terms involving the core-valence and the valence-core matrix elements.

$$
\begin{array}{cc}
 & \text{CORE} \qquad\qquad \text{VALENCE} \\[4pt]
\begin{array}{c} \text{CORE} \\[10pt] \text{VALENCE} \end{array} &
\left(
\begin{array}{c|c}
\text{core}-\text{core} & \text{core}-\text{valence} \\
\hline
\text{valence}-\text{core} & \text{valence}-\text{valence}
\end{array}
\right)
\end{array}
$$

FIG. 2.2   Matrix arrangement for core orthogonalization.

Let us use the superscript v and c to denote the valence (or none-core) and the core portions of the Bloch sums and v' the orthogonalized valence Bloch sum. Expand $b_{i\alpha}^{v'}(\mathbf{k},\mathbf{r})$ as:

$$
b_{i\alpha}^{v'}(\mathbf{k},\mathbf{r}) = b_{i\alpha}^{v}(\mathbf{k},\mathbf{r}) + \sum_{j,\gamma} C_{j\gamma}^{i\alpha} b_{j\gamma}^{c}(\mathbf{k},\mathbf{r}),
\tag{2.31}
$$

and   demanding $\left\langle b_{j\beta}^{c}(\mathbf{k},\mathbf{r}) \middle| b_{i\alpha}^{v'}(\mathbf{k},\mathbf{r}) \right\rangle = \left\langle b_{i\alpha}^{v'}(\mathbf{k},\mathbf{r}) \middle| b_{j\beta}^{c}(\mathbf{k},\mathbf{r}) \right\rangle = 0$,   the   expansion coefficients in (2.31) are given by:

$$
\begin{aligned}
C_{j\gamma}^{i\alpha} &= -\left\langle b_{j\gamma}^{c}(\mathbf{k},\mathbf{r}) \middle| b_{i\alpha}^{v}(\mathbf{k},\mathbf{r}) \right\rangle, \qquad \text{and} \\
C_{j\gamma}^{i\alpha*} &= -\left\langle b_{i\alpha}^{v}(\mathbf{k},\mathbf{r}) \middle| b_{j\gamma}^{c}(\mathbf{k},\mathbf{r}) \right\rangle.
\end{aligned}
\tag{2.32}
$$

Thus after orthogonalization, the matrix elements in the lower-right portion of Fig. 2.2 are given by:

$$\langle b_{i\alpha}^{v'}(\mathbf{k},\mathbf{r})|b_{j\beta}^{v'}(\mathbf{k},\mathbf{r})\rangle = \langle b_{i\alpha}^{v}(\mathbf{k},\mathbf{r})|b_{j\beta}^{v}(\mathbf{k},\mathbf{r})\rangle$$

$$- \sum_{\ell,\gamma} \langle b_{i\alpha}^{v}(\mathbf{k},\mathbf{r})|b_{\ell\gamma}^{c}(\mathbf{k},\mathbf{r})\rangle\langle b_{\ell\gamma}^{c}(\mathbf{k},\mathbf{r})|b_{j\beta}^{v}(\mathbf{k},\mathbf{r})\rangle$$

$$- \sum_{\ell,\gamma} \langle b_{\ell\gamma}^{c}(\mathbf{k},\mathbf{r})|b_{j\beta}^{v}(\mathbf{k},\mathbf{r})\rangle\langle b_{j\beta}^{v}(\mathbf{k},\mathbf{r})|b_{\ell\gamma}^{c}(\mathbf{k},\mathbf{r})\rangle$$

$$+ \sum_{\ell,\gamma}\sum_{m,\delta} \langle b_{i\alpha}^{v}(\mathbf{k},\mathbf{r})|b_{\ell\gamma}^{c}(\mathbf{k},\mathbf{r})\rangle\langle b_{m\delta}^{c}(\mathbf{k},\mathbf{r})|b_{j\beta}^{v}(\mathbf{k},\mathbf{r})\rangle\delta_{\ell m}\delta_{\gamma\delta}.$$

$$(2.33)$$

$$\langle b_{i\alpha}^{v'}(\mathbf{k},\mathbf{r})|H|b_{j\beta}^{v'}(\mathbf{k},\mathbf{r})\rangle = \langle b_{i\alpha}^{v}(\mathbf{k},\mathbf{r})|H|b_{j\beta}^{v}(\mathbf{k},\mathbf{r})\rangle$$

$$- \sum_{\ell,\gamma} \langle b_{i\alpha}^{v}(\mathbf{k},\mathbf{r})|b_{\ell\gamma}^{c}(\mathbf{k},\mathbf{r})\rangle\langle b_{\ell\gamma}^{c}(\mathbf{k},\mathbf{r})|H|b_{j\beta}^{v}(\mathbf{k},\mathbf{r})\rangle$$

$$- \sum_{\ell,\gamma} \langle b_{\ell\gamma}^{c}(\mathbf{k},\mathbf{r})|b_{j\beta}^{v}(\mathbf{k},\mathbf{r})\rangle\langle b_{j\beta}^{v}(\mathbf{k},\mathbf{r})|H|b_{\ell\gamma}^{c}(\mathbf{k},\mathbf{r})\rangle$$

$$+ \sum_{\ell,\gamma}\sum_{m,\delta} \langle b_{i\alpha}^{v}(\mathbf{k},\mathbf{r})|b_{\ell\gamma}^{c}(\mathbf{k},\mathbf{r})\rangle\langle b_{m\delta}^{c}(\mathbf{k},\mathbf{r})|b_{j\beta}^{v}(\mathbf{k},\mathbf{r})\rangle\langle b_{\ell\gamma}^{c}(\mathbf{k},\mathbf{r})|H|b_{m\delta}^{c}(\mathbf{k},\mathbf{r})\rangle.$$

$$(2.34)$$

2.5. The Secular Equation

The new secular equation in the orthogonalized space is

$$\left| \langle b_{i\alpha}^{v'}(\mathbf{k},\mathbf{r})|H|b_{j\beta}^{v'}(\mathbf{k},\mathbf{r})\rangle - \langle b_{i\alpha}^{v'}(\mathbf{k},\mathbf{r})|b_{j\beta}^{v'}(\mathbf{k},\mathbf{r})\rangle E(\mathbf{k}) \right| = 0, \qquad (2.35)$$

which has its dimension drastically reduced. The main numerical error involved in arriving at (2.35) is the assumption of no core-core overlap between the Bloch sums. Test calculations indicate that the eigenvalues for states near the Fermi level or gap obtained from (2.35) differ from those obtained from (2.13) by less than 0.0001 eV. This is of negligible size in most electronic structure studies. One way to improve the accuracy is by treating some high lying core states as valence states in the orthogonalization scheme. It is also possible to retain the core states of one or a few atoms such that the electronic structures of these core states can be studied.

For amorphous calculations, only $\mathbf{k}=0$ case is relevant. The matrix equation is real which can be solved by standard diagonalization method using triangular decomposition.

This is a significant advantage over some other methods since complex matrix equations can take four to eight times more computer time to diagonalize. In general, the matrix in (2.35) is not sparse because the interaction integrals in the first-principles OLCAO method is sufficiently far-ranged. In dealing with complex systems, the dimension of (2.35) may still be formidably large, other time-saving numerical techniques such as iterative solution may be applied, but this has not been tested by us in calculations with metallic glasses.

2.6. Spin-polarized Calculations

For magnetic materials, spin-polarized calculation based on local spin density approximation (LSDA) becomes necessary [34-37]. In the LSDA calculation, spin becomes a variable in $V_{exch}(\mathbf{r})$. The most commonly used form for $V_{exch}(\mathbf{r})$ is the von Barth-Hedin potential [34] as modified by Moruzzi et al. [49] in which the exchange-correlation functional takes the form:

$$\varepsilon_{xc}[\rho_\uparrow,\rho_\downarrow] = \varepsilon_{xc}^p[r_s] + [\varepsilon_{xc}^f(r_s) - \varepsilon_{xc}^p(r_s)]f_{xc}(\rho_\uparrow,\rho_\downarrow) \tag{2.36}$$

The superscripts p and f in (2.36) stand for paramagnetic and ferromagnetic cases and r_s is defined as:

$$(4/3)\pi r_s^3 = 1/\rho,$$

ε_{xc}^f and ε_{xc}^p (2.36) are ferromagnetic and paramagnetic exchange-correlation energies worked out in the parameterized form by fitting results to that of a homogeneous electron gas [49], and f_{xc} is given by:

$$f_{xc}(\rho_\uparrow,\rho_\downarrow) = [(2\rho_\uparrow/\rho)^{4/3} + (2\rho_\downarrow/\rho)^{4/3} - 2]/(2^{4/3} - 2). \tag{2.37}$$

The spin-up states correspond to the case in which the spins are aligned along the direction of internal molecular field (usually taken as the z-direction) and the spin-down states have spins pointed in the opposite direction. The spin-up and spin-down states are coupled together via the exchange potential (2.37). The total charge density $\rho(\mathbf{r})$ in (2.9) is the sum of spin-up and spin-down charge densities: $\rho(\mathbf{r}) = \rho_\uparrow(\mathbf{r}) + \rho_\downarrow(\mathbf{r})$. The difference gives the spin-density function, $\rho_s(\mathbf{r}) = \rho_\uparrow(\mathbf{r}) - \rho_\downarrow(\mathbf{r})$. (For simplicity, we have suppressed the site index $\mathbf{A}$ for $\rho_A(\mathbf{r})$.) To start the self-consistent iterations in the spin-polarized calculation, it is necessary to give the spin-up and spin-down potentials on the magnetic atoms an initial splitting. Equation (2.13) or (2.35) is solved for each spin case and the results are merged to determine the new spin-polarized potential. In general, convergence toward self-consistency in a spin-polarized calculation is much slower because of the additional spin variable and the need for locating the Fermi level accurately.

2.7. Scalar Relativistic Correction

When the speed of an electron is comparable to the velocity of light as in the case of core electrons near the nuclei in heavy elements, relativistic effect must be considered. In this case the relevant equation to be solved is the Dirac equation [50]:

$$c\sigma \cdot \mathbf{p}\psi_v + (mc^2 - E - e\phi)\psi_u = 0, \tag{2.38}$$

$$c\sigma \cdot \mathbf{p}\psi_u - (mc^2 + E + e\phi)\psi_v = 0. \tag{2.39}$$

where c stands for the speed of light, σ are the Pauli spin matrices, $\mathbf{p}$ is the momentum operator, and ϕ is the scalar potential experienced by the electron. ψ_u and ψ_v are the so called "large" and "small" components of a four-component wave function each involving spin-up and spin down components. The ratio of ψ_u to ψ_v can be estimated to be 2c/v. For most problems in condensed matter, we need to concentrate only on the "large" component.

Let $E' = E - mc^2$, equation (2.38) and (2.39) take the form

$$(E'+e\phi)\psi_n = [1/2m](\sigma \cdot \mathbf{P})K(\sigma \cdot \mathbf{P})\psi_n. \tag{2.40}$$

The relativistic kinetic energy term K can be approximated as

$$K = \frac{1}{1+[(E'+e\phi)/2mc^2]} \approx 1 - \frac{E'+e\phi}{2mc^2}, \tag{2.41}$$

and the Dirac equation (2.38) is simplified to:

$$(E'+e\phi)\psi = \left[\frac{1}{2m}\mathbf{p}^2 - \frac{\mathbf{p}^4}{8m^3c^2} - \frac{e\hbar^2}{8m^2c^2}\nabla \cdot \nabla\phi - \frac{e\hbar}{4m^2c^2}\sigma \cdot \nabla\phi \times \mathbf{p}\right]\psi, \tag{2.42}$$

where the subscript u for the "large" component is dropped. Comparing (2.42) with the non-relativistic Schrödinger equation (2.4), three additional terms, mass-velocity, Darwin and spin-orbit coupling appear in the Hamiltonian. Only the spin-orbit term involves spin components, the Darwin and mass velocity terms have no off-diagonal matrix elements between the spin components, thus behave like a scalar. The term "scalar relativistic (SR) correction" is used if only these two terms are taken into account. The separation of spin-dependent and spin-independent parts of relativistic correction is extremely important in terms of practical calculations. In the SR case, the symmetry of the Hamiltonian is unchanged and the mass-velocity and Darwin terms can be absorbed into the one-electron potential and are the same for both spin components. Hence, SR correction can be applied to paramagnetic bands as well as the spin-polarized bands. In most cases, it suffices to consider only the SR effect since the spin-orbit coupling parameters are generally smaller by typically an order of magnitude [51].

In the OLCAO method, the basis functions are expanded in terms of GTO. The matrix elements of the mass-velocity and Darwin terms are easy to compute. For the mass-velocity term, we write:

$$p^4 = (p_x^2 + p_y^2 + p_z^2) \cdot (p_x^2 + p_y^2 + p_z^2) \tag{2.43}$$

where p_x, p_y, p_z are substituted by the corresponding differential operators. It is straightforward to show that differentiations between Bloch sums can be expressed as a series of lattice sums involving differentiations with GTOs. Contribution to the matrix element from the mass-velocity term can be evaluated by repeated differentiation using generalized overlap integrals (2.29).

The Darwin term involves the second derivative of the crystal potential which is expressed as a superposition of atom-centered potential functions $V_A(r)$. The second derivative of the V_A consists of GTO of higher order. With the use of the generalized formula for the three-center integrals similar to (2.29), the contribution from the atom-decomposed Darwin term can be evaluated analytically. However, it is more practical to simply calculate the Darwin correction as a function of r, and numerically fit to the same set of GTO as in $V_A(r)$. The correction then appears as a modification of the coefficient of each term in the original potential. However, problem does arise when repeated differentiations of V_A magnify the numerical error. In this case, it is expedient to separate the total one electron potential into nuclear, Coulomb and exchange-correlation parts:

$$V_A(r) = -\frac{Z}{r} + V_{coul}(r) + V_{xc}(r). \tag{2.44}$$

The correction due to nuclear charge Z is a δ - function of the form:
$$-\nabla \cdot \nabla(-Z/r) = 4\pi Z \delta(r). \tag{2.45}$$

Near the nucleus, only s electrons experience this interaction which is positive. This nuclear term dominates the Darwin contribution in the SR correction. Thus the Darwin correction offsets the mass velocity correction which is always negative. The second order derivative of $V_{Coul}(\mathbf{r})$ is related to the charge density through the Poisson's equation:

$$\nabla^2 V_{coul}(\mathbf{r}) = -4\pi\rho(\mathbf{r}).$$

(2.46)

The $V_{xc}(\mathbf{r})$ term is also related to ρ in the LDA theory. Hence, we may use $\rho_A(\mathbf{r})$ instead of the $V_A(\mathbf{r})$ in the evaluation of the Darwin term. In general, Darwin correction due to $V_{xc}(\mathbf{r})$ is much smaller than that of $V_{coul}(\mathbf{r})$.

2.8 Spin-Orbit Coupling

To account for the full relativistic effect, spin-orbit coupling term must be included in solving the Schrödinger equation and the dimension of the matrix equation is doubled. There are different approaches in treating spin-orbit coupling, depending on the system under study and the physical process involved. In semiconductors, one tends to concentrate on states at the band edges at a high symmetry point in the Brillouin zone. Perturbation theory for the degenerate states is usually applied [52]. The more general approach is to first establish a spin-polarized band structure and then add spin-orbit correction to it [53].

The spin-polarized band structure calculation provides two Hamiltonian matrices, one for the spin-up band (majority spin), the other for the spin-down band (minority spin) which are diagonalized separately. The spin-orbit interaction couples them together and the matrix size is doubled:

$$\begin{pmatrix} \mathbf{U} & 0 \\ 0 & \mathbf{D} \end{pmatrix} \rightarrow \begin{pmatrix} \mathbf{U'} & \Delta \\ \Delta^+ & \mathbf{D'} \end{pmatrix},$$

(2.47)

In (2.47), $\mathbf{U}$ ($\mathbf{U'}$) and $\mathbf{D}$ ($\mathbf{D'}$) are the original (corrected) Hamiltonian matrices for spin-up and spin-down states respectively, $\mathbf{0}$ is the null matrix, and Δ is the coupling matrix. The overlap matrices are the same for both spin cases. It is assumed that SR correction has been included in $\mathbf{U}$ and $\mathbf{D}$. To obtain explicit expressions for Δ, $\mathbf{U'}$ and $\mathbf{D'}$, we need to evaluate the matrix elements of the spin-orbit coupling term (2.42) between spin-polarized Bloch sums. In a truly first-principles approach, explicit expressions can be derived involving integrals of the derivatives of the crystal potential and GTOs [54]. For studying magnetic glasses or alloys with complex structures, such an approach is not practical. A simpler approach is to treat the strength of spin-orbit coupling ξ as a parameter, which can be obtained from either experiments or atomic calculations. In fact, spin-orbit interaction is rather short-ranged, matrix elements between Bloch sums can be well approximated by that between atomic orbitals in the central cell. In the parameterized form, the spin-orbit coupling term is written as:

$$\xi \mathbf{l} \cdot \mathbf{s} = \xi (\mathbf{j}^2 - \mathbf{l}^2 - \mathbf{s}^2)/2 \tag{2.48}$$

where $\mathbf{j}, \mathbf{l}, \mathbf{s}$ are the total, orbital and spin quantum numbers for the atom ξ. is given by the expression:

$$\xi = \frac{\hbar^2}{2m^2 c^2} \int |u_j(\mathbf{r})|^2 \, \frac{1}{r} \frac{dV_A(r)}{dr} \, d\mathbf{r}. \tag{2.49}$$

where V_A, $u_j(\mathbf{r})$ are the atomic potential and the orbital. For the particular state.

The matrix elements of (2.48) between atomic wave functions are quite simple. Due to the spin-polarization, the atomic wave functions are direct products $|\alpha, l, m\rangle |s, m_s\rangle$ where $|\alpha, l, m\rangle$ denotes the spatial parts and $|s, m_s\rangle$ the spin part (with $m_s = \pm 1/2$.). If we separate the angular quantum numbers l,m while absorb all other indices in α, the product functions can be transformed into coupling functions:

$$|l, m\rangle |s, m_s\rangle = \sum_{j, m_j} \langle l, s, j, m_j | l, m, s, m_s \rangle |l, s, j, m_j\rangle \tag{2.50}$$

where $\langle l, s, j, m_j | l, m, s, m_s \rangle$ are obtained from the inverse transformation of Clebsch-Gordon coefficients for two angular momenta coupling [55]. The matrix elements of (2.48) become straightforward because the coupling functions are simultaneous eigenfunctions of $\mathbf{j}, \mathbf{l}, \mathbf{s}$ and we have:

$$\xi \langle \mathbf{l} \cdot \mathbf{s} \rangle = \xi [j(j+1) - l(l+1) - 3/4]/2. \tag{2.51}$$

Test calculations for relativistic corrections in the OLCAO method using the above approach have been carried out for crystalline Ni, Nb, Ce in the SR limit [56]. Full relativistic correction including spin-orbit coupling was tested in the case of ferromagnetic Fe [56]. The results are in good agreement with similar calculations using other approaches.

2.9. Calculation of Physical Observables

From the calculated electronic structures, many physical observables can be obtained and compared with experiments. The most fundamental one is the density of states (DOS), or the number of electronic states per energy unit in the solid. For MG, or amorphous solids in general, the band structure lost its meaning because $\mathbf{k}$ is no longer a good quantum number and the concept of DOS is the most important. In crystalline solids in which sum over BZ and the presence of van-Hove singularities are important, special technique such as linear analytic tetrahedron method has been widely used for accurate determination of DOS. For amorphous solids represented by a large unit cells, sufficient number of energy eigenvalues are at hand and a simple histogram counting method, with appropriate broadening will be quite sufficient. From now on, the $\mathbf{k}$ dependence in the energy and wave function will be dropped, as well as the spin-dependence unless explicitly stated.

It is useful to define a fractional charge for the ith orbital of the αth atom, $\rho_{i\alpha}^m$ of the normalized state $\Psi_m(\mathbf{r})$ with energy E_m according to the Mulliken's population analysis scheme [31]:

$$1 = \int |\Psi_m(\mathbf{r})|^2 \, d\mathbf{r} = \sum_{i,\alpha} \rho_{i\alpha}^m \, , \tag{2.52}$$

$$\rho_{i\alpha}^m = \sum_{j,\beta} C_{i\alpha}^{m*} C_{j\beta}^m S_{i\alpha,j\beta}. \tag{2.53}$$

The fractional charge is a useful quantity and is the natural product with a method in which atomic orbitals are used in the basis expansion. As long as the basis functions are reasonably localized, Mulliken scheme is a simple and effective way of partitioning an electronic state into orbital species. The DOS can be resolved into partial components, or partial DOS (PDOS) by using $\rho_{i\alpha}^m$ as the projection operator. Thus the atom-, orbital- and spin-projected PDOS can be readily obtained which provide a wealth of information about the electronic bonding and interactions in the solid.

A useful quantity of great physical insight for electron states in a noncrystalline material is the localization index which can be defined from the fractional charges (2.53) as:

$$L_m = \sum_{i,\alpha} [\rho_{i\alpha}^m]^2. \tag{2.54}$$

L_m is a measure of the probability density of state m at different sites. L_m lies between 1/N where N is the total number of orbitals in the calculation for a completely delocalized state, to 1 for a completely localized state in which the charge is confined to a single orbital. With a large number of atoms in the model structure for amorphous solids, (2.54) enables us to make realistic estimate of the localization of the one-electron state in a disordered solid across the entire energy spectrum.

The effective charge on each atom can be obtained by summing our occupied orbitals:

$$Q_\alpha^* = \sum_{\substack{m \\ occ.}} \sum_i \rho_{i\alpha}^m \quad . \tag{2.55}$$

In a spin-polarized calculation, the effective charges can be further resolved into spin-up and spin-down components

$$Q_\alpha^* = Q_{\alpha_\uparrow}^* + Q_{\alpha_\downarrow}^* \quad , \tag{2.56}$$

and their difference gives the spin magnetic moment at the site :

$$\langle M_s \rangle_\alpha = Q_{\alpha_\uparrow}^* - Q_{\alpha_\downarrow}^* \quad . \tag{2.57}$$

The site-decomposed effective spin magnetic moment is very important in the study of magnetic properties of materials. The way $\langle M_s \rangle_\alpha$ is evaluated in the OLCAO method does not involve any arbitrary chosen parameters and is highly reliable. It is also straightforward to calculate the susceptibility enhancement in magnetic materials by exploiting the electronic states near the Fermi-level [49].

With spin-polarized bands including spin-orbit interaction, it is possible to estimate the orbital moments $\langle M_l \rangle$ for a crystal. Through spin-orbit coupling, $\langle M_l \rangle$ can be related to the local spin magnetic moments and magnetic anisotropy. In the OLCAO method, $\langle M_l \rangle$ can be expressed as:

$$\langle M_l \rangle = g \sum_{m,k} \sum_{i,\alpha} \sum_{j,\beta}^{occ.} C_{i\alpha}^{m*} C_{j\beta}^m \langle b_{i\alpha}(\mathbf{k},\mathbf{r}) | l_z | b_{j\beta}(\mathbf{k},\mathbf{r}) \rangle \tag{2.58}$$

106

where l_z is the z-component of the angular momentum operator. The matrix elements in (2.58) can be expressed as a lattice sum of overlap integrals since l_z only operates on the angular part of the atomic orbitals in the Bloch sum. The difficulty with orbital moment calculation for a crystal lies in the fact that (2.58) is not invariant under the operation of a wave vector group as in the case of energy eigenstate. Thus the summation over **k** covers the entire BZ instead of the irreducible portion of it. For amorphous materials, wave-vector **k** is irrelevant and the orbital moment may actually be easier to evaluate. To our knowledge, calculation of orbital moments in an amorphous magnetic material has not been attempted. Calculation of $\langle M_l \rangle$ for the hard magnet $Nd_2Fe_{14}B$ shows it to be much smaller than the spin moments on the Fe sites [57].

The transport properties of MG show many interesting physical phenomena unique to disordered alloys [58-62] such as the negative temperature coefficient of resistivity, Mooij correlation for resistivity and thermal power, resistivity saturation, sign reversal of the Hall coefficient, superconductivity, negative magneto-resistance etc.. Several theoretical models were suggested in the past to explain the observed data [63-77]. Most of these models were based on the concept of wave-vector scattering by free electrons. In recent years, the theory based on weak localization and quantum coherence has been quite successful in explaining many transport properties of MG [76,77]. Large scale quantum mechanical calculations of transport properties on MG systems can provide the necessary insight about the scattering process in disordered systems at the microscopic level.

Numerical calculation of transport properties in MG at low temperature starts with the evaluation an energy dependent conductivity function $\sigma(E)$ [78] according to the Kubo-Greenwood formula [79].

$$\sigma(E) = \frac{2\pi\hbar e^2}{3m^2\Omega} \sum_{n,m} \left| \langle n|\mathbf{p}|m \rangle \right|^2 \delta(E_n - E)\delta(E_m - E). \tag{2.59}$$

The double summation in (2.59) is over all energy states and the double delta function describes the scattering of an electron with energy E_n to that of energy E_m. $\langle n|\mathbf{p}|m \rangle$ is the momentum matrix element described in section 2.3. For transport property calculation, only the states near the Fermi level E_F are important.

The temperature dependent d.c. conductivity $\sigma(0,T)$ can be obtained from $\sigma(E)$ through

$$\sigma(O,T) = \int \left[\frac{\partial f(E)}{\partial E} \right] \langle \sigma(E) \rangle dE, \tag{2.60}$$

where f is the Fermi distribution function and $\langle \circ \circ \circ \rangle$ denotes configurational average. Here, $\langle \sigma(E) \rangle$ is interpreted as the average of calculations on as many independent structural models as possible.

The temperature dependence of $\sigma(0,T)$ in (2.60) is strictly through the Fermi distribution function f. At low temperature, f is a step function at E_F and the conductivity of the MG is simply $\sigma(E_F)$ while the resistivity ρ (not to be confused with the electron charge density) is given by $1/\sigma(E_F)$. The resistivity obtained according to (2.60) is purely due to the elastic scattering of the conduction electrons in a disordered solid. The momentum matrix elements and the distribution of energy E_m contain all the information about the quantum coherence and multiple scattering associated with the electron transport. Expression (2.60) is valid only at low temperature. At higher temperature, electron-honon interaction becomes important and must be appropriately accounted for.

In a calculation with a finite number of atoms in the model, the energy spectrum is discrete. The discrete energy spacing implies the δ function condition in (2.59) can never be satisfied at all energies. This difficulty is circumvented by replacing each discrete energy level by a Gaussian of unit area and a finite width. The larger the model, the smaller should be the width of the Gaussian and the more accurate would be the calculation. Hence, the accuracy in determining ρ or $\sigma(E_F)$ depends to a large extent on the state density at E_F which is different for different materials.

The thermopower S(T) can be obtained from $\langle \sigma(E) \rangle$ as :

$$S(T) = -\frac{\pi^2}{3}\frac{k^2 T}{e}\frac{\partial}{\partial E}\log\langle \sigma(E) \rangle \Big|_{E=E_f} \tag{2.61}$$

Because S(T) involves the derivatives of $\ln(\langle \sigma(E) \rangle)$, it is more difficult to evaluate accurately.

The optical properties of MG can also be calculated similar to crystalline metals [7]. The real part of the frequency-dependent optical conductivity $\sigma_1(w)$ can be calculated straightforwardly, using the same momentum matrix elements as in the conductivity function calculation:

$$\sigma_1(\omega) = \frac{2\pi e \hbar^2}{3m^2 \omega \Omega}\sum_{n,m}|\langle n|\mathbf{p}|m \rangle|^2 f_m[1-f_n]\delta(E_n - E_m - \hbar\omega). \tag{2.62}$$

The square of momentum matrix element in (2.62) is averaged over the three Cartesian directions since the optical properties in MG is expected to be isotropic. Because of the absence of the long range order, the optical spectra of MG usually lack rich structures common in crystalline solids.

In crystalline metals, the conductivity at low photon frequency due to intraband transitions is approximated by the Drude formula $\sigma_D = Ne^2\tau / m*(1+\omega^2\tau^2)$ where τ is the relaxation time and $m*$, the effective mass of the conduction electron. In the absence of translational symmetry in disordered solids, the distinction between the interband and intraband transitions no longer exist. The optical conductivity at low frequency can be calculated exactly using (2.62) subject only to the size of the model.

The imaginary part of the dielectric function $\varepsilon_2(\omega)$ can be obtained from σ_1:

$$\varepsilon_2(\omega) = 4\pi\sigma_1(\omega)/\omega, \tag{2.63}$$

and the real part of the dielectric function can be extracted from ε_2 through the usual Kramers-Kronig relation,

$$\varepsilon_1(\omega) = 1 + \frac{2}{\pi} P \int_0^\infty \frac{s\varepsilon_2(\omega)}{s^2 - \omega^2} ds. \tag{2.64}$$

The integration limit in (2.64) is usually replaced by a finite cutoff value, since $\sigma_1(\omega)$ is calculated only for a finite range of photon energy. It is also expedient to express $\varepsilon_2(\omega)$ as the optical absorption power $\alpha(\omega)$ involving wave length λ of the photon.

$$\alpha(\omega) = \varepsilon_2(\omega)/\lambda. \tag{2.65}$$

The transport and optical properties of several MG have been calculated within the general framework of the OLCAO-LDA theory [25-27]. In principle, it is possible to extend such calculation to magnetic glasses and to study magneto-optical properties using the spin-polarized electronic structures including spin-orbit interactions. Such complicated calculation is very challenging and has not been attempted.

3. Application to Metallic Glasses

In this section, we summarize some of the results on the electronic structures of MG using the OLCAO method. In the past two decades, significant progress has been made in the realistic calculations of electronic structures of amorphous alloys [2,6,26,27]. In this this area of endeavor, it is necessary to distinguish two structurally different disordered alloys: the randomly substituted binary or other multicomponent alloys and MG. The disordered nature of the former comes from potential fluctuations due to random substitution on a crystalline lattice. The system thus retains at least in part the concept of reciprocal lattice. For MG, additional topological disorder associated with the structure of the glass is present. Short range chemical order as well as intermediate range order are present in all MG. The second case is more difficult to treat, and the use of direct space OLCAO-LDA method for MG system is most effective.

There are several special features in the OLCAO method discussed in section 2 that make computation for MG very efficient. First, the minimal atomic basis set provides the best balance between practicality and accuracy. By adopting the same exponential set $\{\alpha_j\}$ in the basis function for all the atoms and all orbitals, the number of integrals that need to be calculated can be greatly reduced with little or no loss in accuracy. Second, the orthogonalization to the core technique reduces the effective dimension of the matrix equation such that much larger models can be studied. Third, all the interaction integrals are calculated in the real space with no restrictions on the symmetry of the system. Fourth, there is no artificial boundary on the potential functions. This is a very important point for topologically disordered solids, a point frequently ignored by many researchers. Fifth, there is no restriction on the types of elements involved in MG. And, sixth, spin-polarized version of the method enables us to study magnetic glasses. In what follows, we briefly summarize the calculated results on four representative glasses:(1) a-Ni, a single-component, metastable, weakly ferromagnetic transition metal glass; (2) a-$Ni_{1-x}P_x$, a transition metal-metalloid glass; (3) a-$Mg_{1-x}Zn_x$, a free electron-like glass; and (4) a-$Cu_{1-x}Zr_x$, an early transition metal (Zr) and a late transition metal (Cu) glass.

3.1. Periodic Structure Models

Molecular dynamics has been a popular simulational technique for the structures of both crystalline and non crystalline solids [80,81]. In those studies, focus has been on the dynamic properties and temporal behavior of the system. Properties are evaluated in a statistically averaged sense with no individual atom playing an essential role. This is different from the simulation and construction of periodic models for electronic structure studies of glasses. In the later case, the bonding for all atoms have to be reasonable and the calculation is similar to that for a very complex crystal. A highly distorted local environment for one atom that is inconsequential in molecular dynamic simulation may turns out to be fatal in the electronic structure calculation. We construct the structural models for MG by means of Monte-Carlo relaxation method [82]. For each glass considered, a fixed number of atoms are randomly distributed in a cubic cell of volume L^3. The cell dimension L is determined by the mass density of the glass which is usually obtained from the experiment. The total number of atoms N in the cell is mainly dictated by the type of atoms in the glass and the computational resource available. With the OLCAO method and a reasonable computer facility (for example, a topline Vax α-station), N can be as large as 1200 for atoms involving no d electrons and 500 for transitional metals with d-electrons. Pair-wise Leonard Jones type of potentials are used to describe the effective interactions between atoms. The parameters for the pair potentials are generally obtained from cohesive energy data and interatomic distances of relevant crystalline compounds with small adjustments so as to reproduce the first peak position in the calculated radial distribution function (RDF) of the relaxed model the same as in the experiment. It is also possible to obtain effective pair potentials by fitting to data obtained from total energy calculation in the LDA approximation on smaller crystals and molecules. At each Monte-Carlo step, interactions between pairs of atoms up to a distance of L/2 are included. Periodic boundary condition is imposed by assigning displacement vectors to each atom when considering its interactions with other atoms. For atoms in the interior of the cell, the displacement vectors are zero vectors. Initially, the atoms are randomly distributed in the cell, corresponding to a high temperature configuration. Following the standard Monte-Carlo algorithm [82] according to a Boltzman distribution, the elastic energy and the force on each atom are calculated, and the atoms are moved in a sequential order. The fictitious temperature 'T' is then successively reduced, accompanied by a graduate reduction in the average elastic energy of the cell. After a sufficiently large number of Monte-Carlo steps, the total elastic energy stabilizes and an equilibrium low temperature model structure for the glass is obtained. The quality of the model is checked by comparing the calculated RDF of the model with the measured one. A typical RDF for a-$Fe_{80}B_{20}$ glass with 160 Fe atoms and 40 B atoms is illustrated in the Fig. 2.3.

In spite of a rather modest size of the model, the periodic boundary condition and the exact mass density ensure the model representing an infinite solid. In a cluster type of model with a free surface, the density cannot be fixed exactly and the relaxation process can result in inhomogeneous and porous regions in the model. There is a common misconception that the periodicity in the model implies the use of Bloch theorem and is therefore inappropriate for amorphous systems. The fact is that for a sufficiently large model, the periodic boundary condition is a simple mathematical 'trick' to circumvent the problem of free surface in a finite calculation.

The differences between a periodic model with substitutional disorder, a periodic model with topological disorder and a cluster model are illustrated in Fig. 2.4.

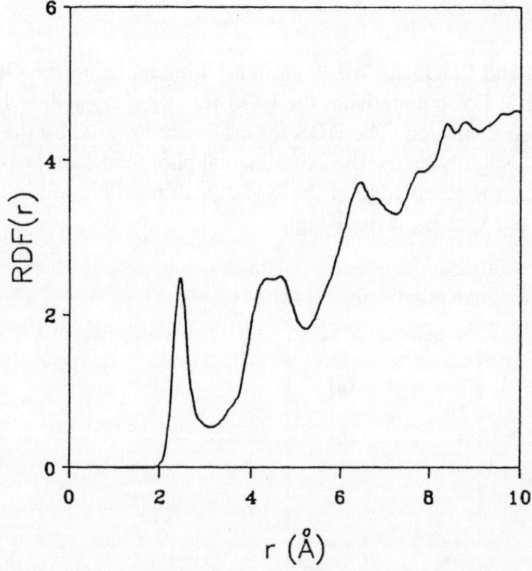

FIG. 2.3:   The RDF of the a-$Fe_{80}B_{20}$ model with 160 Fe atoms and 40 B atoms.

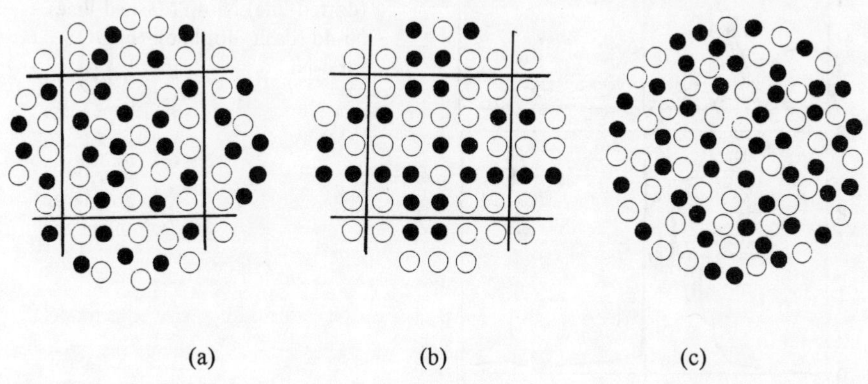

FIG. 2.4:   (a) Model structure for: (a) a two-component MG with periodic boundary; (b) random substitutional alloy with periodic boundary; (c) a cluster model for MGwith free surface.

112

3.2. Amorphous Ni and Ni-P Glasses

The calculated DOS and PDOS for a 200 atom a-Ni model using the OLCAO method [26] is shown in Fig. 2.5. For comparison, the DOS for f.c.c crystalline Ni (c-Ni) using the same potential is also displayed. The DOS for a-Ni is dominated by the Ni-3d peak at about -1.8 eV below E_F, slightly below the experimental photoemission peak for c-Ni [83]. The Fermi energy cuts at the steep edge of the Ni-3d peak resulting in a d-band hole. The DOS value of 0.47 states/eV at E_F, is quite high.

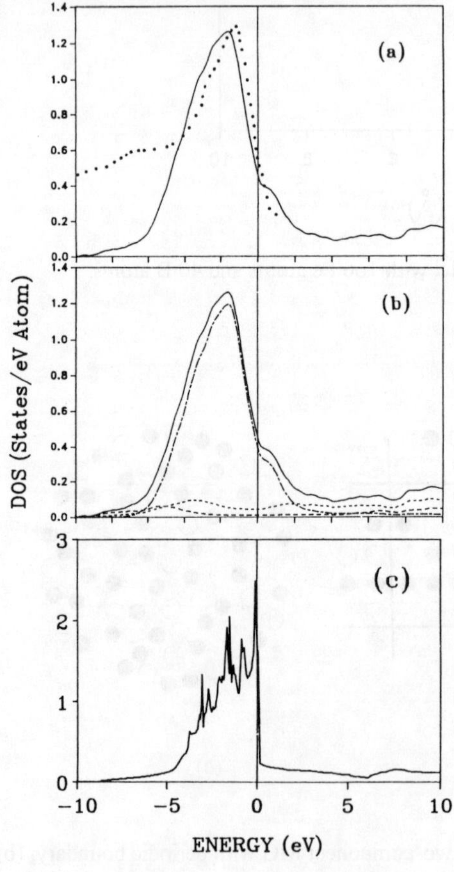

FIG. 2.5: (a) Total DOS of a-Ni. The dots are data forc-Ni from ref. 83. (b) PDOS for Ni-4s (dotted-line);Ni-4p (dashed-lline), Ni-4d (dash -dot line). (c) DOS for c-Ni.

In Fig. 2.6, we show the PDOS for a-$Ni_{1-x}P_x$ glasses with x=0.15, 0.20, and 0.25 using smaller models of only 100 atoms per cell [21]. The DOS are resolved for different types of Ni atoms depending on the number N of P atoms as nearest neighbors. (In MG, a nearest neighbor atom is defined as one whose distance of separation is smaller than the first minimum in the RDF). The PDOS of Ni-N is sensitive to the short range order. They range from a very sharp peak for Ni atoms with no P as a nearest neighbor, to rather broadened profiles with lower peak position for Ni atoms with increasing number of P atoms as nearest neighbors. The PDOS of P atoms at the bottom of Fig. 2.6 show very little variation with x because states for P are deeper and more localized, hence are less sensitive to local potential fluctuations.

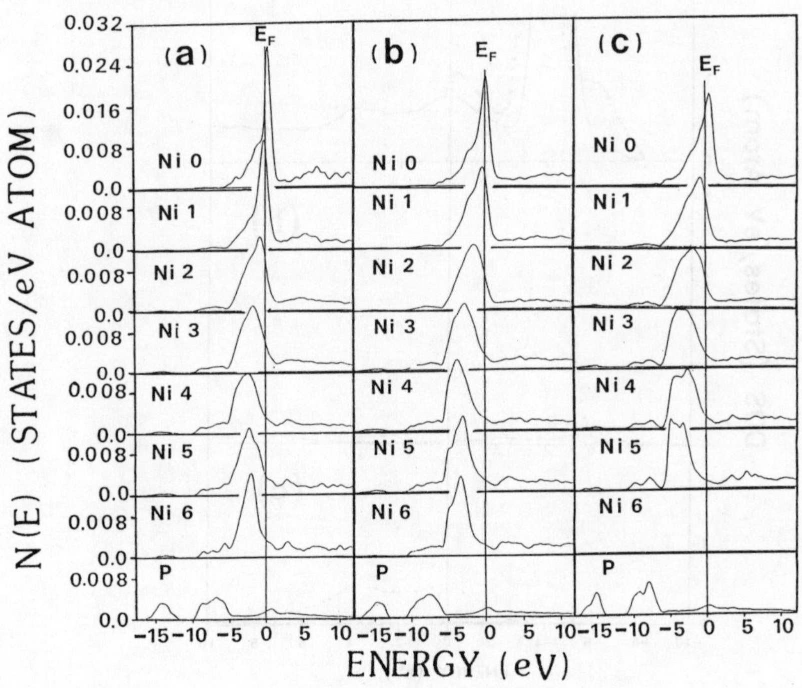

FIG. 2.6: DOS per atom for Ni-N where N denotes the NO of P atoms as nearest neighbors, and P: (a) a-$Ni_{75}P_{25}$; (b) a-$Ni_{80}P_{20}$; (c) a-$Ni_{85}P_{15}$.

3.3. Amorphous Cu-Zr Glasses

A-$Cu_{1-x}Zr_x$ is probably the most well-studied metallic glass of all. It serves as a paradigm case in many instances. The electronic structure for a-$Cu_{1-x}Zr_x$ with x = 0.33, 0.50 and 0.67 has been studied using the OLCAO method with 90-atom models [23] and later with 200-atom models [26]. The later calculation used atomic-like potentials derived from a fully self-consistent pseudo-crystal alloy calculation and is more accurate. The DOS and the PDOS for a-$Cu_{60}Zr_{40}$ are shown in Fig. 2.7.

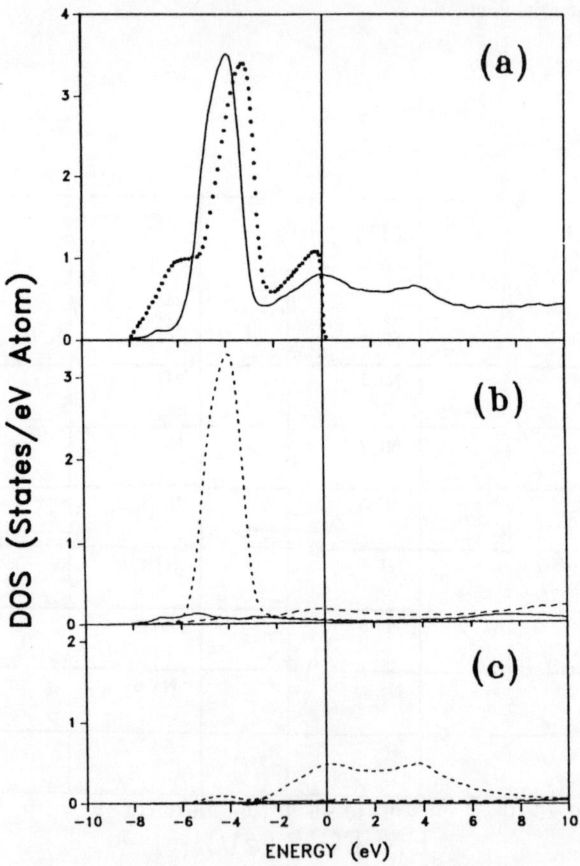

FIG. 2.7:   DOS of a-$Cu_{60}Zr_{40}$: (a) total; dots are data from ref. 86. (b) PDOS for Cu-4s (solid-line), Cu-4p (dashed-line) and Cu-3d (dotted-line). (c) PDOS for Zr-5s (solid-line), Zr-5p (dashed line) and Zr-4d (dotted line).

The DOS spectrum can be simply interpreted as the overlap of a narrowly peaked Cu-3d band (at -3.91 eV) and a more broad Zr-4d band at a higher energy. The DOS at the Fermi level is 0.80 states/eV-atom which is in close agreement with the experimental measurements [84,85]. The overlap of the Cu-3d band and the Zr-4d band results in a minimum in the DOS at -2.30 eV. These features are in good agreement with the photoemission data [86,87] reproduced in Fig. 2.7. The calculated Cu-3d peak is about 0.5 eV lower than the measured one.

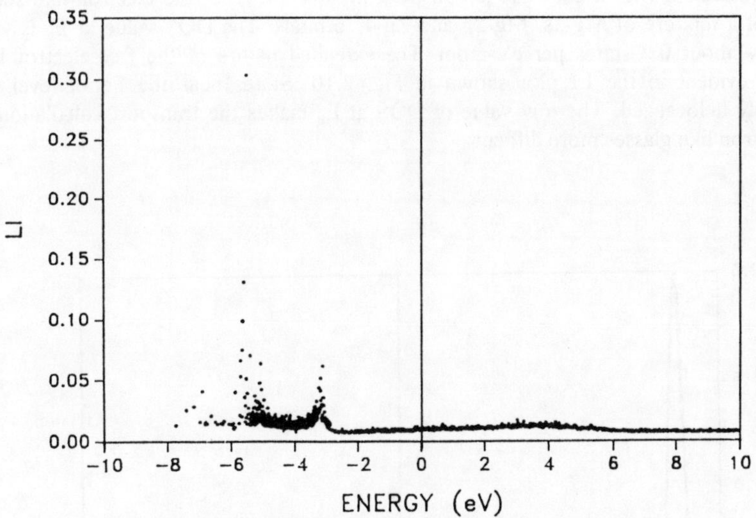

FIG. 2.8:   Localization index of a-$Cu_{60}Zr_{40}$.

The LI for states in a-$Cu_{60}Zr_{40}$ are plotted in Fig. 2.8. States at the edge of the Cu-3d band are relatively localized and states at the edge of the Zr-4d band are not. This is because the Zr-3d band is much wider than the Cu-3d band. Since the Fermi level is near the broad peak of the Zr DOS, states near E_F in a-$Cu_{60}Zr_{40}$ glass are not highly localized contrary to what was generally envisioned for a TM glass. On the other hand, they are not as delocalized as those in the free-electron-like MG such as a-$Mg_{75}Zn_{25}$.

116

3.4. Amorphous Mg-Zn Glasses

The electronic structure of tne free electron-like glass a-$Mg_{1-x}Zn_x$ has been studied by the OLCAO method using a 200-atom model [26]. This MG is not relevant to magnetic systems and is included here for comparative purpose. The DOS for a-$Mg_{75}Zn_{25}$ is shown in Fig. 2.9. The dashed line in the same figure is obtained by diagonalizing the Hamiltonian at '$\mathbf{k}$' = $(1,1,1)[2\pi/a]$ and is compared with the solid line for '$\mathbf{k}$' =0. The difference of the two is an indicator of the adequacy of the cell size. As can be seen, a 200-atom model for a free-electron-like glass is barely sufficient. For transition metal glasses such as a-Ni or a-$Cu_{1-x}Zr_x$, similar tests show no discernible differences between the two curves.

The DOS of a-$Mg_{75}Zn_{25}$ is completely different from that of a-Ni or a-$Cu_{1-x}Zr_x$. It is almost featureless with a core-like Zn-3d peak at -8.4 eV. The free electron-like states consist of a mixture of Mg-3s, Mg-3p and Zn-4s orbitals. The DOS value at E_F is very low, only about 0.2 states per eV-atom. The extended nature of the free-electron-like states is evident in the LI plot shown in Fig. 2.10. States near the Fermi level are completely delocalized. The low value of DOS at E_F makes the transport calculation in free-electron like glasses more difficult.

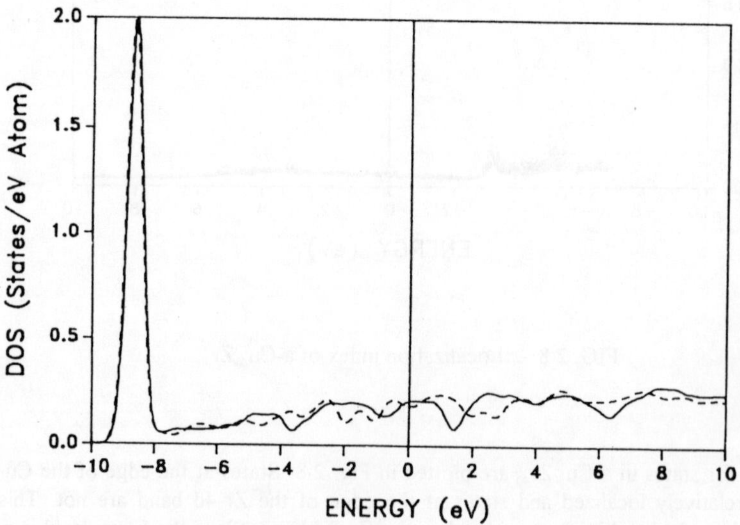

FIG. 2.9:   Calculated DOS for a-$Mg_{75}Zn_{25}$ obtained at $(0,0,0)[2\pi/a]$ and $(1,1,1)[2\pi/a]$ points of the quasi Brillouin zone.

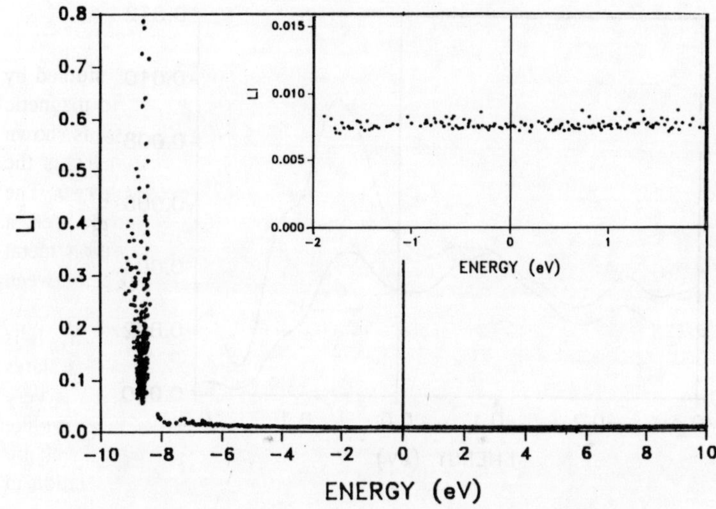

FIG. 2.10:   Localization index for a-Mg$_{75}$Zn$_{25}$. Inset shows the states near E$_F$ on a finer scale.

3.5. Transport Properties

The transport properties of several MG discussed above have been studied using rather modest size models [25,27].

The calculated energy-dependent conductivity function $\langle \sigma_E \rangle$ for a-Ni is shown in Fig. 2.11. Because the size of the model is small (N=100), the configuration average in $\langle \sigma_E \rangle$ is taken over 8 'k' points of the small quasi-Brillouin zone. This is less desirable than configuration average over independent models. A distinct feature of $\langle \sigma_E \rangle$ is that there exits a minimum in $\langle \sigma_E \rangle$ near E$_F$ which is important in understanding transport properties in a-Ni. The low conductivity value or (high resistivity) at E$_F$ is consistent with the rather localized nature of the electron wave functions near and above E$_F$. Also shown in Fig. 2.12. are the calculated temperature-dependent resistivity ρ(T) and thermopower S(T). At low temperature (T < 200K), the temperature coefficient of resistivity α is negative in agreement with experimental observation. The calculated thermopower is also negative and decreases rather rapidly as temperature is increased. The measured resistivity at room temperature for metastable a-Ni film is about 100 $\mu\Omega$-cm which is close to the calculated ρ of 111 $\mu\Omega$-cm at 200 K. The corresponding data for liquid Ni metal is about 85-87 $\mu\Omega$-cm [88].

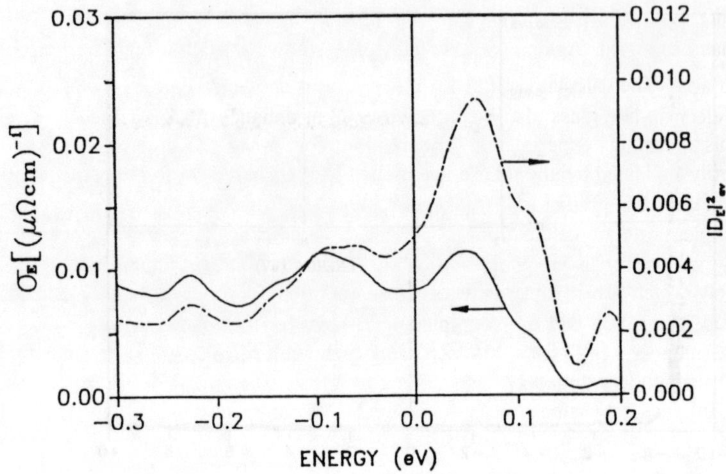

FIG. 2.11: Calculated $\sigma(E)$ for a 100-atom a-Ni model.

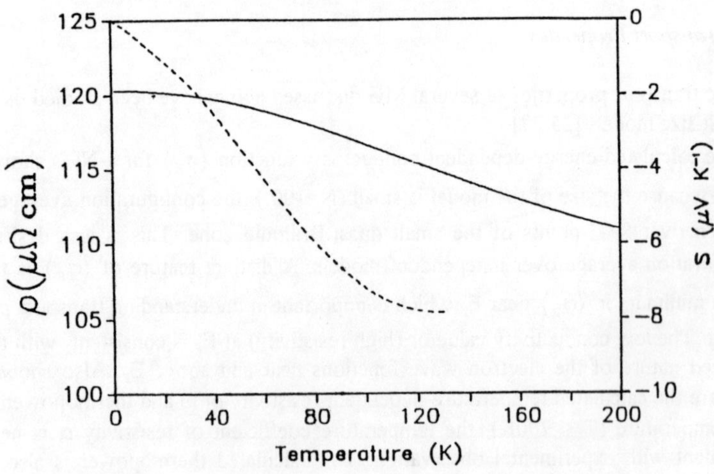

FIG. 2.12: Calculated $\rho(T)$ and $S(T)$ for the 100-atom a-Ni model.

Similar results for a-Mg$_{75}$Zn$_{25}$ is shown in Figs. 2.13. and 2.14. In this case, a 200-atom models has been used. Again, $\langle\sigma_E\rangle$ in Fig. 2.13 sthows a shallow minimum around E$_F$ similar to a-Ni. The calculated ρ(T) for a-Mg$_{1-x}$Zn$_x$ is shown in Fig. 2.14 which indicates this free electron-like glass also has a negative α in defiance with the Mooij correlation [89]. This puzzling experimental observation for a-Mg$_{1-x}$Zn$_x$ glass has not been satisfactorily explained by any theoretical model. Models based on free-electron scattering mechanism generally predict a positive α for a free-electron-like MG. It is satisfying that a first-principles type of calculation is capable of predicting this result. The calculated resistivity values for a-Mg$_{70}$Zn$_{30}$ and a-Mg$_{75}$Zn$_{25}$ at 2K^0 are 68 and 67.8 $\mu\Omega$ cm respectively which are in the range of those reported for coevaporated a-MgZn films (ρ=76-85 $\mu\Omega$ cm) [90], and for melt spinning samples (ρ=55 $\pm$ 5$\mu\Omega$ cm) [91].

The thermopower S(T) for a-Mg$_{1-x}$Zn$_x$ is more difficult to calculate accurately. Only the order of magnitude estimate is possible. The calculated value of -0.95 mVK^{-1} for S(T) in a-Mg$_{70}$Zn$_{30}$ has the same sign as the reported experimental value of -0.20 mVK^{-1} [92].

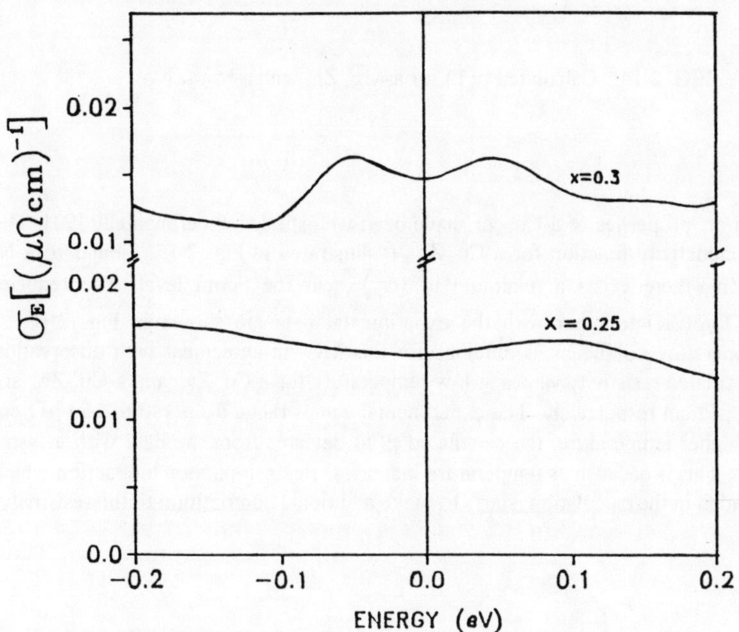

Fig. 2.13 : Ca'culated σ(E) for a-Mg$_{70}$Zn$_{30}$ and a-Mg$_{75}$Zn$_{25}$.

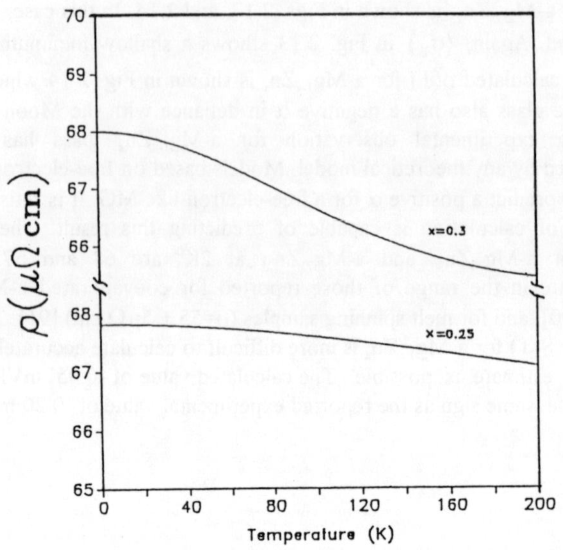

FIG. 2.14: Calculated $\rho(T)$ for a-Mg$_{70}$Zn$_{30}$ and a-Mg$_{75}$Zn$_{25}$.

The transport properties of a-Cu$_{1-x}$Zr$_x$ have been well studied experimentally [93]. The calculated conductivity function for a-Cu$_{60}$Zr$_{40}$ is illustrated in Fig. 2.15. Similar to a-Ni and a-Mg$_{1-x}$Zn$_x$, there exists a minimum in $\langle\sigma_E\rangle$ near the Fermi level. The reduced resistivity $\rho(T)/\rho(2K)$ together with the experimental data are shown in Fig. 2.16. A negative temperature coefficient is obtained for this MG in agreement with observation [94]. The calculated resistivity values at low temperature for a-Cu$_{60}$Zr$_{40}$ and a-Cu$_{50}$Zr$_{50}$ are 197 and 192 $\mu\Omega$ cm respectively. The experimental values range from 180 to 250 $\mu\Omega$ cm [93,95]. At higher temperature, the calculated $\rho(T)$ deviates from the data with a faster decrease in ρ. This is because as temperature increases, electron-phonon interaction which was not included in the calculation, starts to make additional contribution to the resistivity.

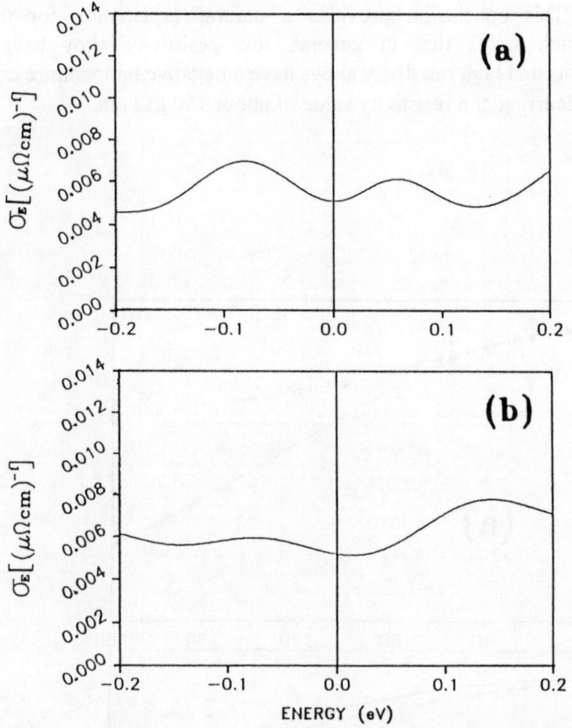

FIG. 2.15: Calculated $\sigma(E)$ near E_F for: (a) a-$Cu_{60}Zr_{40}$ and (b) a-$Cu_{50}Zr_{50}$.

The transport properties calculations for several MG described above take into account the short and intermediate range order of the glass. Hence, the effects of multiple scattering, localization and quantum interference are implicitly included. The elastic electronic scattering in a disordered medium, irrespective of the strengthof the scatterers, is treated exactly within the limitations of the local density theory and the adequacy of the structural model. On the other hand, electron-phonon interaction is totally neglected. It is gratifying to see that the calculated low temperature resistivities of all these MG are close to experimentally measured values. All three MG are predicted to have negative temperature coefficients. At a higher temperature, electron-phonon interaction becomes increasingly important. If electron-phonon interaction is not strong enough to overcome the effect of elastic disorder scattering, as is in the case of a-$Cu_{60}Zr_{40}$, the temperature coefficient remains negative. For a metallic glass where the elastic disordered scattering is weak, electron-phonon interaction may be sufficiently strong to make the temperature

122

coefficient positive. This conclusion provides a natural explanation for the Mooij correlation [89], which states that in general, low resistivity alloy have positive temperature coefficients and high resistivity alloys have a negative temperature coefficients with the cross over occurring at a resistivity value of about 150 $\mu\Omega$ cm.

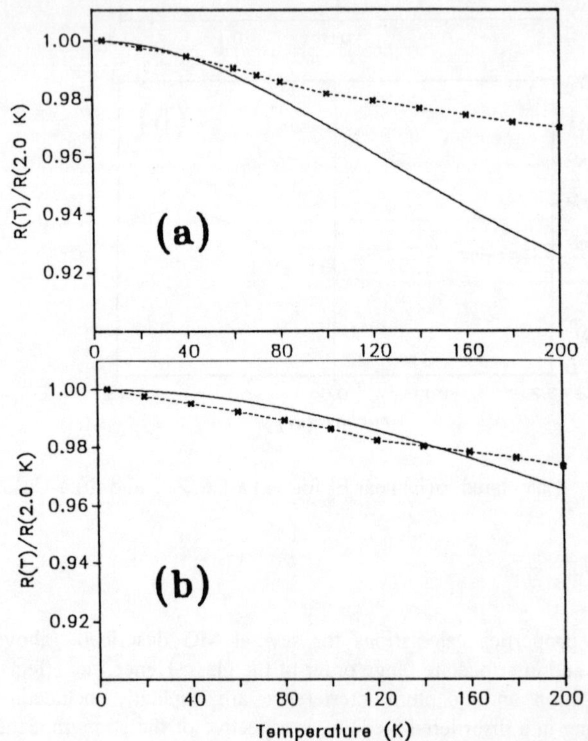

FIG. 2.16: Calculated $\rho(T)$ for: (a) a-$Cu_{60}Zr_{40}$ and (b) a-$Cu_{50}Zr_{50}$. Experimental data from ref. 95.

3.6. Optical Properties

The optical properties calculations of MG have also been attempted. Fig. 2.17 shows the calculated real part of optical conductivity $\sigma_1(w)$ for a-Ni using the 100-atom model [27]. Below 0.5 eV, the optical absorption increases dramatically, characteristic of a Drude-type behavior. The experimental conductivity for crystalline Ni [95] is also plotted for comparison.

The calculated optical absorption coefficient of a-$Mg_{70}Zn_{30}$ is shown in Fig. 2.18 using the 200-atom model[27]. At low energy, a Drude-type of absorption is even more evident as would be expected for a free-electron-like metal. Also shown is one set of experimental data [96]. Above 2 eV, the agreement with experiment is excellent. Both the calculation and the data show a featureless, graduate decreasing absorption spectrum. This result is consistent with a rather featureless DOS above and below E_F for a-$Mg_{70}Zn_{30}$.

Fig. 2.19 shows the calculated optical absorption for a-$Cu_{50}Zr_{50}$ using a 150-atom model [27]. At low frequency (< 0.3 eV), similar Drude-type of behavior is again quite evident. The main features are the existence of some minima in the background of a gradually decreasing absorption profile up to 8 eV. Also shown is one set of experimental data [97] for comparison. The agreement is reasonably satisfactory.

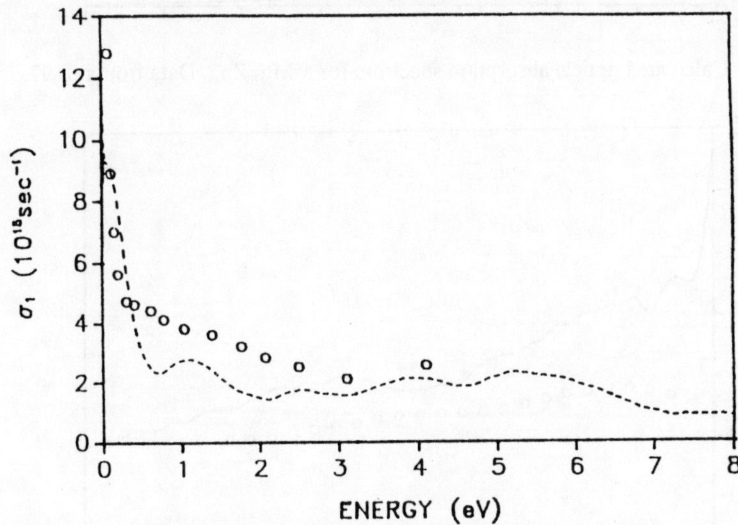

FIG. 2.17: Calculated optical conductivity for a-Ni.

The optical properties calculationsfor MG show general agreement with experiments. Of particular significance is the fact that the Drude term can be directly evaluated in a first-pinciples manner. This is because in MG, one need not to differentiate between the inter-band the intraband transitions. The information on the scattering process in a disordered environment and the accompanying relaxation process of the conduction electrons are all contained in the electron wave functions.

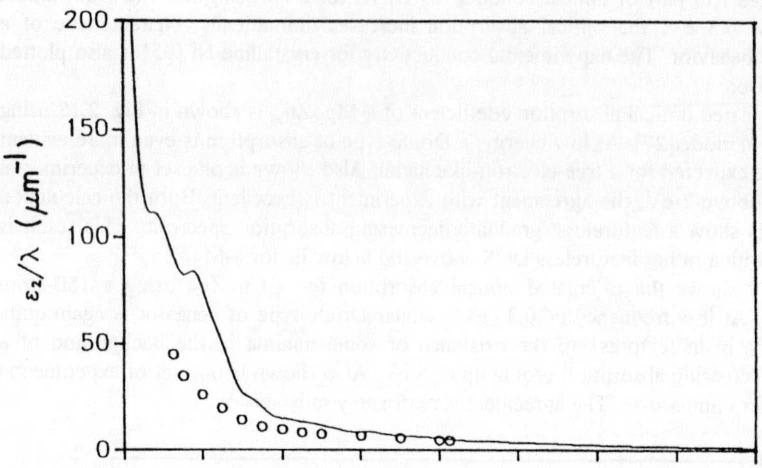

FIG. 2.18: Calculated optical absorption spectrum for a-$Mg_{70}Zn_{30}$. Data from ref. 97.

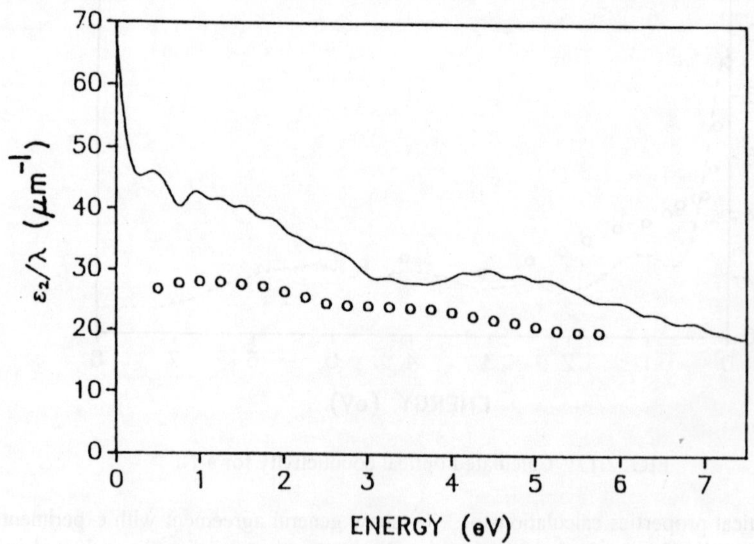

FIG. 2.19: Calculated absorption power for a-$u_{50}Zr_{50}$. Data points from ref. 98.

4. Application to Magnetic Glasses

4.1. Crystalline FeB, Fe₂B and Fe₃B Compounds

In this section, we discuss applications of the self-consistent OLCAO method to the magnetic materials by means of spin-polarized calculations. The intermetallic iron boron series FeB, Fe_2B and Fe_3B constitute an ideal testing ground because it allows us to investigate the change in magnetic properties with different B concentrations with precisce local atomic environments, thereby providing valuable insights on the electronic and magnetic structures of $Fe_{1-x}B_x$ glasses [99].

The three FeB crystals have different crystal structures. FeB has an orthorhombic unit cell containing 4 formula units [100] (space group *Pnma.*). Fe_2B has a body-centered tetragonal cell (space group *I4/mcm*) with two formula units per cell [101]. For Fe_3B, two phases exist, a body-centered tetragonal (bct) phase and the metastable orthorhombic phase (space group *Pbnm*) which has four formula units in the cell [102]. We only consider the simpler orthorhombic phase of Fe_3B. The three crystals differ in the short-range-order for Fe atoms due to different B concentrations. The Fe-Fe separations can be larger or smaller than the Fe-Fe separation of 2.482 Å in bcc Fe. The Fe-B and B-B distances also vary. The short range order in FeB, Fe_2B and Fe_3B does not necessarily correlate with B concentration, but depends on the actual crystal symmetry of each compound. In Fe_3B, the B-B distances are much larger because of the low B concentration and the local atomic arrangement may already resembles that of a-$Fe_{75}B_{25}$ glass in which close metalloid-metalloid contact is not expected. Recent Mössbauer isomer shift data clearly show the dependence on the local environment of Fe at different sites [103].

Fig. 2.20 shows the spin-projected partial DOS for Fe_3B, F_2B and FeB. For comparison, result for bcc Fe is also included. Even with 50% of B concentration in FeB, the compound is still ferromagnetic. Low lying states in the range of -10. to -14 eV are derived from B atoms. Effective charge calculation shows B atoms to be electron acceptors in the F-B compounds. The calculated spin magnetic moments (M_s) at the Fe sites are 1.26 and 1.95 μ_B for FeB and Fe_2B respectively, which compare favorably with the measured values of 0.9 and 1.6 μ_B respectively [104]. For Fe_3B, the calculated M_s values for the two Fe sites are 1.91 and 2.01 μ_B. The difference in the local moments at two different Fe sites indicates the sensitivity of the Fe moment to the local short range order in the crystal. For the pure bcc Fe, the calculated M_s value is 2.15 μ_B, the same as the experimental value. The B atoms are oppositely polarized in all three compounds, in agreement with neutron scattering measurements [104]. The calculated moments are -0.10, -0.23 and -0.29 μ_B per B atom in FeB, Fe_2B and Fe_3B respectively. In Fig. 2.21, the negative polarizations of B atoms are more clearly shown in the spin density maps in crystalline planes containing both Fe and B atoms. It also shows the opposite spin density around the B atom increases as B concentration decreases.

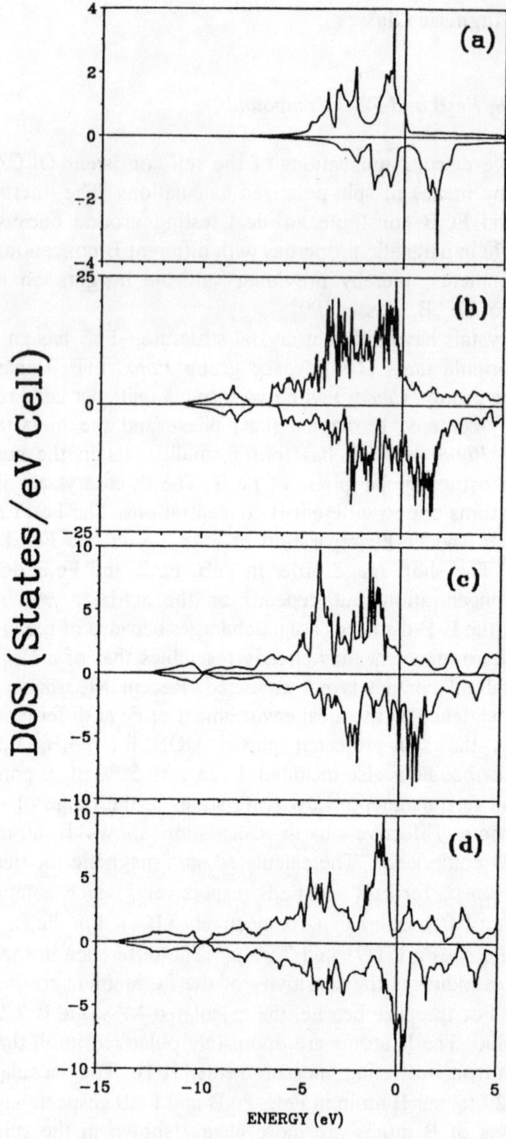

FIG. 2.20: Spin projected PDOS for (a) bcc Fe; (b) Fe$_3$B; (c) Fe$_2$B and (d) FeB. Positive (negative) for spin-up (down) bands.

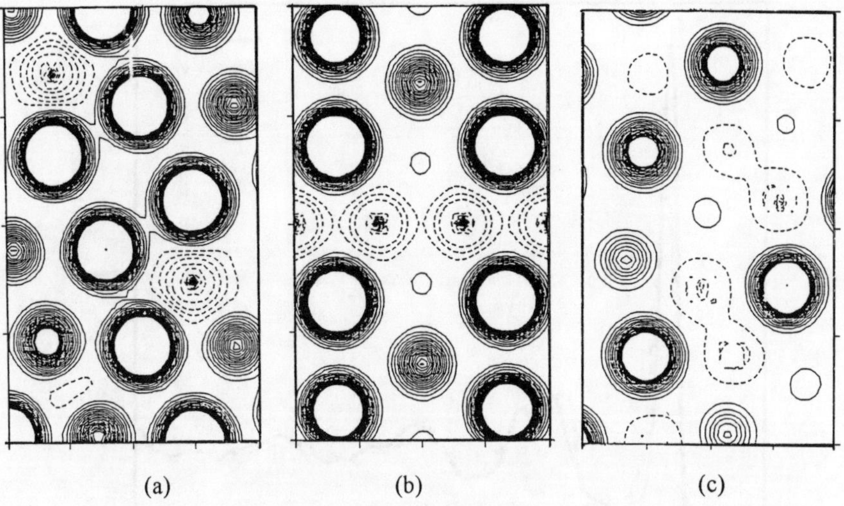

FIG. 2.21: Calculated spin density contour for (a) FeB_3; (b) FeB_2 and (c) FeB. The contours are from 0.002 to 0.04 a.u.$^{-3}$ in units of 0.001. a.u.$^{-3}$. Dashed lines indicate negative values

4.2. Amorphous Fe Model

Pure amorphous Fe is not easy to prepare in laboratories [105] and its low temperature magnetic behavior may be pertinent to that of a spin glass [106]. Nevertheless, it provides a good model to investigate the effect of disorder on the magnetic properties by comparing the results with that for crystalline Fe. The electronic structure of a-Fe has been studied with a 200-atom model [107] with a cell dimension of 13.808 Å corresponding to a mass density of 7.04 gm/c.c.. This model has a larger density deficit than the usual few percent lower than the crystalline phase. The model gives a pair distribution function G(r) in good agreement with electron diffraction data [108] which is shown in Fig. 2.22.

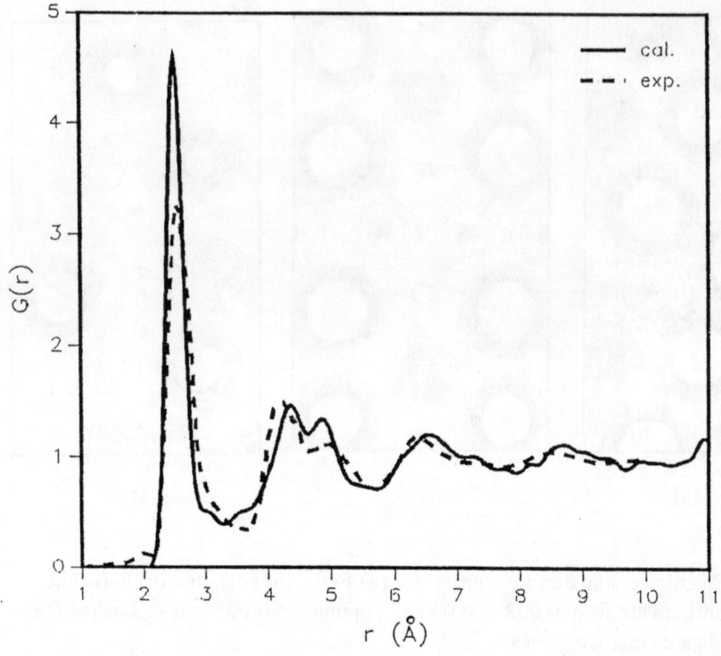

FIG. 2.22:   PDF for the 200-atom a-Fe model. Dashed line is the data from ref. 108.

The calculated DOS for the spin-up and spin-down states for this model is shown in Fig. 2.23 and is quite different from that of bcc Fe shown in Fig. 2.20. The majority spin band has the major peak at -0.5 eV while the minority spin band has the major peak at 1.8 eV above E_F. The DOS at E_F for the spin-down state is three times as large as that for the spin-up states. In bcc-Fe, Fermi level resides in a deep valley in the spin-down DOS such that the value of DOS at E_F is larger for the spin-up states. No such deep minimum in the DOS exists in the amorphous case.

The calculated effective charges and spin magnetic moments for the Fe atoms in the a-Fe model are shown in Fig. 2.24. The effective charge plot shows a larger dispersion for the spin-down states than for the spin-up states. This is because the Fermi level cuts at the middle of the spin-down band. The effective charges are very sensitive to the energy levels of individual Fe atoms relative to E_F, resulting in a large scattering for the spin-

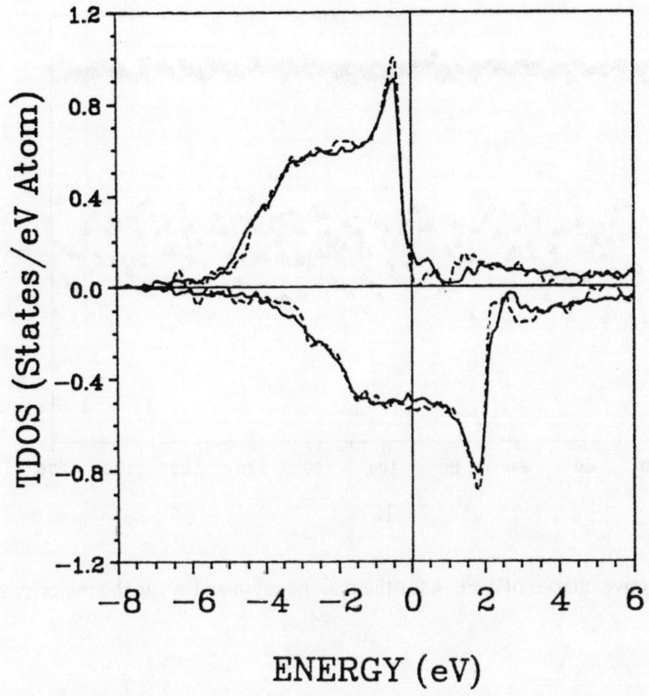

FIG. 2.23: Calculated DOS for the 200 atom a-Fe model. Upper (lower) panel for the spin-up (down) states. The dashed line is the calculated result at another 'k' point.

down effective charges. For the spin-up states, E_F is in a region of low DOS and no large variation in effective charges is expected. The M_s values obtained from $Q_\uparrow^* - Q_\downarrow^*$ for the 200 Fe atoms in the model vary from $1.68\mu_B$ to $3.56\mu_B$. The average value is $2.46\mu_B$ which is larger than $2.15\ \mu_B$ for bcc Fe. The increase in M_s value in a-Fe is related to the larger Fe-Fe separations in the model and the large difference between the spin-up and spin-down DOS at E_F. A histogram distribution of the 200 Fe moments in this model is shown in Fig. 2.25 which peaks at approximately $2.25\ \mu_B$. This distribution is not symmetric but skews towards the lower values with the maximum below the mean value.

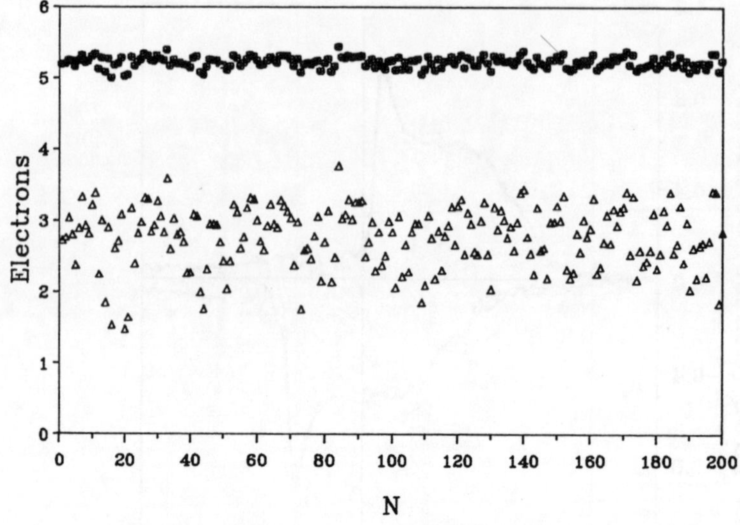

Fig. 2.24:   Effective charge of a-Fe. ● for the majority spin and Δ for the minority spin.

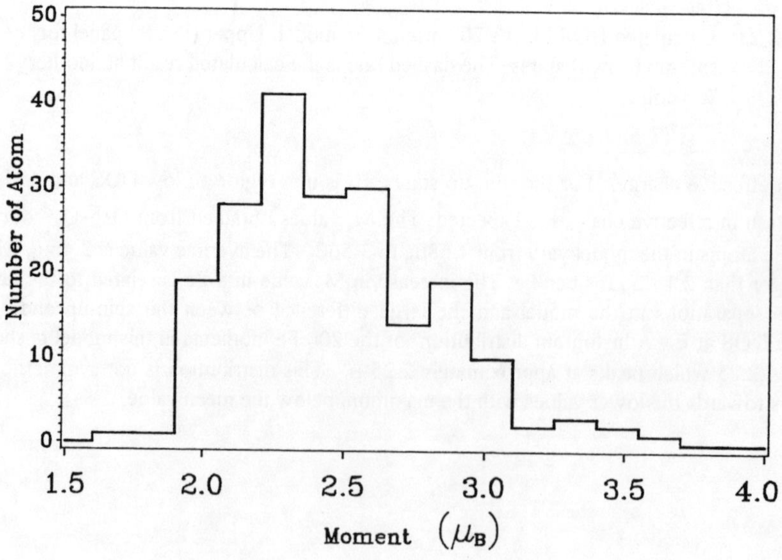

FIG. 2.25 Distribution of local moments in a-Fe.

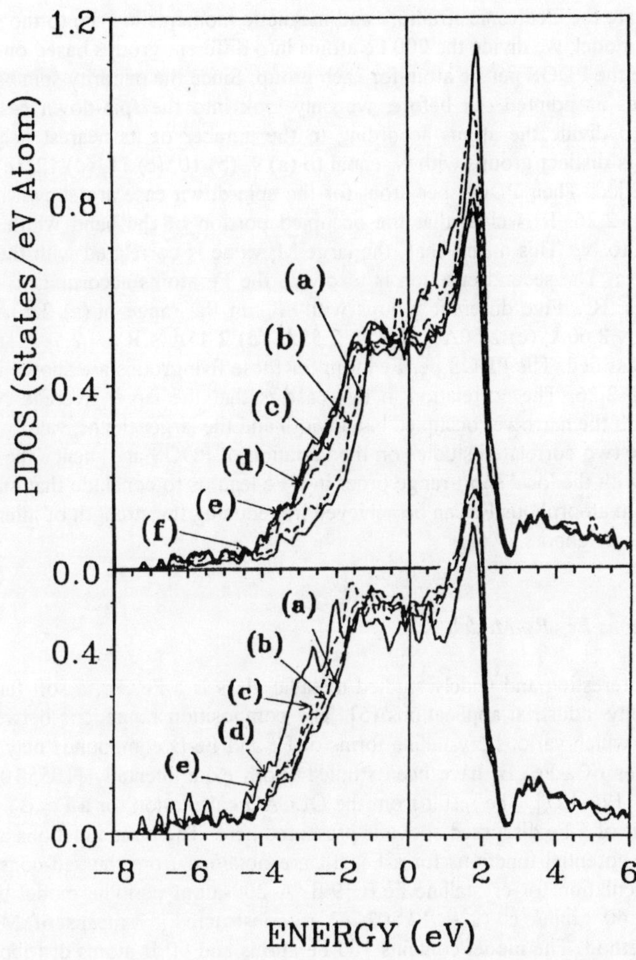

FIG. 2.26: Partial DOS in a-Fe according to local environment. Upper panel, according to number of nearest neighbors; lower panel according to the average nearest neighbor distances. See text for detail.

132

To correlate the electronic structure and magnetic moments in a-Fe to the short-range-order of the model, we divide the 200 Fe atoms into different groups based on two criteria and compare the PDOS per Fe atom for each group. Since the minority spin band controls the M_s values as pointed out before, we only look into the spin-down case. The first criterion is to divide the atoms according to the number of its nearest neighbor (NN) atoms, N_n. Six distinct groups with N_n equal to (a) 9, (b) 10, (c) 11, (d) 12, (e) 13, and (f) 14 are identified. Their PDOS per atom for the spin-down case are shown in the upper panel of Fig. 2.26. It is clear that the occupied portion of the band width is inversely proportional to N_n. This means that the large M_s value is correlated with the smaller N_n and vice versa. The second criterion is to divide the Fe atoms according to the average NN distances, R_{av}. Five different groups with R_{av} in the range of:(a) 2.60Å < R_{av}; (b) 2.55Å < R_{av} < 2.60Å; (c) 2.50Å < R_{av} < 2.55Å; (d) 2.45Å < R_{av} < 2.50Å and (e) R_{av} < 2.45Å are classified. The PDOS per Fe atom for these five groups are shown in the lower panel of Fig. 2.26. The correlation in this case is that the larger average NN distance correlates with the narrower occupied band width and the larger the M_s value.

The above two correlated studies on the variation of PDOS and hence the Fe moment distribution with the local short range order in a-Fe lead us to conclude that an increase in Fe moments in amorphous Fe can be achieved by reducing the strength of interaction with its immediate neighbors.

4.3. Amorphous $Fe_{80}B_{20}$ Model

A more interesting and widely studied metallic glass is a-$Fe_{1-x}B_x$, a soft ferromagnetic MG with many industrial applications [5]. The composition range x is between 0.10 to 0.30 outside which various crystalline forms of Fe and Fe-B compounds may precipitate. The properties of a-$Fe_{1-x}B_x$ have been studied both experimentally [105, 109-123] and theoretically [124-127]. The result from the OLCAO calculation for a-$Fe_{80}B_{20}$ glass [128] similar to that of a-Fe discussed above is presented here. The basis functions and the site-decomposed potential functions for a-$Fe_{80}B_{20}$ are obtained from the self-consistent spin-polarized calculation for crystalline Fe_3B [99]. A 200-atom periodic model with a mass density of 7.40 gram/c.c. (a = 12.802 Å) is constructed by means of Monte Carlo relaxation method. The model contains 160 Fe atoms and 40 B atoms distributed initially at random. No B-B contact is assumed as is expected for a transition metal-metalloid glass. The RDF of the model is shown in Fig. 2.3.

The spin-projected DOS for a-$Fe_{80}B_{20}$ is shown in Fig. 2.27. For the majority spin band, the Fermi level cuts at the steep edge of the d band while in the minority band, it resides in a shallow valley of the d-band splitting. Contrary to the a-Fe case, the DOS at the Fermi level for both spin cases are comparable, namely, 0.34 and 0.33 state per eV-atom-spin respectively. The main effect of introducing B atoms is to reduce the height of the Fe 3d peak and in pushing them closer to the E_F level. The calculated DOS for a-$Fe_{80}B_{20}$ is consistent with the results of x-ray photoemission [105] and UPS [121] experiments.

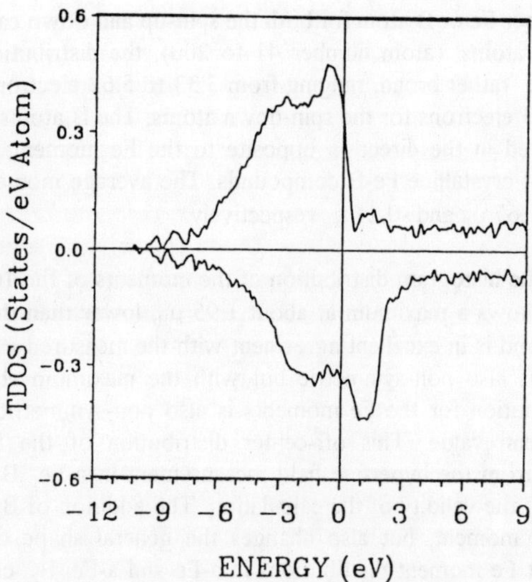

FIG. 2.27: Calculated DOS of a-$Fe_{80}B_{20}$. Upper (lower) panel for the majority (minority) spin.

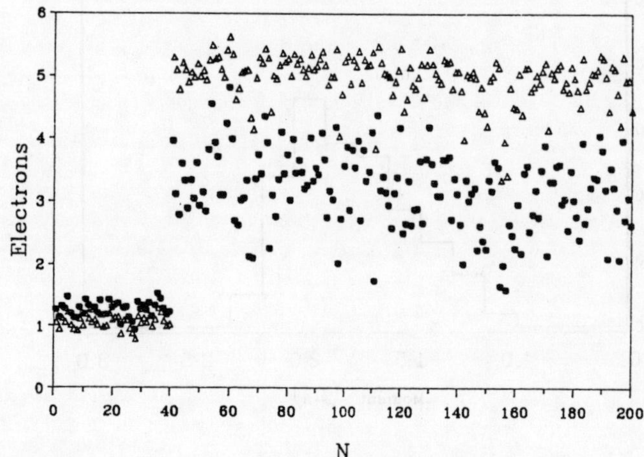

FIG. 2.28: Effective charge of a-$Fe_{80}B_{20}$. * for the majority spin and ^ for the minority spin.

134

The effective charges per Fe or B atom for both the spin-up and down cases are plotted in Fig. 2.28. For the Fe atoms, (atom number 41 to 200), the distribution of effective charges for both spins are rather broad, ranging from 3.33 to 5.62 electrons for the spin-up atoms and 1.60 to 4.83 electrons for the spin-down atoms. The B atoms (atom number 1-40) are slightly polarized in the direction opposite to the Fe moment, a fact already obvious from the study of crystalline Fe-B compounds. The average moments for Fe and B atoms in a-$Fe_{80}B_{20}$ are 1.67 μ_B and -0.21 μ_B respectively.

In Fig. 2.29, we plot the histogram distribution of the moments of the 160 Fe atoms in the a-$Fe_{80}B_{20}$ model. It shows a maximum at about 1.95 μ_B, lower than the maximum at 2.25 μ_B in the a-Fe case and is in excellent agreement with the measured value of 1.90 μ_B [120]. This distribution is also non-symmetric but with the maximum above the mean value. In a-Fe, the distribution for the Fe moments is also non-symmetric, but with the maximum below the mean value. This off-center distribution of the Fe moment is consistent with the data from the hyperfine field measurement in a-$Fe_{1-x}B_x$ films [5,113] and thus directly confirms the validity of the calculation. The addition of B atoms to a-Fe not only reduces the Fe moment, but also changes the general shape of the moment distribution. The different Fe moment distributions in a-Fe and a-$Fe_{80}B_{20}$ clearly indicates the effect of different local atomic scale short-range-order on their magnetic properties. The reduction of Fe moment in a-$Fe_{80}B_{20}$ glass is mainly due to the narrowing of the Fe-3d band and shifting of the center of gravity towards E_F.

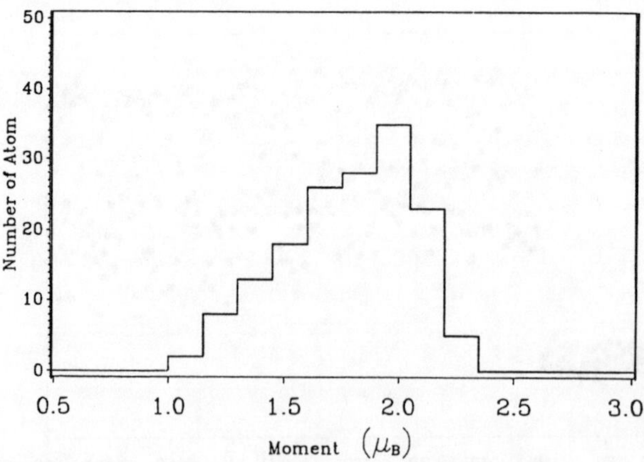

FIG. 2.29: Local Fe moments distribution in a-$Fe_{80}B_{20}$.

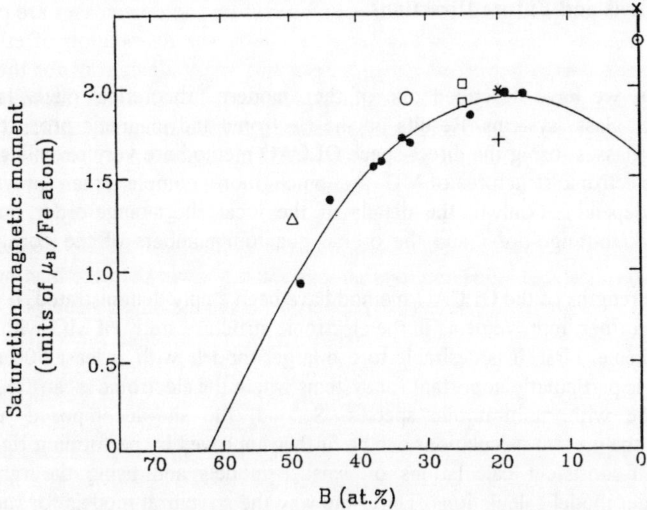

FIG. 2.30: Average Fe moments in a-Fe-B films: •,experimental data from ref. 120; open symbols, data from crystalline calculations (ref. 99); + (average Ms) and × (maximum in distribution) form a-$Fe_{80}B_{20}$ calculation; × and ⊙ are data points from a-Fe calculations.

In Fig. 2.30, we reproduce the Fig. 15 of ref. [120], showing the experimental data on the dependence of average Fe moment in a-$Fe_{1-x}B_x$ films as a function of B concentration. Hyperfine field measurement by Xiao and Chien [122] show very similar dependence. On top of it, we add the calculated data from crystalline FeB, Fe_2B and Fe_3B compounds discussed in section 3.1, the x=0.20 data from a-$Fe_{80}B_{20}$ and the x=0 data from pure a-Fe. In the amorphous cases, both the average M_s and the position of maximum in the distribution curve are plotted. For x=0.20, the maximum in the distribution falls right on the experimental curve. While for x=0, the calculated data points are above what is to be extrapolated from the x≠0 data. The subject of functional dependence of M_s on x in a-$Fe_{1-x}B_x$ is a complicated and unsettled issue [105]. Experimental data [120] seem to suggest a maximum M_s at x=0.20. However, there is very little data for x < 0.10 below which the Fe atoms in the glass begin to crystalline [122]. It has been shown that for a-$Fe_{1-x}B_x$ glass, there is a critical value for x ≅ 0.62 above which the glass is no longer magnetic [116]. This is consistent with the trend of the DOS diagram in Fig. 2.20 for Fe-B alloys. Microscopic calculations for a-$Fe_{1-x}B_x$ models for x near 0.62 will be most revealing.

5. Some Conclusions and Future Directions

In this chapter, we have described one of the modern theoretical methods for studying metallic glass systems. Results on the electronic and magnetic properties of several metallic glasses using the direct-space OLCAO method are very revealing. It is shown that the electronic structures of MG are much more complex than previously envisioned. They depend not only on the details of the local short-range-order, but also on the intermediate-range-order and the orbital quantum numbers of the constituent atoms.

Although the strengths of the OLCAO method have been amply demonstrated, it is still possible to make further improvements in the electronic structure study of MG within the OLCAO-LDA scheme. First, it is desirable to use larger models with at least 500 atoms per unit cell. This is particularly important for systems where the electronic mean-free-path is long and those with multi-atomic species. Second, the site-decomposed atomic potentials used in the present calculations can be further improved by performing rigorous first-principles self-consistent calculations on smaller models and using the resulting potentials for bigger model calculations. Third, the way the structural models for the MG have been generated can be refined by employing more realistic pair potentials. Extraction of accurate and effective pair-potentials (with possible many-body corrections) for simulational studies in metallic systems is currently a hotly pursued area of theoretical condensed matter physics. Fourth, the computational speed may be improved by further optimization of the basis set and the potential fitting functions. This is particularly important for transport properties calculations where configurational average may require repeated calculations on many model structures in order to obtain better statistical average, or in the case of spin-polarized calculations where convergence to self-consistency is rather slow. With a rapid increase in computing power in recent years, the first-principles approach to the study of disordered systems is far more promising than empirical or semi-empirical methods.

While the results presented in this chapter are quite limited, especially on magnetic systems, they demonstrate the effectiveness and capabilities of the method. It is therefore appropriate to speculate on the type of problems that can be meaningfully studied using the present approach. Several topics naturally come to mind as an extension of present studies:

(1) The magnetism of amorphous Fe-based alloys and glasses is a subject of immense complexity. Much insight can be obtained by more detailed and systematic calculations on models of different constituents, concentrations and local structural characteristics. Characterization of local short-range-order and its correlation to the magnetic properties can be carried out at a more refined level.

(2) It is desirable to extend the present first-principles calculation to include temperature dependent properties in a computationally tractable way. A different strategy involving the use of electronic structure results may be pursued.

(3) It is obvious that the OLCAO method can be applied to multi-component glasses such as metaglass (a-Ni-Fe-Si-P). Careful analysis of PDOS and effective charges may reveal the secret of fine mixing in MG to achieve certain specific properties.

(4) The problem of metal-insulator transition in disordered alloys is still not fully resolved. With the feasibility of calculating the zero temperature conductivity of a glass, the conductivity as a function of glass composition x in systems such as a-Nb$_{1-x}$Si$_x$ can be studied. Information on the microscopic origin of the metal-insulator transition and its relationship to electron localization can be obtained.

(5) The study of transport properties in various MG systems can be further improved by using larger models, better configuration averages and more accurate wave functions. The calculation can also be extended to thermo-power and Hall effects.

(6) It will be extremely challenging to include the effect of electron-phonon interactions into the transport property calculations. In the OLCAO method where the atomic basis consists of GTO, the electron-phonon matrix elements which involve the gradient of the site-decomposed one-electron potential are not too difficult to evaluate.

(7) It is possible to study the orbital moments of magnetic atoms in MG by spin-polarized calculations including the effect of spin-orbit coupling. The distribution of local orbital moments and its relationship to local short-range-order of the MG will be highly interesting. Such a calculation has been demonstrated recently for a-Fe [129].

(8) It is further possible to couple the orbital moments and the local spin moments together such that the local spin directions of magnetic atoms in disordered magnetic systems can be elucidated from the standing point of realistic computation. In the past, such problems were invariably treated by phenomenological Hamiltonians whose parameters cannot always be easily justified.

(9) With the demonstration of the calculation of optical properties of MG and the possibility of including the spin-orbit interactions, it is highly promising that magneto-optical properties of some technologically important glasses can be studied.

Acknowledgments: I am indebted to several of my colleagues, especially G.-L. Zhao, M.-Z. Huang and Y.-N. Xu for most of the work presented in this chapter. The work at the University of Missouri-Kansas City has been supported by the U.S. Department of Energy.

6. References:

1. N.F. Mott and E.A. Davis, *Electronic Process in Non-crystalline Materials* , Clarendon Press. Oxford, 1979.
2. *Glassy Metals I and II*, edited by H.-J. Güntherodt and H. Bec (Springer-Verlag, New York, 1981 and 1983).
3. *Magnetism of Metals and Alloys*, edited by M. Cyrot, North Holland, Amsterdam, 1982.
4. K. Moorjani and J.M.D. Coey, *Magnetic Glasses*, Elsevier, 1984.

5. *Glassy Metals Magnetic Chemical and Structural Properties*, CRC Press, Boca Raton, Florida. edited by R. Hasegawa.

6. Y. Kakehahsi and H. Tanaka, Chapter 1 of this book.

7. W.Y. Ching, J. Amer, Ceram. Soc. **71**, 3135 (1990) and references cited therein.

8. P. Hohenberg and W. Kohn, Phys. Rev. **136**, 138 (1964).

9. W. Kohn and L. J. Sham, Phys. Rev. **140**, A1133 (1965); L.J. Sham and W. Kohn, Phys. Rev. **145**, 561 (1966);

10. W.Y. Ching and C.C. Lin,   Phys. Rev. Lett. **34**, 1223 (1975).

11. W.Y. Ching and C.C. Lin, and D.L. Huber, Phys. Rev. **B14**, 620 (1976).

12. W.Y. Ching, C.C. Lin, and L. Guttman, Phys. Rev.**B16**, 5488 (1977).

13. W.Y. Ching, and C.C. Lin, Phys. Rev. **B18**, 6829 (1978).

14. W.Y. Ching, D.J. Lam, and C.C. Lin, Phys. Rev. **B21,** 2378 (1980).

15. W.Y. Ching, J. Non. Cryst. Solids, **35** and **36**, 61 (1980); W.Y.Ching, Phys. Rev. **B22,** 2816 (1980); W.Y. Ching and C.C. Lin, in A.I.P. Conference Proceedings, NO **73** *Tetrahedrally Bonded Semiconductors,* p153, (1987). Edited by R.A. Street, D.K. Biegelsen, and J.C. Knight.

16. W.Y.Ching, Phys. Rev. Lett. **46**, 607 (1981).

17. W.Y.Ching, Phys. Rev. **B26**, 6610 (1982).

18. W.Y.Ching, Phys. Rev. **B26**, 6622 (1982).

19. W.Y.Ching, Phys. Rev. **B26**, 6633 (1982).

20. R.A. Murray, L.W. Song and W.Y. Ching, J. Non-Cryst. Solid **94**, 133 (1987); **94**, 144 (1987).

21. R.A. Murray and W.Y. Ching, Phys. Rev. **B39**, 1320(1989).

22. S.S. Jaswal and W.Y. Ching, Phys. Rev. **B26**, 1064 (1982).

23. W.Y. Ching, L.W. Song and S.S. Jaswal, Phys. Rev. **B30**, 544 (1984); J. Non-Cryst. Solid **61-62**, 1207 (1984).

24. W.Y.Ching, Phys. Rev. **B34**, 2080 (1986).

25. G.-L. Zhao and W.Y. Ching, Phys. Rev. Lett. **62**, 2511 (1989).

26. W.Y. Ching, G.-L. Zhao and Yi He, Phys. Rev. **B42**, 10878 (1990).

27. G.-L. Zhao, Yi. He and W.Y. Ching, Phys. Rev. **B42**, 10887 (1990).

28. W.Y. Ching, M.-Z. Huang, and Y.-N. Xu, Phys. Rev. **B46**, 9910 (1992).

29. M.-Z. Huang, W.Y. Ching, and T.L. Lenosky, Phys. Rev. **B47**, 1593 (1993).

30. E. Clementi and C. Roetti, At. Nucl. Data Tables, **14**, 177-478, (1974); ibid **26**, 197-385, (1981).

31. R.S. Mulliken, J. Am. Chem. Soc. **23**, 1833 (1955).

32. See for example, *Density Functional Methods in Chemistry*, edited by J.K. Labanowski and J.W. Andzelm, Springer-Verlag, New York (1991).

33. L. Hedin and B.I. Lundqv st, J. Phys. **C4** 2064 (1971).

34. U. Von Barth and L. Hedin, J. Phys. C **5**, 1629 (1972).

35. O. Gunnarsson and B.I. Lundqvist, Phys. Rev. **B13**, 4274 (1976).

36. D.M. Ceperly and B.J. Alder, Phys. Rev. Lett., **46,** 556 (1980).

37. S.H. Vosko, L. Wilk and M. Nusair, Can. J. Phys. **58**, 1200 (1980).

38. R.G. Parr and W. Yang, *Density Functional Theory of Atoms and Molecules*, Oxford University Press, (1989).

39. E.P. Wigner, Phy. Rev. **46**, 1002 (1934).

40. H. Sambe and R.H. Felton, J. Chem. Phys. **61**, 3862 (1974); **62**, 1122 (1974).

41. W.Y. Ching and B.N. Harmon, Phys. Rev. **B34** 5305 (1986).

42. F. Zandiehnadem and W.Y. Ching, Phys. Rev. **B41**, 12162 (1990).

43. I. Shavitt, *Methods in Computational Phys.*, Vol.2, edited by B. Aldler, S. Fernbach and M. Rotenberg. Academic Press, New York, 1963, p1.

44. R.C. Chaney, T.K. Tung, C.C. Lin, and E.E. Lafon, J. Chem. Phys. **52**, 361 (1970).

45. L.J. Schaad and G.O. Morrell, J. Chem. Phys., **54**, 1965 (1971).

46. Y.P. Li, Z-Q. Gu, and W.Y. Ching, Phys. Rev. **B32**, 8377 (1985).

47. E.E. Lafon, J. Comput. Phys. **83**, 185 (1989).

48. W.Y. Ching and C.C. Lin, Phys. Rev. **B12**, 5536 (1975); W.Y. Ching and C.C. Lin, Phys. Rev. **B16**, 2989 (1977).

49. V.L. Moruzzi, J.F. Janak, and A.R. Williams, *Calculated Electronic Properties of Metals* (Pergamon, New York, 1978).

50. See, for example, M.Weissbluth, *Atoms and Molecules* (Academic, New York), 1978, p 315.

51. D.D. Koelling and A.H. MacDonald, in *Relativistic Effects in Atoms Molecules and Solids*, NATO ASI Series, B: Physics., Vol. 87 Edited by G.L. Malli, Plunum, New York., 1983.

52. J. Callaway, *Quantum Theory of the Solid State*, Academic Press, New York, 1991. p38.

53. D.D. Koelling and B.N. Harmon, J. Phys. **C10**, 3107 (1977).

54. C.S. Wang and J. Callaway, Phys. Rev. **B9**, 4897 (1974).

55. See, for example, D.M. Brink and G.R. Satchler, *Atomic Momentum* (Clarendon, Oxford, 1968).

56. X.-F. Zhong, Y.-N. Xu, and W.Y. Ching, Phys. Rev. **B41,** 10545 (1990).

57. X.-F. Zhong and and W.Y. Ching, J. Appl. Phys., **67**(9), 4768 (1990).

58. S. R. Nagel, Adv. Chem. Phys. **51**, 227 (1982).

59. U. Mizutani, Prog. Mater. Sci. **28**, 97 (1983).

60. D.G. Nagle, J. Phys. Chem. Solids **45**, 367 (1984).

61. J.S. Dugdale, Contemp. Phys. **28**, 547 (1987).

62. M.A. Howson and B. L. Gallagher, Physics Rep. **170**, 265 (1988).

63. M. Ziman, Philos. Mag. **6**, 1031 (1961).

64. T.E. Faber and J. M. Ziman, Philos. Mag. **11**, 153 (1965).

65. R. Evans, D.A. Greenwood, and P. Lioyd, Phys. Lett. **A35**, 57 (1971).

66. N.F. Mott. Philos. Mag. **27**, 1249 (1972).

67. G.F. Weir, M. A. Howson, B. L. Gallagher and G. J. Morgan, Philos. Mag. **47**, 163 (1983); G. J. Morgan and G. F. Weir, Philos. Mag. **47**, 177 (1983).

68. P.W. Anderson, B. I. Halperin and C. M. Varma, Philos. Mag. **25**, 1 (1972).

69. R. Cochrane, R. Harris, J. Ström-Olsem and M. Zuckerman, Phys. Rev. Lett. **35**, 676 (1975).

70. C.C. Tsuei, Solid State Commun. **27**, 691 (1978).

71. S.M. Girvin and M. Jonson, Phys. Rev. **B22**, 3583 (1980).

72. Y. Imry, Phys. Rev. Lett. **44**, 469 (1980).

73. D. Belitz and W. Götze, J. Phys. **C15**, 981 (1982).

74. E. Abrahams, P. W. Anderson, D. C. Licciardello and T. V. Ramakrishnan, Phys. Rev. Lett, **47**, 1617 (1981).

75. P.A. Lee and T.V. Ramakrishnam, Rev. Mod. Phys. **57**, 287 (1985).

76. B.L. Alt'shuler and A.G.Aronov, in *Electron-Electron Interaction in Disordered Conductors*, edited by A.L. Efros and M. Pollak (North-Holland, Amesterdam, 1985), p.4.

77. M.A. Howson, J. Phys. **F14**, L25 (1984); **F16**, 984 (1986).

78. L.E. Ballentine and J.E. Hammerberg, Canadian J. Phys., **62**, 692 (1984).

79. D. Greenwood, Proc. Phys. Soc. London **71**, 585-96 (1958).

80. A. Raman and F.H. Stillinger, J. Chem. Phys., **53**, 3336 (1971).

81. R. Car and M. Parrinello, Phys. Rev. Lett. **55**, 2471 (1985).

82. N. Metropolis, A. W. Rosenbluth, M. N. Rosenbluth, A.H. Teller and E. Teller, J. Chem. Phys. **21**, 1087 (1953).

83. A. Amamon, D. Aliaga-Guema, P.Panissod, G. Krill, and R.Kuentzler, J. of Phys. Paris C8, **41**, 396 (1980).

84. Z. Altounian and J. O. Strom-Olsen, Phys. Rev. **B27**, 4149 (1983).

85. K.Samwerand H. V. Löhneysen, Phys. Rev. B **26**, 107 (1982).

86. P. Oelhafen, E. Hauser, H.-J. Güntherodt, and K. H. Bennemann, Phys. Rev. Lett. **43**, 1134 (1979); P. Oelhafen, E. Hauser, and H.-J. Güntherodt, Solid State Commun. **35**, 765 (1980).

87. P. Oelhafen in: Glassy Metals II, eds. H. J. Güntherodt and H. Beck (Springer, New York, 1983)P. 283.

88. T. Iida and R.I.L. Guthrie, in *The Physical Properties of Liquid Metals* (Claredon, Oxford), 1988, p232-233.

89. J. H. Mooij, Phys. Stauts Solidi **A17**, 521 (1973).

90. V. Nguyen-Van, S. Fisson and M.L. Theye, J. Non-Cryst. Solids, **61-62**, 1325 (1984).

91. Y. Calvayrac, et al. , Philos. Mag. **B48**, 323 (1983).

92. M.N. Baibich, W.B. Muir, Z. Altounian, and Tu Guo-Hua, Phys. Rev. **B26**, 2963 (1982).

93. See ref. 27 for a large number of experimental references cited.

94. E. Babic, R. Ristic, M.Miljak, and M.G. Scott, in *Proceedings of the Fourth International Conference on Rapidly Quenched Metals, Sendai*, 1981, edited by T. Masumoto nd K. Suzuki (The Japan Institute of Metals, Sendai, 1981). Vol. 2, p. 1079.

95. T. Murata, S. Tomizawa, T. Fukae, and T. Masumoto, Scr. Metsall. **10**, 181 (1976).

96. M.M. Kirillova, Zh. Eksp. Teor. Fiz. **61**, 336 (1971) [Sov. Phys. -JETP **34**, 178 (1972)].

97. M.L. Theye, Van Nguyen-Van, and S.Fisson, Phys. Rev. **B31**, 6447 (1985).

98. J. Rioory and J.M. Frigerio, in *Proceedings of the Fourth International Conference on Rapidly Quenched Metals*, edited by T. Masumoto and

K. Suzuki (The Japan Institute of Metals, Sendai, 1981). Vol. 2, p. 1287.

99. W.Y. Ching, Y.N. Xu, B.N. Harmon, J. Ye and T.C. Leung, Phys. Rev. B **42**, 4460 (1990).

100. T. Bjurstrom and H. Arnfelt, Z. Phys. Chem. B**4**, 469 (1929).

101. F. Wever and A. Muller, Mitt. K. Wilhelm-Inst. Eisenforsch, Dusseldorf 11, 193 (1930).

102. R.W.G. Wyckoff, *Crystal Structures* (Interscience, New York, 1965), p114.

103. F.H. Sánchez and M.N. Fernandez Van Raap, Phys. Rev. B**46**, 9013 (1992).

104. P.J. Brown and J.L. Cox, Phil. Mag. **23**, 705 (1970); R.S. Perkins and P.J. Brown, J.Phys. F: **4**, 906, (1974).

105. C.L. Chien, Chapter 4 of this book.

106. Y. Kakehashi, Phys. Rev. B**43**, 10820 (1991).

107. Y.-N. Xu, Yi He and W.Y. Ching, J. Appl. Phys. **69**(8), 5460 (1991).

108. T. Ichikawa, Phys. Status Solidi, (a) **19**, 707 (1973).

109. H. Hiroyoshi, K. Fukamichi, M. Kikuchi, A. Hoshino, T. Masumoto, Phys. Lett. 65, 163 (1978).

110. F.E. Luborsky,and H.H. Liebermann, Appl. Phys. Lett., **33**, 233 (1978).

111. R. Hasegawa and R. Ray, J. Appl. Phys. **49**, 4174 (1978).

112. M. Matsuura, Solid State Commun. **30**, 231 (1979).

113. I. Vincze, D.S. Boudreaux, M.Tegze, Phys. Rev. B**19**, 4896 (1979).

114. I. Vincze and E. Babic, Solid State Commun., **27**, 1425 (1978).

115. M. Matsuura, T. Nomoto, F.Itoh, and K. Suzuki, Solid State. Commun. 33, 895 (1980).

116. C.L. Chien and K.M. Unruh, Phys. Rev B**24**, 1556 (1981).

117. Y.D. Yao, A. Sigurds and S.T. Lin, *Proc. of 4th Int. Conf. on Rapidly Quenched Metals (Sendai),* Edited by T.Masumoto and K. Suzuki, Vol. 2, p839 (1981).

118. R. Ray and A.K. Majumdar, Phys. Rev. B**31**, 2033 (1985).

119. W. Matz, H. Herman and N. Mattern, J. Non-ryst. Solids, **93**, 217 (1987).

120. N. Cowlam and G.E. Carr, J. Phys. F: Met. Phys. **15**, 1109, (1985); ibid, 1117 (1985).

121. Th. Paul and H. Neddermeyer, J. Phys. F:Met. Phys. **15**, 79 (1985).

122. G. Xiao and C.L. Chien, J. Appl. Phys. **61**, 3246 (1987).

123. F.H. Sánchez, M.N. Fernandez Van Raap, and J.I. Budnick, Phys. Rev. B**46**, 13881(1992).

124. T. Fujiwara, J. Phys. F: Met. Phys. **12**, 661 (1982).

125. U. Krey, H. Ostermeier and J. Zweck, Phys. Status Solidi b, **144**, 203 (1987).

126. M. Krojci and P. Mrafko, J. Phys. Met. Phys. **18**, 2137 (1988).

127. S. Varga and J. Krempasky, J. Phys.: Conden. Mat. **1**, 7851 (1989).

128. W.Y. Ching and Y.-N. Xu, J. Appl. Phys. **70** (10) 6305 (1991).

129. X.-F. Zhong and W.Y. Ching, J. Appl. Phys. **75** (8), 6834 (1994).

CHAPTER THREE

RANDOM ANISOTROPY IN AMORPHOUS ALLOYS

Eugene M. CHUDNOVSKY

Physics Department, Lehman College, CUNY, Bronx, NY 10468-1589

1. The Random Anisotropy Model

In this Chapter I will review the random-anisotropy real-space model which turned out to be quite powerful in interpreting the experimental data on amorphous ferromagnets. I start with the discussion of the physical picture underlying this model.

Amorphous magnetic alloys are typically obtained by the rapid freezing of the melt. While there are many models of amorphous disorder, one picture is commonly kept in mind. This picture describes the process of solidification as a diffusion-driven rearrangement of atoms towards the minimum energy state. The length of the short range structural order in a solid, r_0, is then determined by the average size of the volume in which the atoms can successfully rearrange before the diffusion coefficients become exponentially small as the temperature drops. Depending on the rate of cooling, one obtains solids ranging from monocrystals to disordered networks of atoms. The diffusion-driven rearrangement of atoms also determines the properties of amorphous alloys obtained by the deposition of very thin layers of different elements, using the evaporation technique.

Having this picture in mind, let us now turn to the magnetic properties of amorphous alloys. The model that we are going to employ is a direct generalization of the common approach to the theory of ferromagnetic crystals. As it is known, the magnetic order in crystalline solids is determined by two major factors: the exchange interaction and the magnetic anisotropy. The ferromagnetic exchange leads to the parallel orientation of neighboring magnetic moments, while the anisotropy aligns the

143

resulting magnetization along some preferred directions determined by the symmetry of the crystal. As one turns to amorphous solids, the concept of ferromagnetic exchange remains almost unperturbed by the structural disorder. The concept of magnetic anisotropy, however, becomes somewhat less obvious in the absence of the global crystallographic anisotropy. In what follows we will analyze both concepts in more detail.

Let us start with the exchange interaction. It is described by the Heisenberg Hamiltonian,

$$\mathcal{H}_{ex} = -\sum_{\alpha\beta} J_{\alpha\beta}\hat{\mathbf{J}}_\alpha\cdot\hat{\mathbf{J}}_\beta \quad , \tag{3.1}$$

where $\hat{\mathbf{J}}_\alpha$ is the operator of the angular momentum of the α-th magnetic ion, and the coefficients $J_{\alpha\beta}$ rapidly go to zero as the distance between the ions, $|\mathbf{r}_\alpha - \mathbf{r}_\beta|$, increases. Eq.(3.1) is mathematically equivalent to

$$\mathcal{H}_{ex} = -\frac{1}{(g\mu_B)^2}\int d^3r \int d^3r' \, J(\mathbf{r} - \mathbf{r}')\hat{\mathbf{M}}(\mathbf{r}) \cdot \hat{\mathbf{M}}(\mathbf{r}') \quad , \tag{3.2}$$

where $\hat{\mathbf{M}}(\mathbf{r})$ is the operator of the magnetization,

$$\hat{\mathbf{M}}(\mathbf{r}) = g\mu_B \sum_\alpha \hat{\mathbf{J}}_\alpha\delta(\mathbf{r} - \mathbf{r}_\alpha) \quad , \tag{3.3}$$

g is the gyromagnetic factor, and $J_{\alpha\beta} = J(\mathbf{r}_\alpha - \mathbf{r}_\beta)$. For problems involving the coherent behavior of a large number of elementary magnetic moments, $\hat{\mathbf{M}}(\mathbf{r})$ in Eq.(3.2) may be treated as a classical vector:

$$\mathbf{M}(\mathbf{r}) = \frac{1}{v} < \int_v d^3r' \, \hat{\mathbf{M}}(\mathbf{r}') > \quad , \tag{3.4}$$

where v is a microscopic volume centered at $\mathbf{r}$. The dimensions of v are chosen from the condition that all individual magnetic moments of ions are aligned inside that volume due to the strong exchange coupling. The size of v is therefore defined by the range of the ferromagnetic exchange, r_{ex}. The value of r_{ex} depends on the nature of the exchange interaction (RKKY, direct exchange, etc.) and is typically smaller than the correlation length of amorphous alloys r_0.

Ferromagnetism exists in amorphous alloys with a sufficiently high concentration of magnetic ions. More rigorously, the average distance between the ions should be smaller than (or comparable to) r_{ex}. In the presence of short-range order, the atoms of an amorphous alloy have a regular arrangement on the scale of r_0. This arrangement most likely (but not necessarily) resembles the crystalline counterpart of the alloy. When $r_0 > r_{ex}$, the number of magnetic ions inside the volume v is fixed by the short-range order. It then follows that in a homogeneous amorphous solid with $r_0 > r_{ex}$, the length of the magnetization vector is constant throughout the sample, $[\mathbf{M}(\mathbf{r})]^2 = M_0^2$, as in its crystalline counterpart. Thus, the possible configurations of the $\mathbf{M}(\mathbf{r})$-field inside the solid can be obtained by local rotations of the $\mathbf{M}(\mathbf{r})$-vector. This determines the behavior of the magnet in a magnetic field. As we will see in Sec.3.3, a large number of amorphous ferromagnets exhibits magnetization curves consistent with the above picture. It should be noted that the condition $[\mathbf{M}(\mathbf{r})]^2 = constant$ may also be approximately valid for $r_{ex} > r_0$, if the range of the exchange is sufficiently large (due to, e.g., RKKY interactions) to provide a large number of magnetic ions, $N >> 1$, inside the exchange volume v. This would give $\delta M_0/M_0 \sim 1/\sqrt{N} << 1$ for the average fluctuation of the length of the magnetization vector.

As we will see in Section 3.2, the rotation of $\mathbf{M}(\mathbf{r})$ in an amorphous ferromagnet typically occurs on the scale $R_f >> r_{ex}$. This, with the condition $J(r >> r_{ex}) \to 0$, allows one to replace $\mathbf{M}(\mathbf{r}')$ in Eq.(3.2) with the first three terms of the Taylor series expansion near $\mathbf{r}$,

$$\mathbf{M}(\mathbf{r}') = \mathbf{M}(\mathbf{r}) + (r'_i - r_i)\frac{\partial \mathbf{M}(\mathbf{r})}{\partial r_i} + \frac{1}{2}(r'_i - r_i)(r'_k - r_k)\frac{\partial^2 \mathbf{M}(\mathbf{r})}{\partial r_i \partial r_k} \quad . \tag{3.5}$$

When evaluating the integral in Eq.(3.2) the first term becomes a constant, $M_0^2 V \int d^3r\, J(r)$. The second term vanishes due to the homogeneity of the system and the third term, after integration by parts, gives the following expression for the energy of the ferromagnet, corresponding to a given static $\mathbf{M}(\mathbf{r})$-configuration:

$$E_{ex} = \frac{A}{M_0^2}\int d^3r\, \frac{\partial \mathbf{M}}{\partial r_i}\cdot\frac{\partial \mathbf{M}}{\partial r_i} \quad , \tag{3.6}$$

where $A(erg/cm)$ is the exchange constant,

$$A = \frac{2\pi M_0^2}{3(g\mu_B)^2}\int_0^\infty dr\, r^4 J(r) \quad . \tag{3.7}$$

It should be emphasized that the possibility of describing an amorphous ferromagnet by a coordinate-independent exchange constant arises from
  1) the condition $r_0 > r_{ex}$ which leads to $\mathbf{M}^2(\mathbf{r}) = M_0^2$, and
  2) the global spatial isotropy of the system.
For a crystalline ferromagnet the same calculation leads to

$$E_{ex} = \frac{1}{M_0^2}\int d^3r\, A_{ik}\frac{\partial \mathbf{M}}{\partial r_i}\cdot\frac{\partial \mathbf{M}}{\partial r_k} \tag{3.8}$$

with

$$A_{ik} = \frac{M_0^2}{2(g\mu_B)^2}\int d^3r\, r_i r_k J(\mathbf{r}) \quad . \tag{3.9}$$

Only for a cubic crystal $A_{ik} = A\delta_{ik}$; e.g., an uniaxial crystal is characterized by two exchange constants, $A_{11} = A_{22} \neq A_{33}$. If the exchange constants A_{ik} are known for the crystalline counterpart of the amorphous alloy, one may expect that the exchange constant A of the amorphous ferromagnet is $\frac{1}{3}\sum_i A_{ii}$.

Let us now turn to the question of the magnetic anisotropy. Eq.(3.6), while suggesting that the minimum energy corresponds to the uniform magnetization state, does not specify the direction of $\mathbf{M}$. The degeneracy of the energy with respect to the global orientation of $\mathbf{M}$ is removed in the presence of the crystalline field. For a uniaxial crystal, for example, one must add to Eq.(3.8) the anisotropy term,

$$E_{an} = -\frac{K_c}{M_0^2} \int d^3 r \, (\mathbf{n} \cdot \mathbf{M})^2 \quad , \tag{3.10}$$

where $\mathbf{n}$ is a unit vector along the anisotropy axis of the crystal, and $K_c(erg/cm^3)$ is the anisotropy constant. The total energy,

$$E = \frac{1}{M_0^2} \int d^3 r \left[A_{ik} \frac{\partial \mathbf{M}}{\partial r_i} \cdot \frac{\partial \mathbf{M}}{\partial r_k} - K_c(\mathbf{n} \cdot \mathbf{M})^2 \right] \quad , \tag{3.11}$$

is now minimized by $\mathbf{M}$ being along one of two directions parallel to $\mathbf{n}$. Let N_0 be the concentration of magnetic ions, then K_c/N_0 is the energy per ion needed for a global rotation of $\mathbf{M}$ away from the easy direction. In magnetic crystals this energy is typically small compared to the energy $A/N_0 r_{ex}^2$ needed to destroy the ferromagnetic alignment of the magnetic moment of a single ion with the moments of other ions. The ratio of the two energies, $K_c r_{ex}^2/A$ is a small parameter of the conventional ferromagnetic theory. As we will see below, the key parameter of the theory of the amorphous ferromagnetism is [1] $K_r R_a^2/A$, with $R_a \geq r_0 > r_{ex}$, K_r being the strength of the random anisotropy. This ratio can be either small or large, depending on the length R_a, and its physical meaning is discussed below.

Before turning to the consideration of an amorphous magnet, it is useful to remind the reader that for a crystalline ferromagnet the equations that minimize Eq.(3.11) have also non-uniform solutions for $\mathbf{M}(\mathbf{r})$, which correspond to domain walls separating uniformly magnetized domains. The static domain structure with a zero net magnetic moment becomes the minimum of the total energy if one takes into account the magnetic dipole interaction between elementary magnetic moments, which is not included in Eq.(3.11). Simply speaking, the magnet breaks into domains in order to decrease the energy of the magnetic field generated by the global magnetic moment. The width of the domain wall, $\delta = (A/K_c)^{1/2}$ is typically in the range of $50 - 500$Å, while the average size of domains is much greater, $10^{-4} - 10^{-2} cm$, depending on the geometry of the sample.

Let us now turn to the consideration of the magnetic anisotropy in amorphous magnetic alloys. Based upon our picture of an amorphous ferromagnet, it is clear that there should be local preferential directions of the magnetization, $\mathbf{n}(\mathbf{r})$, determined by the local arrangement of atoms. Since the strength of the magnetic anisotropy, K_r, is determined by the crystalline field on the scale r_0 :

1) K_r should not differ significantly from the strength of the anisotropy
in the crystalline counterpart, and

2) K_r should not significantly fluctuate from one point to another. Consequently, the energy functional [1]

$$E = \frac{1}{M_0^2} \int d^3r \left\{ A \frac{\partial \mathbf{M}}{\partial r_i} \cdot \frac{\partial \mathbf{M}}{\partial r_i} - K_r[\mathbf{n}(\mathbf{r}) \cdot \mathbf{M}(\mathbf{r})]^2 \right\} \qquad (3.12)$$

should be a good approximation in the case of an uniaxial local anisotropy. In this formula, $\mathbf{n}(\mathbf{r})$ is a static random vector field $(\mathbf{n}^2 = 1)$ of properties which depend on the type of the structural disorder. Note that the second term in Eq.(3.12) can be written as $-K_r n_i n_j M_i M_j$, with $n_i n_j$ having a non-zero average, $< n_i n_j >= \frac{1}{3}\delta_{ij}$. It is more convenient, therefore, to write the energy in the form

$$E = \frac{1}{M_0^2} \int d^3r \left[A \frac{\partial \mathbf{M}}{\partial r_i} \cdot \frac{\partial \mathbf{M}}{\partial r_i} - K_r R_{ij}(\mathbf{r}) M_i M_j \right] \quad , \qquad (3.13)$$

where

$$R_{ij} = n_i n_j - \frac{1}{3}\delta_{ij} \qquad (3.14)$$

is a random tensor with zero average. The energy functionals (3.12) and (3.13) are equivalent because they differ by a constant, $\frac{1}{3}K_r V$.

The random anisotropy is described by the correlation function

$$< \hat{R}(\mathbf{r})\hat{R}(\mathbf{r}') > \propto \Gamma(|\mathbf{r} - \mathbf{r}'|) \quad . \qquad (3.15)$$

It is tempting to relate $\Gamma(r)$ to the pair correlation function of the amorphous structure, which can be obtained by X-ray or electron microscopy methods. In this case $\Gamma(r)$ would rapidly go to zero at $r >> r_0$. While admitting such a possibility in principle, we would like to emphasize that this is not necessarily true for all amorphous magnetic materials. The reason for this is based on a very general argument [2,3]. When a solid with a short-range structural order is formed, two distinct local symmetries become violated: the symmetry with respect to local translations and the symmetry with respect to local rotations. Correspondingly, two kinds of short-range order are established: the short-range order in the arrangement of atoms (the translational order) and the short-ranged order in the orientation of locally defined crystallographic axes (the orientational order). In a perfect crystal, both are long-ranged and related to each other. In disordered systems, however, the translational and the orientational order may have different correlation lengths. The range of the translational order can be visualized as the average size of a perfect lattice which approximately matches the local arrangement of atoms. Regions separated by $r > r_0$ must be approximated by different pieces of the perfect lattice. Each such a piece has a certain orientation in space, and the length at which these orientations are correlated determines the range of the orientational order, R_a. In recent years a few disordered systems of different nature have been discovered which show a very short length of the translational order but extended orientational correlations, $R_a >> r_0$

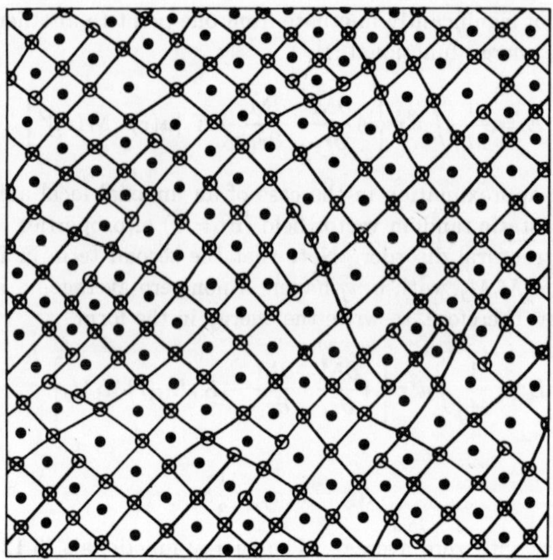

Figure 3.1: Amorphous structure with extended orientational order, formed by two kinds of atoms.

[4] (Fig. 3.1). This is in agreement with theoretical suggestions that the orientational order is more robust than the structural order. It has also been argued [5] that a similar situation may occur in amorphous solids. The position of the first peak of the pair correlation function of an amorphous system, obtained from diffraction experiments, determines the length of the translational order r_0. It is obvious, however, that the correlation function of the magnetic anisotropy axes, $\Gamma(r)$, is related to the orientational order. Consequently, correlations in the orientation of easy axes may decay on a scale $R_a > r_0$. Thus R_a (the length of the orientational order), and not r_0, (the parameter of the amorphous structure), is responsible for the magnetic properties of amorphous solids. This means that magnetic measurements may uncover hidden features of the amorphous state, which cannot be uncovered by diffraction methods. We will return to this question in Section 3.3.

Let us now go back to Eq.(3.12) and ask what magnetic state follows from this equation. In the limit of strong anisotropy the answer is more or less obvious. The strong anisotropy simply aligns $\mathbf{M(r)}$ along the local easy axes $\mathbf{n(r)}$ (see Fig. 3.2a). We will see in Section 3.2 that this happens when $K_r R_a^2 / A > 1$. Note that this condition is often misinterpreted as the condition that the anisotropy energy is greater than the exchange energy. This latter condition would require $K_r r_{ex}^2 / A > 1$, which is much more difficult to satisfy since the anisotropy energy has a relativistic $(v/c)^2$-smallness compared to the exchange. Typically, the limit of the large anisotropy occurs because of the condition $R_a \gg r_{ex}$. In that limit, the magnetic properties

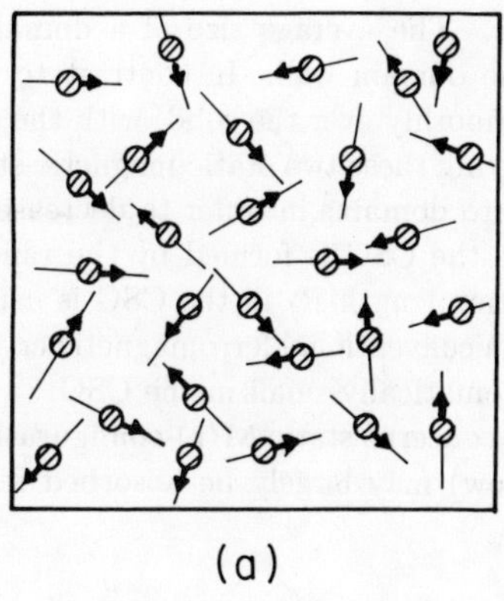

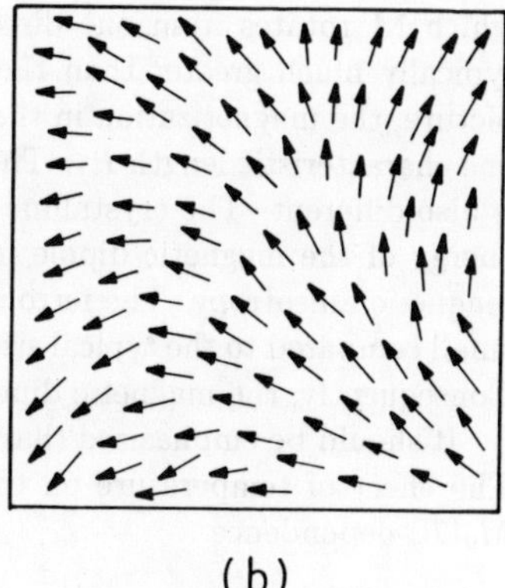

Figure 3.2: (a) Random orientation of local moments in an amorphous magnet with strong local anisotropy. (b) Correlated spin glass state in an amorphous ferromagnet with weak local anisotropy.

of the amorphous solid are similar to the magnetic properties of a solid formed by randomly packed ferromagnetic crystallites of the average size R_a. Such systems have been extensively studied long before the era of amorphous ferromagnets. The magnetic state shown in Fig. 3.2a is sometimes called a speromagnetic (SM) state.

Let us now turn to the more subtle case of weak anisotropy, which provides an example of a continuous model with frustration. On one hand, the local random anisotropy favors the direction of the local magnetization along $\mathbf{n}(\mathbf{r})$. On the other hand, if this was the case, $\mathbf{M}(\mathbf{r})$ would rotate on a small scale R_a, which would be unfavored by the strong exchange interaction. The ferromagnetic alignment of spins should certainly prevail in the limit of $K_r \to 0$. Even this obvious case, however, leaves us with the uncertainty about the global orientation of $\mathbf{M}$. The solution of the weak anisotropy problem is rather non-trivial and will be discussed in the next Section. This solution shows that in the limit of weak anisotropy, the magnetization chaotically rotates over the sample (Fig. 3.2b) with a characteristic length $R_f \sim R_a/\lambda^2$, where $\lambda \sim K_r R_a^2/A$ is a small parameter of the theory. This magnetic state of an amorphous ferromagnet has been called the correlated spin glass (CSG), and is sometimes referred to as an state of Imry-Ma domains. In this respect it may be useful to mention that the CSG magnetic state is essentially different from the domain structure of a crystalline ferromagnet. The domain structure consists of well defined uniformly magnetized domains separated by domain walls, inside of

150

which $\mathbf{M}$ rotates from one direction to another. The average size of a domain is typically much greater than the thickness of the domain wall. In contrast to that picture, the magnetization in the CSG rotates smoothly over the solid, with the only one characteristic length R_f. The physics underlying these two static magnetic states is also different. The crystalline magnet splits into domains in order to decrease the energy of the magnetic dipole interaction, while the CSG is formed by the random magnetic anisotropy. The ferromagnetic correlation length R_f in the CSG is usually small compared to the typical size of a domain in a conventional ferromagnetic crystal. Consequently, the magnetic dipole energy is automatically small in the CSG.

It should be emphasized that all of the above concerns static $\mathbf{M(r)}$-configurations. The effect of temperature on the CSG (see below) may largely be absorbed in the $M_0(T)$-dependence.

2.   The Correlated Spin Glass

The exact solution of the random anisotropy model in three dimensions is not known. In what follows I will review several approximations, all of which lead to the CSG picture described in the previous Section.

2.1.   The Imry-Ma arguments

Consider the random anisotropy model in d dimensions, described by the fictitious random field $\mathbf{h(r)}$ instead of the random anisotropy $\mathbf{n(r)}$. The corresponding energy functional is

$$E = \int d^d r \left[\frac{A}{M_0^2} \left(\frac{\partial \mathbf{M}}{\partial r_i} \cdot \frac{\partial \mathbf{M}}{\partial r_i} \right) - \mathbf{h} \cdot \mathbf{M} \right] \ . \tag{3.16}$$

We assume that $\mathbf{h(r)}$ randomly rotates over the sample on a scale R_a, as $\mathbf{n(r)}$ does. One may expect that this model leads to an $\mathbf{M(r)}$-behavior similar to that of the random anisotropy model, if the strength of the random field is chosen as

$$h = H_r \equiv \frac{2K_r}{M_0} \ . \tag{3.17}$$

It is also convenient to introduce the field

$$H_{ex} = \frac{2A}{M_0 R_a^2} \ . \tag{3.18}$$

These two fields, H_r and H_{ex}, are important parameters of the theory of amorphous ferromagnetism. We will call them the anisotropy field and the exchange field respectively. It should be noted that while H_r of Eq.(3.17) coincides with the definition of the anisotropy field for a crystalline ferromagnet, H_{ex} of Eq.(3.18) is different from the exchange field of the conventional ferromagnetic theory. The latter is defined as $2A/M_0 r_{ex}^2$ and is typically large compared to the experimentally accessible magnetic fields. As to H_{ex} of Eq.(3.18), it may be much smaller due to the possibility that $R_a >> r_{ex}$, so that the magnetic field $H \sim H_{ex}$ may well be within the experimental range.

For small H_r, the ferromagnetic order favored by the first term in Eq.(3.16) cannot be disrupted on a small scale R_a. Let $R_f >> R_a$ be the ferromagnetic correlation length. The average energy of the exchange interaction is

$$E_{ex} \sim \frac{A}{R_f} V \sim H_{ex} M_0 V \left(\frac{R_a}{R_f}\right)^2 , \tag{3.19}$$

where V is the d-dimensional volume of the system. If $< \mathbf{h} >$ was exactly zero on the scale R_f, the average anisotropy energy would be also zero. Due to statistical fluctuations, however, the average anisotropy field on the scale R_f is $H_r(R_a/R_f)^{d/2}$. This gives the average anisotropy energy

$$E_{an} \sim -H_r M_0 V \left(\frac{R_a}{R_f}\right)^{d/2} . \tag{3.20}$$

Minimizing the total energy, $E = E_{ex} + E_{an}$ with respect to R_f, one obtains

$$R_f \sim R_a \left(\frac{H_{ex}}{H_r}\right)^{\frac{2}{4-d}} \tag{3.21}$$

as a rough estimate of R_f for $d = 1, 2, 3$. Eq.(3.21) is applied when $R_f >> R_a$, i.e. in the limit of weak anisotropy, $H_r << H_{ex}$. At $H_r \sim H_{ex}$ this equation gives $R_f \sim R_a$. In the limit of large anisotropy, $H_r >> H_{ex}$, $\mathbf{M}(\mathbf{r})$ simply follows the direction of the local anisotropy field.

The important observation that follows from the above arguments is that R_f remains finite, no matter how weak the random anisotropy is [6]. This non-trivial fact can be easily understood in terms of the random walk. When moving along some path through an amorphous magnet, the magnetization vector "feels" the random interactions of the local anisotropy, like a particle in the Brownian motion experiences random "pushes" from the surrounding molecules. The random walk of the direction of $\mathbf{M}$ leads to the destruction of the long-range ferromagnetic order. This analogy is illustrated by the numerical simulation [7] of the one-dimensional random magnet shown in figures 3.3 and 3.4. The Hamiltonian of the system is [8]

$$\mathcal{H} = -\sum_{\alpha}[J\mathbf{S}_\alpha \cdot \mathbf{S}_{\alpha+1} + D(\mathbf{n}_\alpha \cdot \mathbf{S}_\alpha)^2] . \tag{3.22}$$

The spins are interacting ferromagnetically ($J > 0$), while the direction of the anisotropy $\mathbf{n}_\alpha$ at any site α is random (R_a is the distance between the spins). Fig.3.3 shows the actual orientation of spins in a small segment of a chain that consists of

Figure 3.3: One-dimensional correlated spin glass, D/J=0.3

1000 spins. Fig. 3.4 shows the entire chain by choosing the following representation. When moving along the chain, the direction of spin randomly rotates in the plane normal to the chain. As the chain is rolled up into a circle, the ends of spins form a random spiral on the surface of a torus. The average distance between the turns is the ferromagnetic correlation length. The strength of the anisotropy in Fig. 3.4b is ten times greater than in Fig. 3.4a, leading to a smaller R_f. Note that although Fig. 3.4 is in qualitative agreement with Eq.(3.21), there is a pronounced quantitative difference between the numerical results and the scaling law $R_f \propto D^{-2/3}$ that follows from this equation. There might be a few reasons for this: First, the simulations were performed for the random-anisotropy and not for the random-field model. Second, they may reflect average magnetic configurations of numerous metastable states, common to disordered systems, rather than the property of the ground state. This may be especially important in low dimensions where the system has less freedom to escape from metastable states. Third, both the Imry-Ma arguments and the numerical simulations are likely to miss the long-range fluctuations responsible for the fast decay of correlations in low dimensions. It would be nice, therefore, to have something better than the Imry-Ma arguments. Because physics is sometimes better understood through exact mathematics, I will study here two exactly solvable models: the random-anisotropy XY model, and the random-anisotropy mean-spherical model.

2.2. The random-anisotropy XY model

Consider the model represented by Eq.(3.12) in d dimensions, with $\mathbf{M}$ and $\mathbf{n}$ having only two components, along the X and Y axes. The orientations of $\mathbf{M}$ and $\mathbf{n}$ can be parameterized by polar angles in the XY-plane, $\theta(\mathbf{r})$ and $\phi(\mathbf{r})$ respectively. In this subsection we will also allow for the dependence of the magnitude of the anisotropy on the coordinates, $K_r = K_r(\mathbf{r})$.

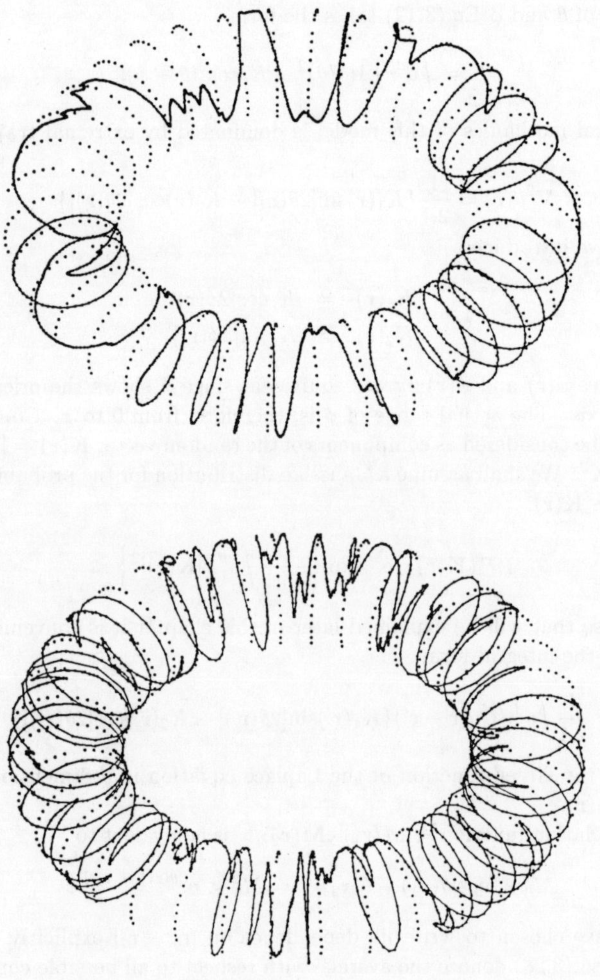

Figure 3.4: Toroidal representation of the CSG state in a chain of 1000 spins. (a) $D/J = 0.02$ (b) $D/J = 0.2$ (from Dickman and Chudnovsky, Ref.7)

In terms of θ and ϕ Eq.(3.12) takes the form

$$E = \int d^d r \left[A(\nabla\theta)^2 - K_r \cos^2(\theta - \phi) \right] \quad . \tag{3.23}$$

The statistical mechanics of this model is dominated by extremal trajectories satisfying

$$\nabla^2 \theta(\mathbf{r}) = \frac{1}{2A}\{ K_1(\mathbf{r})\sin[2\theta(\mathbf{r})] - K_2(\mathbf{r})\cos[2\theta(\mathbf{r})] \} \quad , \tag{3.24}$$

where we have introduced

$$\begin{aligned} K_1(\mathbf{r}) &= K_r \cos[2\phi(\mathbf{r})] \\ K_2(\mathbf{r}) &= K_r \sin[2\phi(\mathbf{r})] \end{aligned} \tag{3.25}$$

The angles $\phi(\mathbf{r})$ and $\phi(\mathbf{r})\pm\pi$ are equivalent since ϕ shows the orientation of the anisotropy axis. The actual range of ϕ is, therefore, from 0 to π. Consequently, K_1 and K_2 may be considered as components of the random vector $\mathbf{K}(\mathbf{r}) = [K_1(\mathbf{r}), K_2(\mathbf{r})]$, $K_1^2 + K_2^2 = \mathbf{K}^2$. We shall assume a Gaussian distribution for the probability of a given configuration $\mathbf{K}(\mathbf{r})$,

$$P\{\mathbf{K}(\mathbf{r})\} \propto \exp\left\{ -\frac{1}{w_d}\int d^d r \, [\mathbf{K}(\mathbf{r})]^2 \right\} \quad . \tag{3.26}$$

For a purpose that will be explained later in this chapter it is convenient to present Eq.(3.24) in the integral form

$$\theta(\mathbf{r}) = \frac{1}{2A}\int d^d r' G_d(\mathbf{r} - \mathbf{r}')\{ K_1(\mathbf{r}')\sin[2\theta(\mathbf{r}')] - K_2(\mathbf{r}')\cos[2\theta(\mathbf{r}')] \} \quad , \tag{3.27}$$

where G_d is the Green function of the Laplace equation in d-dimensions, satisfying, $\nabla^2 G_d(\mathbf{r}) = \delta(\mathbf{r})$.

The correlation function $< \mathbf{M}(\mathbf{r}_1) \cdot \mathbf{M}(\mathbf{r}_2) >$ is equivalent to

$$M_0^2 < \cos[\theta(\mathbf{r}_1) - \theta(\mathbf{r}_2)] > = M_0^2 < e^{i[\theta(\mathbf{r}_1) - \theta(\mathbf{r}_2)]} > \quad , \tag{3.28}$$

where we have chosen to write its dependence on $|\mathbf{r}_1 - \mathbf{r}_2|$ explicitly. The angular brackets in Eq.(3.28) denote the average with respect to all possible configurations of $\mathbf{K}(\mathbf{r})$ occurring with probabilities given by Eq.(3.26). Thus, the correlation function (3.28) is determined by the functional integral

$$\begin{aligned} < \mathbf{M}(\mathbf{r}_1) \cdot \mathbf{M}(\mathbf{r}_2) > = \ & M_0^2 \\ \times \ & \frac{\int D\{\mathbf{K}\}\exp\left\{\int d^d r \left[\frac{i}{2A}(G_d^{(1)} - G_d^{(2)})(K_1\sin2\theta - K_2\cos2\theta) - \frac{1}{w_d}\mathbf{K}^2 \right]\right\}}{\int D\{\mathbf{K}(\mathbf{r})\} \exp\left\{ -\frac{1}{w_d}\int d^d r \, [\mathbf{K}(\mathbf{r})]^2 \right\}} \end{aligned} \quad , \tag{3.29}$$

where $G_d^{(n)} \equiv G_d(\mathbf{r}_n - \mathbf{r})$. Generally speaking, it is impossible to evaluate this integral because $\theta(\mathbf{r})$ depends on $\mathbf{K}(\mathbf{r})$ via the non-linear Eq.(3.24). It should be recalled, however, (this statement is also justified by the following calculation) that in the limit of weak anisotropy, $H_r^2 = 4 < \mathbf{K}_r^2 > /M_0^2 << H_{ex}^2$, the local magnetization is formed by a random walk on the scale $R_f >> R_a$. Consequently, the local value of $\theta(\mathbf{r})$ is almost independent of $\mathbf{K}(\mathbf{r})$ at the same point. Taking into account that the integration over $\mathbf{K}(\mathbf{r})$ in Eq.(3.29) is local in spatial coordinates, this suggests that the integral is Gaussian with high accuracy. Performing such an integration and noticing that after the integration $\sin[2\theta(\mathbf{r})]$ and $\cos[2\theta(\mathbf{r})]$ combine into $\sin^2[2\theta] + \cos^2[2\theta] = 1$, one obtains a simple formula which is valid for any d in the limit of $H_r << H_{ex}$,

$$< \mathbf{M}(\mathbf{r}_1) \cdot \mathbf{M}(\mathbf{r}_2) > = M_0^2 \exp\left\{ -\frac{w_d}{8A^2} \int d^d r \, [G_d(\mathbf{r}_1 - \mathbf{r}) - G_d(\mathbf{r}_2 - \mathbf{r})]^2 \right\} \; . \quad (3.30)$$

The Fourier transform of the integral in the exponent then gives

$$< \mathbf{M}(\mathbf{r}_1) \cdot \mathbf{M}(\mathbf{r}_2) > = M_0^2 \exp\left\{ -\frac{w_d}{4A^2} \int \frac{d^d q}{(2\pi)^d} \frac{1 - \cos[\mathbf{q} \cdot (\mathbf{r}_1 - \mathbf{r}_2)]}{q^4} \right\} \; , \quad (3.31)$$

where we have substituted $G_d(q) = -1/q^2$. The integral must be cut off at $1/R_a$ at large q and at $1/L$ at small q, where L is the size of the system. Our final result at $R_a << |\mathbf{r}_1 - \mathbf{r}_2| << L$ is

$$< \mathbf{M}(\mathbf{r}_1) \cdot \mathbf{M}(\mathbf{r}_2) >_{d=2} = M_0^2 \exp\left\{ -\frac{w_2}{32\pi A^2} |\mathbf{r}_1 - \mathbf{r}_2|^2 \ln\left[\frac{L}{|\mathbf{r}_1 - \mathbf{r}_2|}\right] \right\} \quad (3.32)$$

$$< \mathbf{M}(\mathbf{r}_1) \cdot \mathbf{M}(\mathbf{r}_2) >_{d=3} = M_0^2 \exp\left\{ -\frac{w_3}{32\pi A^2} |\mathbf{r}_1 - \mathbf{r}_2| \right\} \quad (3.33)$$

$$< \mathbf{M}(\mathbf{r}_1) \cdot \mathbf{M}(\mathbf{r}_2) >_{d=4} = M_0^2 \left(\frac{R_a}{|\mathbf{r}_1 - \mathbf{r}_2|}\right)^{w_4/32\pi^2 A^2} \quad (3.34)$$

The parameter w_d can be estimated as $w_d \sim < \mathbf{K}^2 > \Omega_d$, where $\Omega_d \propto R_a^d$ is the d-dimensional volume inside which anisotropy axes are correlated. Presenting Eq.(3.33) as

$$< \mathbf{M}(\mathbf{r}_1) \cdot \mathbf{M}(\mathbf{r}_2) >_{d=3} = M_0^2 \exp\left(-\frac{|\mathbf{r}_1 - \mathbf{r}_2|}{R_f}\right) \; , \quad (3.35)$$

one obtains $R_f \sim R_a(H_{ex}/H_r)^2$, in agreement with Eq.(3.21) at $d = 3$. As follows from Eq.(3.32), the Imry-Ma arguments also provide a reasonably good answer in two dimensions, $R_f \sim R_a(H_{ex}/H_r)$, if one neglects the $ln(L)$-dependence of the correlation function. At $d = 4$ the ferromagnetic correlations decay as a power law. This is the marginal dimensionality for the random anisotropy ferromagnet. The weak random anisotropy does not affect the long range ferromagnetic order at $d \geq 5$.

2.3. *The random-anisotropy mean-spherical model*

In this Subsection we will give up the fixed length condition, $\mathbf{M}^2 = M_0^2$. Instead, we let M_0 be the average value of the local magnetization,

$$< \mathbf{M}^2(\mathbf{r}) > = \frac{1}{V} \int d^3 r \, \mathbf{M}^2(\mathbf{r}) = M_0^2 \; . \quad (3.36)$$

With this condition Eq.(3.12) is still equivalent to Eq.(3.13). Our problem is to minimize the functional

$$E' = \frac{1}{M_0^2} \int d^3r \left(A \frac{\partial M_i}{\partial r_j} \frac{\partial M_i}{\partial r_j} - K_r R_{ij} M_i M_j \right) - \lambda_M \int d^3r \, \mathbf{M}^2(\mathbf{r}) \quad , \tag{3.37}$$

where λ_M is a (constant) Lagrange multiplier. Extremal $\mathbf{M}(\mathbf{r})$-trajectories satisfy

$$\frac{A}{M_0^2} \nabla^2 M_i + \frac{K_r}{M_0^2} R_{ij} M_j + \lambda_M M_i = 0 \tag{3.38}$$

or

$$(\nabla^2 - k^2) M_i = -\frac{K_r}{A} R_{ij} M_j \quad , \tag{3.39}$$

where $\lambda_M \equiv - A k^2 / M_0^2$ $(k^2 > 0)$. The formal solution of Eq.(3.38) is

$$M_i(\mathbf{r}) = -\frac{K_r}{A} \int d^3r' G_k(\mathbf{r} - \mathbf{r}') R_{ij}(\mathbf{r}') M_j(\mathbf{r}') \quad , \tag{3.40}$$

where the Green function is

$$G_k(\mathbf{r}) = -\frac{e^{-k|\mathbf{r}|}}{4\pi|\mathbf{r}|} \quad . \tag{3.41}$$

From Eq.(3.40) we have a formal expression for the correlation function

$$< \mathbf{M}(\mathbf{r}_1) \cdot \mathbf{M}(\mathbf{r}_2) > = \left(\frac{K_r}{A} \right)^2$$
$$\times \int d^3r' \int d^3r'' G_k(\mathbf{r}_1 - \mathbf{r}') G_k(\mathbf{r}_2 - \mathbf{r}'') < R_{ij}(\mathbf{r}') R_{ik}(\mathbf{r}'') M_j(\mathbf{r}') M_k(\mathbf{r}'') > \tag{3.42}$$

Now for weak anisotropy, $\mathbf{M}(\mathbf{r})$ is only weakly correlated with $\mathbf{n}(\mathbf{r})$ and we can replace $< R_{ij}(\mathbf{r}') R_{ik}(\mathbf{r}'') M_j(\mathbf{r}') M_k(\mathbf{r}'') >$ with $< R_{ij}(\mathbf{r}') R_{ik}(\mathbf{r}'') > < M_j(\mathbf{r}') M_k(\mathbf{r}'') >$. The random tensor $\hat{R}$ has only short-range correlations, so that

$$< R_{ij}(\mathbf{r}') R_{ik}(\mathbf{r}'') > = \frac{2}{9} \delta_{kj} \Gamma(|\mathbf{r}' - \mathbf{r}''|) \quad . \tag{3.43}$$

The coefficient 2/9 is obtained at $\mathbf{r}' = \mathbf{r}''$, by using the explicit form of R_{ij}, Eq.(3.14), to satisfy the condition $\Gamma(0) = 1$. Eq.(3.42) then reduces to

$$< \mathbf{M}(\mathbf{r}_1) \cdot \mathbf{M}(\mathbf{r}_2) > = \frac{2}{9} \left(\frac{K_r}{A} \right)^2$$
$$\times \int d^3r' \int d^3r'' \Gamma(|\mathbf{r}' - \mathbf{r}''|) G_k(\mathbf{r}_1 - \mathbf{r}') G_k(\mathbf{r}_2 - \mathbf{r}'') < \mathbf{M}(\mathbf{r}') \cdot \mathbf{M}(\mathbf{r}'') > \quad . \tag{3.44}$$

$\Gamma(|\mathbf{r}' - \mathbf{r}''|)$ rapidly goes to zero at $|\mathbf{r}' - \mathbf{r}''| >> R_a$. Correspondingly, only the values of $\mathbf{r}'$ which are close to $\mathbf{r}''$ contribute to the integral. This allows one to put $\mathbf{r}' = \mathbf{r}''$ in the Green functions and in the correlation function, $< \mathbf{M}(\mathbf{r}') \cdot \mathbf{M}(\mathbf{r}'') >$, under the integral, assuming that $R_f = k^{-1} >> R_a$, as is justified below. Using Eq.(3.36) we obtain

$$< \mathbf{M}(\mathbf{r_1}) \cdot \mathbf{M}(\mathbf{r_2}) >= \frac{2}{9} M_0^2 \left(\frac{K_r}{A}\right)^2 \Omega \int d^3 r \, G_k(\mathbf{r_1} - \mathbf{r}) G_k(\mathbf{r_2} - \mathbf{r}) \quad , \qquad (3.45)$$

where $\Omega = \int d^3 r \Gamma(r)$. is the volume inside which the anisotropy axes are correlated due to the short-range order. Further integration gives

$$< \mathbf{M}(\mathbf{r_1}) \cdot \mathbf{M}(\mathbf{r_2}) >= M_0^2 \left(\frac{K_r}{A}\right)^2 \frac{\Omega}{36\pi k} e^{-k|\mathbf{r_1}-\mathbf{r_2}|} \quad . \qquad (3.46)$$

The parameter k is related to the Lagrange multiplier and can be evaluated from Eq.(3.46) at $\mathbf{r_1} = \mathbf{r_2}$ with the condition (3.36). Finally we have

$$< \mathbf{M}(\mathbf{r_1}) \cdot \mathbf{M}(\mathbf{r_2}) > \; = \; M_0^2 \exp\left(-\frac{|\mathbf{r_1} - \mathbf{r_2}|}{R_f}\right)$$

$$R_f = k^{-1} \; = \; \frac{36\pi}{\Omega} \left(\frac{A}{K_r}\right)^2 \quad . \qquad (3.47)$$

A similar calculation in two dimensions gives $R_f \propto A/K_r$. This is again in a qualitative agreement with the Imry-Ma arguments and the results of the XY model.

2.4. Experimental evidence of the correlated spin glass

As follows from the above discussion, the CSG shown in figures 3.2-3.4 is the most likely magnetic state of an amorphous ferromagnet in zero magnetic field. In three dimensions it is characterized by the exponential decay of correlations, $< \mathbf{M}(\mathbf{r_1}) \cdot \mathbf{M}(\mathbf{r_2}) >= M_0^2 \exp(-|\mathbf{r_1} - \mathbf{r_2}|/R_f)$, with $R_f \sim R_a(H_{ex}/H_r)^2 >> R_a$. This magnetic state can be revealed by neutron scattering. The intensity, $I(Q)$ of the small Q neutron scattering by the static CSG structure is defined by the Fourier transform of $< \mathbf{M}(\mathbf{r_1}) \cdot \mathbf{M}(\mathbf{r_2}) >$,

$$I_{csg}(Q) = \frac{B}{(Q^2 + R_f^{-2})^2} \quad , \qquad (3.48)$$

where B is proportional to M_0^2/R_f.

Within spatial regions smaller than R_f, the magnetic moments are ferromagnetically correlated. In these regions, well-defined spin waves with momenta $Q > R_f^{-1}$ must exist, so that there must be a spin-wave contribution to the total intensity of neutron scattering,

$$I_{sw} = \frac{C}{Q^2 + k^2} \quad . \qquad (3.49)$$

The parameter k can be obtained from the following argument. The $(Q^2 + k^2)$ in the denominator of Eq.(3.49) originates from the spin-wave energy $\epsilon(Q)$. In a crystalline

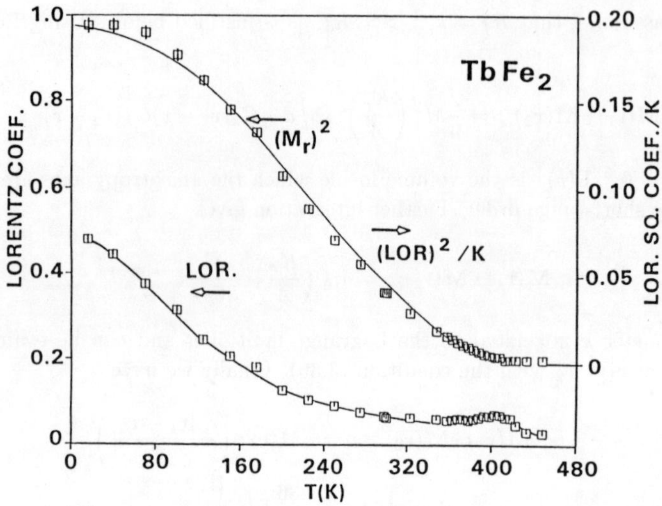

Figure 3.5: Temperature dependence of the Lorentzian squared coefficient B divided by $k = R_f^{-1}$. The solid line in the B/k curve is the temperature dependence of the zero-field bulk magnetic moment.(From Rhyne, Ref.9)

ferromagnet $\epsilon(Q) \propto AQ^2 + K_c$ where A and K_c are exchange and anisotropy constants. In an amorphous ferromagnet the effective anisotropy, evident at large scales (Section 3), is $K_{eff} \sim A/R_f^2$. Consequently, one should expect that k of Eq.(3.49) is of the order of R_f^{-1}.

A total intensity of the form

$$I(Q) = \frac{B}{(Q^2 + R_f^{-1})^2} + \frac{C}{Q^2 + R_f^{-1}} \tag{3.50}$$

has been used for the analysis of experimental results on neutron scattering in several amorphous systems. For $R_x Fe_{1-x}$ alloys (R=rare earth) a good agreement with the theoretical prediction, $B \propto M_0^2(T)/R_f(T)$, has been obtained [9] (see Fig. 3.5). In $R_x Fe_{1-x}$ amorphous alloys $R_f \sim 50 - 80$Åat low temperature and increases with increasing temperature. Since R_a in these systems is of the order of 5Å, the CSG picture seems quite plausible. The behavior of the neutron cross-section given by Eq.(3.50) also has been observed, and interpreted as a random field effect, in amorphous $Fe - Mn$ alloys [10]. It is likely that Eq.(3.50) provides a natural explanation for many experimental features common to a wide variety of amorphous and spin-glass systems. It should also be noted that the problem of spin-waves in amorphous ferromagnets is more complicated than its short-wavelength aspect discussed here (see Chapters 5 and 6 of this book).

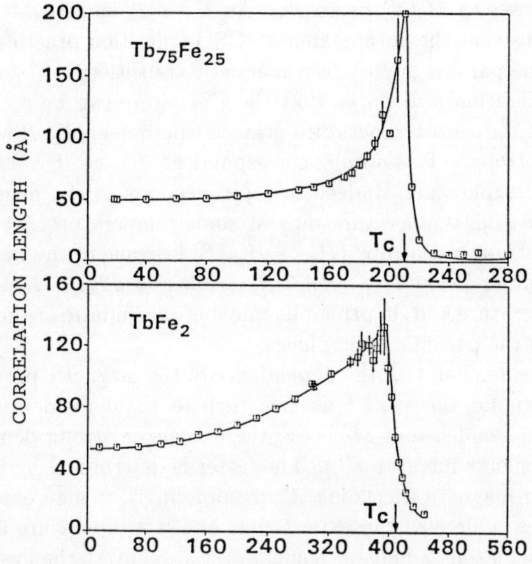

Figure 3.6: Temperature dependence of the ferromagnetic correlation length. (From Rhyne, Ref.9)

2.5. The phase diagram

Experimental data show that R_f increases as the Curie temperature is approached (Fig. 3.6). This behavior of R_f can be understood in terms of its temperature dependence on the parameters of the theory. According to Eq.(3.47), $R_f \propto A^2/K_r^2\Omega$. The volume Ω inside of which the easy axes are correlated must be practically independent of temperature, unless some annealing effects (i.e., significant diffusion of atoms) are involved. The latter would lead to the irreversible temperature dependence of R_f which has not been commonly observed. The exchange interaction parameter, A, depends on the temperature through $M_0(T)$ according to Eq.(3.7). Note that the exchange function $J(r)$ is temperature independent. In our model K_r is related, through the short-range structural order, to the magnetic anisotropy of the crystalline counterpart of an amorphous magnet. It is well known that the crystalline anisotropy can depend on higher powers of the magnetization. If p is the power characterizing the angular dependence of the anisotropy, $K_r(\mathbf{n} \cdot \mathbf{M})^p$, then $K_r(T) \propto [M_0(T)]^{p(p+1)/2}$. The predictions of the random anisotropy model may be not sensitive to the value of p until the temperature effects are studied. We consider here the lowest possible value of p, $p = 2$, which should apply to rare-earth-based compounds with a strong single-ion anisotropy. Even in this case the dependence of K_r on M_0 is rather strong, $K_r \propto M_0^3$, and gives $R_f \propto M_0^{-2}(T)$ for the temperature dependence of the ferromagnetic correlation length. In general, R_f must be proportional to a some inverse power of M_0 because the dependence of A on M_0 is weaker than $K_r(M_0)$. Correspondingly, R_f increases as $M_0(T)$ decreases. As $T \to T_c$ and $M_0 \to 0$, R_f goes to infinity. This means that the paramagnetic - CSG transition practically cannot be distinguished from the paramagnetic - ferromagnetic transition.

The above consideration also shows that the CSG state can be realized at finite temperatures even if the zero temperature state is speromagnetic (SM) due to the strong random anisotropy. This would correspond to $H_r > H_{ex}$ at $T = 0$ but $H_r < H_{ex}$ at finite temperature. Indeed, a rapid decrease in the magnitude of the anisotropy with increasing temperature may at some temperature $T = T_{csg}$ convert the system from the strong anisotropy ($H_r > H_{ex}$) SM regime to the weak anisotropy ($H_r < H_{ex}$) CSG regime. Below T_{csg}, $R_f = R_a$, while above T_{csg}, R_f starts growing with increasing temperature and, in principle, may become infinite at the temperature of the transition into the paramagnetic phase.

As many experiments deal with the dependence of the magnetic properties on the concentration of magnetic ions, x, let us now turn to the dependence of R_f on x. Besides the obvious dependence of M_0 on x, there is also a strong dependence of A on x through the exchange integral $J(r)$. The latter is determined by the overlap of the wave functions of magnetic electrons. Correspondingly, it may exponentially go to zero as x decreases. One also can expect that at low temperature an increase in the exchange with increasing x may, at some $x = x_{csg}$, convert the system from the strong anisotropy ($H_r > H_{ex}$) SM regime to the weak anisotropy ($H_r < H_{ex}$) CSG regime.

The dependence of R_f on x and T leads to the phase diagram [11] shown in Fig. 3.7. At large x, as the temperature decreases, the system first enters the CSG phase. As discussed above, the corresponding phase transition must resemble the paramagnetic - ferromagnetic transition. As the temperature continues to decrease, the system enters a highly disordered speromagnetic state. Such a behavior, often referred as a reentrant behavior, has been observed in a number of Eu and Gd based amorphous materials [12]. In accordance with Fig. 3.7, it has also been observed that the temperature of the reentrant (CSG - SM) transition decreases with increasing x [13].

2.6. Coherent anisotropy

Up to this moment we have assumed that the anisotropy is completely random. Meanwhile, in real samples the presence of a some coherent anisotropy, K_c, is inevitable. This can be caused by elastic stresses, weak structural anisotropy arising from the technology of the preparation of the amorphous alloy, etc.

The limit of strong coherent anisotropy corresponds to a crystalline ferromagnet. The ground state of such a ferromagnet is the domain structure. This must be the case even in the presence of a weak random component of the anisotropy, which is always present in ferromagnetic crystals due to defects. The domain structure is characterized by two lengths: the average size of a domain, which depends on the

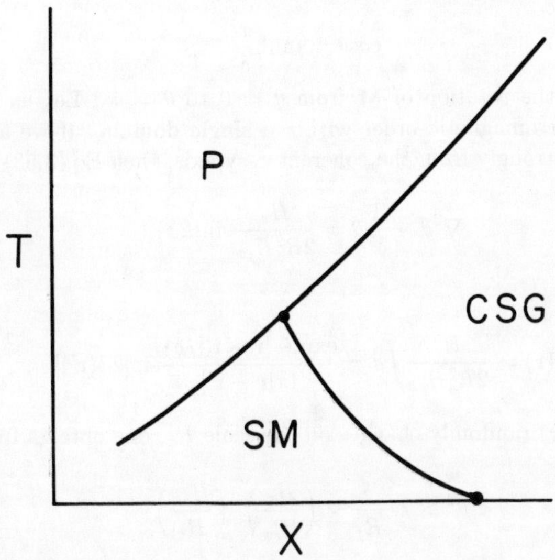

Figure 3.7: Schematic phase diagram for random-anisotropy ferromagnet.

geometry of the sample, and the domain wall thickness, $\delta = (H_{ex}/H_c)^{1/2} R_a$. Here we have introduced the coherent anisotropy field $H_c = 2K_c/M_0$. The question is how large should K_c be to transform the CSG into an usual ferromagnetic domain structure.

In the presence of both random and coherent anisotropy the energy becomes

$$E = \frac{1}{M_0^2} \int d^3r \left\{ A\frac{\partial \mathbf{M}}{\partial r_i} \cdot \frac{\partial \mathbf{M}}{\partial r_i} - K_r[\mathbf{n}(\mathbf{r}) \cdot \mathbf{M}(\mathbf{r})]^2 - K_c[\mathbf{n}_c(\mathbf{r}) \cdot \mathbf{M}(\mathbf{r})]^2 \right\} \quad , \qquad (3.51)$$

where $\mathbf{n}_c$ is the direction of the coherent easy axis. It is much easier to obtain the result we are looking for from the XY version of Eq.(3.51). Assuming all vectors to be in the XY plane, with $\mathbf{n}_c$ along the X axis, we get

$$E = \int d^3r \left[A(\nabla\theta)^2 - K_r\cos^2(\theta - \phi) - K_c\cos^2\theta\right] \quad , \qquad (3.52)$$

where $\theta(\mathbf{r})$ and $\phi(\mathbf{r})$ are the angles that $\mathbf{M}(\mathbf{r})$ and $\mathbf{n}(\mathbf{r})$ form with the X axis respectively. The angle $\phi(\mathbf{r})$ is random, while $\theta(\mathbf{r})$ adjusts to the minimum energy configuration satisfying

$$A\nabla^2\theta - \frac{1}{2}K_c\sin(2\theta) = \frac{1}{2}K_r\sin[2(\theta - \phi)] \quad . \qquad (3.53)$$

In the limit of weak random anisotropy ($K_r \to 0$), the coherent anisotropy aligns $\mathbf{M}$ along the X axis, $\theta = 0$ or $\theta = \pi$. There is also a domain wall solution of Eq.(3.53) at $K_r = 0$,

$$\cos\theta = \tanh\frac{x}{\delta} \quad , \tag{3.54}$$

corresponding to the rotation of $\mathbf{M}$ from $\theta = 0$ to $\theta = \pi$. Let us investigate the stability of the ferromagnetic order within a single domain. If we assume that $\mathbf{M}$ does not deviate strongly from the coherent easy axis, then Eq.(3.53) reduces to

$$\nabla^2\theta - \frac{1}{\delta^2}\theta = \frac{H_r}{2R_a^2 H_{ex}}\sin(2\phi) \quad . \tag{3.55}$$

The solution is

$$\theta(\mathbf{r}) = \frac{H_r}{2R_a^2 H_{ex}}\int d^3r' \frac{\exp(-|\mathbf{r} - \mathbf{r}'|/\delta)}{4\pi|\mathbf{r} - \mathbf{r}'|}\sin[2\phi(\mathbf{r}')] \quad . \tag{3.56}$$

Assuming that $\phi(\mathbf{r})$ randomly changes on the scale R_a, one obtains from Eq.(3.56)

$$< \theta^2 > \sim \frac{\delta}{R_f} \sim \left(\frac{H_r}{H_{ex}}\right)^2 \left(\frac{H_{ex}}{H_c}\right)^{1/2} \quad , \tag{3.57}$$

where $R_f \sim R_a(H_{ex}/H_r)^2$ is the ferromagnetic correlation length of the CSG state. $< \theta^2 >$ is small if $R_f >> \delta$, i.e. $H_r/H_{ex} << (H_c/H_{ex})^{1/4}$. Consequently, the coherent anisotropy [14]

$$K_c \sim K_r \left(\frac{H_r}{H_{ex}}\right)^3 \tag{3.58}$$

transforms the CSG into the ferromagnetic domain structure. This means that in the limit of weak random anisotropy, $H_r << H_{ex}$, the coherent anisotropy which is small compared to K_r will destroy the CSG. For that reason it is more likely to observe the CSG in materials where the ratio H_r/H_{ex} is not very small.

3.   Behavior in a magnetic field

In the presence of a magnetic field the energy functional of the random-anisotropy magnet becomes

$$E = \int d^3r \left[\frac{A}{M_0^2}\frac{\partial\mathbf{M}}{\partial r_i} \cdot \frac{\partial\mathbf{M}}{\partial r_i} - \frac{K_r}{M_0^2}(\mathbf{n} \cdot \mathbf{M})^2 - \mathbf{M} \cdot \mathbf{H}\right] \quad . \tag{3.59}$$

We will be interested in the component of the average magnetization parallel to the direction of the field, $< M_H >$. The total magnetic moment of the sample is $< M_H > V$.

In the case of a strong anisotropy, $H_r >> H_{ex}$, the problem is essentially the same as for a polycrystalline ferromagnetic solid. At $H = 0$ each moment is aligned with the local anisotropy axis but the application of a magnetic field breaks this alignment. A non-trivial aspect of the problem comes from the fact that at $H <$ H_r each moment may have two equilibrium orientations with respect to the field and the local anisotropy axis. One direction corresponds to the absolute energy minimum, and the other direction corresponds to a metastable state. At high fields, $H >> H_r >> H_{ex}$, when local metastable states disappear, one can show [15] that $< M_H >$ approaches saturation as $1/H^2$. At low fields the magnetization curve can be obtained by computer simulation. The temperature effects become important at a temperature comparable to the anisotropy energy per correlated volume, $\Omega \sim R_a^3$. At low temperature, amorphous materials with large H_r are hard magnets with small zero-field susceptibility.

In this section we will be concerned with the field behavior of the CSG. The process of the magnetization of such a system will be determined by the rotation of $\mathbf{M(r)}$ on a large scale (comparable to R_f at small H). In that sense it is analogous to the process of the magnetization of a crystalline ferromagnet which occurs through the motion of domain walls. Because of the large scales involved, thermal effects should manifest through the temperature dependence of the local magnetization, $M_0(T)$. This, of course, does not apply at temperatures close to T_c, where long-range thermal fluctuations become important.

3.1. Zero-field susceptibility of the CSG

Amorphous materials with weak random anisotropy, $H_r << H_{ex}$, may be extremely soft magnets. After some initial confusion, it has been realized that the zero-field susceptibility of the CSG may be very large, but finite [14,16], $\chi \propto (H_{ex}/H_r)^4$ in three dimensions. In order to avoid complications, we will demonstrate this for the $3d$ XY model.

In terms of the angles $\theta(\mathbf{r})$ and $\phi(\mathbf{r})$, which $\mathbf{M}$ and $\mathbf{n}$ make with respect to $\mathbf{H}$, Eq.(3.59) becomes

$$E = \int d^3r \left[A(\nabla\theta)^2 - K_r\cos^2(\theta - \phi) - M_0 H\cos\theta \right] \quad . \tag{3.60}$$

This expression is minimized by $\theta(\mathbf{r})$ satisfying

$$\nabla^2\theta - R_H^{-2}\sin\theta = \frac{H_r}{2R_a^2 H_{ex}}\sin[2(\theta - \phi)] \quad , \tag{3.61}$$

where we have introduced a characteristic length

$$R_H = \left(\frac{H_{ex}}{H}\right)^{1/2} R_a \quad . \tag{3.62}$$

Let $\theta_0(\mathbf{r})$ be the equilibrium configuration at $H = 0$. A very weak field transforms this configuration into

$$\theta(\mathbf{r}) = \theta_0(\mathbf{r}) + \delta\theta(\mathbf{r}) \quad . \tag{3.63}$$

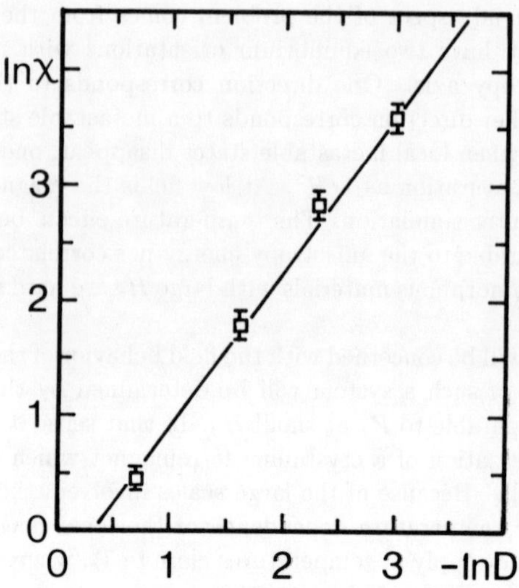

Figure 3.8: Dependence of the magnetic susceptibility of the $1d$ random-anisotropy ferromagnet on the exchange constant D (see Hamiltonian (3.22)). The slope of the straight line is 4/3. (From Dickman and Chudnovsky, Ref.7)

To the lowest order in H, $\delta\theta(\mathbf{r})$ satisfies

$$\nabla^2\delta\theta(\mathbf{r}) = R_H^{-2}\sin\theta_0(\mathbf{r}) \tag{3.64}$$

or

$$\delta\theta(\mathbf{r}) = -\frac{1}{R_H^2}\int d^3r'\,\frac{\sin[\theta_0(\mathbf{r}')]}{4\pi|\mathbf{r}-\mathbf{r}'|} \quad . \tag{3.65}$$

The susceptibility is

$$\chi = \frac{<M_H>}{H} = \frac{M_0}{H}<\cos\theta> = -\frac{M_0}{H}<\sin(\theta_0)\delta\theta> \quad . \tag{3.66}$$

Substituting here $\delta\theta$ of Eq.(3.65) we obtain

$$\chi = \frac{M_0}{R_a^2 H_{ex}}\int d^3r'\,\frac{<\sin[\theta_0(\mathbf{r})]\sin[\theta_0(\mathbf{r}')]>}{4\pi|\mathbf{r}-\mathbf{r}'|} \quad . \tag{3.67}$$

According to Eq.(3.35), the correlator under the integral is

$$<\sin[\theta_0(\mathbf{r})]\sin[\theta_0(\mathbf{r}')]> = \frac{1}{M_0^2}<M_y(\mathbf{r})M_y(\mathbf{r}')> = \frac{1}{2M_0^2}<\mathbf{M}(\mathbf{r})\cdot\mathbf{M}(\mathbf{r}')>$$

$$= \frac{1}{2}\exp\left(-\frac{|\mathbf{r}-\mathbf{r}'|}{R_f}\right) \quad . \tag{3.68}$$

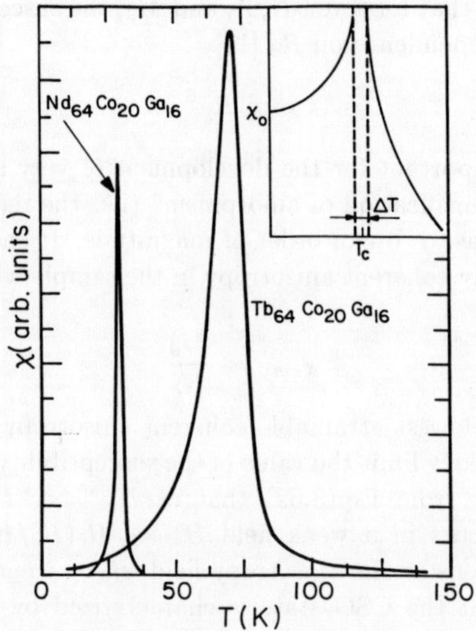

Figure 3.9: Temperature dependence of susceptibility for strong anisotropy magnets. $\chi(T)$ becomes large at $T \to T_c$, while $\chi(0)$ can be very small.(From Sellmyer and Nafis, Ref.18)

This gives

$$\chi = \frac{M_0}{H_{ex}}\left(\frac{R_f}{R_a}\right)^2 \sim \frac{M_0}{H_{ex}}\left(\frac{H_{ex}}{H_r}\right)^4. \tag{3.69}$$

Note that a similar calculation for an arbitrary number of dimensions, $d < 4$, yields [14] $\chi \propto H_r^{4/(d-4)}$. For the one dimensional chain of spins described in Section 3.2 this gives $\chi \propto K_r^{-4/3}$. This result is in good agreement with numerical simulations [7] (Fig. 3.8).

In the limit $H_r << H_{ex}$ the susceptibility can be very large; in three dimensions it is proportional to K_r^{-4}. This seems to be in agreement with the experimental result [17], $\chi \propto 1/K_r^{3.8\pm0.2}$, obtained in $Dy_xGd_{1-x}Ni$. The temperature dependence of χ in the CSG phase is dominated by the temperature dependence of R_f. As R_f rapidly grows at $T \to T_c$ (see Section 2.5), the susceptibility must have a sharp maximum at T_c. It has been discussed in Section 2.5 that strong anisotropy SM systems should turn into the CSG state when temperature is sufficiently close to the PM transition. Consequently, the prediction of the susceptibility cusp, based upon the CSG theory, actually has more generality and is applied to any random anisotropy system including strong anisotropy magnets. This is in agreement with experiment [18] (Fig. 3.9). Note that the vicinity of the CSG-PM transition of course cannot be described in terms of temperature-dependent phenomenological constants.

Eq.(3.69) also gives the dependence of χ on the parameters of the amorphous structure. It turns out that for given K, A, and M_0, the susceptibility of the CSG has an extremely strong dependence on R_a [19]

$$\chi \propto R_a^{-6} \ .$$

(3.70)

This observation is important for the development of very soft magnetic materials. It shows that the "intensification of amorphism" (i.e. the reduction of R_a) by only a factor of 1.5 can increase χ by an order of magnitude. It should be noted, however, that the presence of any coherent anisotropy in the sample will restrict the maximum value of χ by [20]

$$\chi_{max} \sim \frac{M_0}{K_c} \ .$$

(3.71)

Thus, in practice, the lowest attainable coherent anisotropy, not the lowest random anisotropy, will most likely limit the value of the susceptibility in amorphous magnets.

One should expect from Eq.(3.69) that at $H_r << H_{ex}$ the saturation knee ($< M_H > \sim M_0$) occurs in a weak field $H_s \sim H_r(H_r/H_{ex})^3$. For a crystalline ferromagnet this happens at the anisotropy field, $H_s \sim H_c = 2K_c/M_0$. Correspondingly, one can say that the CSG state is characterized by the effective anisotropy $K_{eff} \sim K_r(H_r/H_{ex})^3$ which is a statistical average of the random anisotropy on the scale R_f.

3.2. Magnetic disorder in the presence of a field

As discussed above, the CSG is a soft magnet which develops a large magnetic moment in the presence of a weak field. In such materials the major part of the magnetization curve is the approach to saturation, $< M_H > \rightarrow M_0$. This regime turns out to be quite important for technological applications, as well as for the understanding of the structure of amorphous magnets. The field $H_s \sim H_r(H_r/H_{ex})^3$ removes most of metastable states in random anisotropy systems. From the theoretical point of view, this means that in the saturation regime one may hope to obtain a rigorous solution of a $3d$ random-anisotropy model. This turns out to be true.

Consider a functional

$$E = \int d^3r \left[\frac{A}{M_0^2} \frac{\partial \mathbf{M}}{\partial r_i} \cdot \frac{\partial \mathbf{M}}{\partial r_i} - \frac{K_r}{M_0^2}(\mathbf{n} \cdot \mathbf{M})^2 - \mathbf{M} \cdot \mathbf{H} - \gamma(\mathbf{r})\mathbf{M}^2 \right] \ ,$$

(3.72)

where $\gamma(\mathbf{r})$ is the Lagrange multiplier responsible for the condition $\mathbf{M}^2 = M_0^2$. The equation for $\mathbf{M}(\mathbf{r})$ that follows from Eq.(3.72) is

$$\frac{A}{M_0^2}\nabla^2\mathbf{M} + \frac{K}{M_0^2}\mathbf{n}(\mathbf{n} \cdot \mathbf{M}) + \frac{1}{2}\mathbf{H} + \gamma(\mathbf{r})\mathbf{M} = 0 \ .$$

(3.73)

Multiplying this equation by $\mathbf{M}$ and expressing $\gamma(\mathbf{r})$ via $\mathbf{M}(\mathbf{r})$, we get

$$\nabla^2\mathbf{m} + \lambda\mathbf{n}(\mathbf{n} \cdot \mathbf{m}) + \mathbf{h} = \mathbf{m}[\mathbf{m}\nabla^2\mathbf{m} + \lambda(\mathbf{n} \cdot \mathbf{m})^2 + \mathbf{h} \cdot \mathbf{m}] \ ,$$

(3.74)

where we have introduced

$$\mathbf{m} = \frac{\mathbf{M}}{M_0}, \quad \mathbf{h} = \frac{\mathbf{H}}{H_{ex}}, \quad \lambda = \frac{H_r}{H_{ex}} \quad , \tag{3.75}$$

and the dimensionless coordinates $\mathbf{x} = \mathbf{r}/R_a$.

The equation (3.74) gives $\mathbf{m}(\mathbf{x})$ for the fixed $\mathbf{h}$ and the fixed configuration of the random field $\mathbf{n}(\mathbf{x})$. The CSG corresponds to the zero net magnetization in the absence of the applied field. In the saturation region one can write $\mathbf{m}$ in the form

$$\mathbf{m} = m_\parallel \frac{\mathbf{h}}{h} + \mathbf{m}_\perp \tag{3.76}$$

with $|\mathbf{m}_\perp| << 1$, $\quad m_\parallel = 1 - \frac{1}{2}m_\perp^2$. The component of the magnetization $\mathbf{m}_\perp$ is a characteristic of the local disorder. We will be interested in the correlation function $< \mathbf{m}_\perp(\mathbf{r}_1) \cdot \mathbf{m}_\perp(\mathbf{r}_2) >$. To obtain this correlation function let us write the equation for $\mathbf{m}_\perp$ which follows from Eq.(3.74) to the first order in $\lambda << 1$:

$$(\nabla^2 - p^2)\mathbf{m}_\perp = -\lambda n_\parallel \mathbf{n}_\perp \quad . \tag{3.77}$$

In this equation $p^2 = h$, $\quad n_\parallel$ and $\mathbf{n}_\perp$ are components of the vector

$$\mathbf{n} = n_\parallel \frac{\mathbf{h}}{h} + \mathbf{n}_\perp \quad . \tag{3.78}$$

The solution of Eq.(3.76) is

$$\mathbf{m}_\perp(\mathbf{x}) = \lambda \int d^3 x' \frac{\exp(-p|\mathbf{x} - \mathbf{x}'|)}{4\pi|\mathbf{x} - \mathbf{x}'|} n_\parallel(\mathbf{x}')\mathbf{n}_\perp(\mathbf{x}') \quad . \tag{3.79}$$

Correspondingly,

$$< \mathbf{m}_\perp(\mathbf{r}_1) \cdot \mathbf{m}_\perp(\mathbf{r}_2) >=$$
$$\left(\frac{\lambda}{4\pi}\right)^2 \int d^3 x' \int d^3 x'' \frac{e^{[-p|\mathbf{x}_1-\mathbf{x}'|-p|\mathbf{x}_2-\mathbf{x}''|]}}{|\mathbf{x}_1 - \mathbf{x}'||\mathbf{x}_2 - \mathbf{x}''|} < n_\parallel(\mathbf{x}')n_\parallel(\mathbf{x}'')\mathbf{n}_\perp(\mathbf{x}') \cdot \mathbf{n}_\perp(\mathbf{x}'') > \tag{3.80}$$

The correlation function for anisotropy axes can be written as

$$< n_\parallel(\mathbf{x}')n_\parallel(\mathbf{x}'')\mathbf{n}_\perp(\mathbf{x}') \cdot \mathbf{n}_\perp(\mathbf{x}'') >= \frac{2}{15}C(|\mathbf{x}' - \mathbf{x}''|) \tag{3.81}$$

which is the definition of $C(x)$. The coefficient $2/15$ is obtained at $\mathbf{x}' = \mathbf{x}''$, to satisfy $C(0) = 1$. One should expect that $C(x)$ rapidly goes to zero at $x >> 1$ $(r >> R_a)$.

Let us compute (3.80) in the range of the field $H_s << H << H_{ex}$, which is typically the experimental range. In this case $p << 1$ and the exponential function under the integral changes slowly on the scale $x \sim 1$. This reduces the integral to

$$< \mathbf{m}_\perp(\mathbf{r}_1) \cdot \mathbf{m}_\perp(\mathbf{r}_2) >= \frac{\lambda^2 \omega}{30\pi} \int d^3 x \frac{\exp[-p|\mathbf{x}_1 - \mathbf{x}| - p|\mathbf{x}_2 - \mathbf{x}|]}{|\mathbf{x}_1 - \mathbf{x}||\mathbf{x}_2 - \mathbf{x}|} \tag{3.82}$$

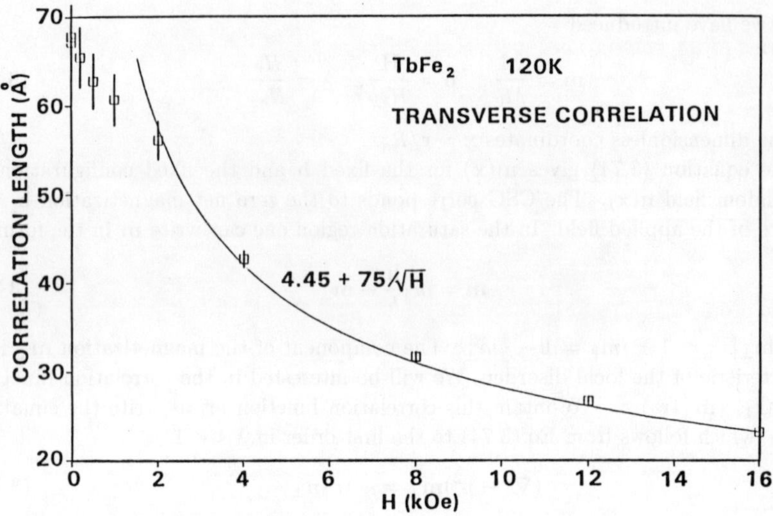

Figure 3.10: Field dependence of the correlation length R_H corresponding to the transverse part of the neutron scattering cross-section. (From Rhyne, Ref.9)

with $\omega = \frac{1}{4\pi}\int d^3 x C(x) \sim 1$. Further integration gives

$$< \mathbf{m}_\perp(\mathbf{r}_1) \cdot \mathbf{m}_\perp(\mathbf{r}_2) > = \frac{\lambda^2 \omega}{15 p} e^{-p|\mathbf{x}_1 - \mathbf{x}_2|} \quad . \tag{3.83}$$

Eq.(3.83) shows that there is a field dependent correlation length [11],

$$R_H = R_a \left(\frac{H_{ex}}{H}\right)^{1/2} \quad , \tag{3.84}$$

which determines the scale of transverse correlations. In a weak field, i.e. $H \sim H_s = H_r (H_r/H_{ex})^3$, one obtains from Eq.(3.84) $R_H \sim R_f \sim R_a (H_{ex}/H_r)^2$, that is the zero-field CSG state. As the field grows, the transverse correlations decay on a shorter scale, yielding the minimal length $R_H \sim R_a$ at $H \sim H_{ex}$. This behavior of the random anisotropy magnet is in agreement with the experiments [9]. Fig. 3.10 shows the dependence of R_H on H obtained for amorphous $TbFe_2$ from neutron scattering data.

3.3. The magnetization curve

In the saturation region

$$\frac{\delta M}{M_0} = \frac{1}{2} < \mathbf{m}_\perp^2 > \quad . \tag{3.85}$$

From Eq.(3.80) at $\mathbf{x}_1 = \mathbf{x}_2$ and Eq.(3.81), we can obtain an expression for $< \mathbf{m}_\perp^2 >$ in terms of the correlation function of the easy axes $C(x)$,

$$< \mathbf{m}_\perp^2 >= \frac{\lambda^2}{120\pi^2} \int d^3x \int d^3y \, \frac{\exp[-p(x+y)]}{x\,y} C(|\mathbf{x}-\mathbf{y}|) \quad . \tag{3.86}$$

After a few transformations this integral can be reduced to [21]

$$< \mathbf{m}_\perp^2 >= \frac{\lambda^2}{15p} \int_0^\infty dx \, e^{-px} x^2 C(x) \quad . \tag{3.87}$$

A nice property of this formula is that the dependence of $< \mathbf{m}_\perp^2 >$ on p is not sensitive to the form of $C(x)$ in both limits $p << 1$ and $p >> 1$. For these two field regimes one obtains from equations (3.85) and (3.87) [14]

$$\frac{\delta M}{M_0} = \frac{\omega}{30} \left(\frac{H_r}{H_{ex}}\right)^2 \left(\frac{H_{ex}}{H}\right)^{1/2} \quad at \quad H << H_{ex} \tag{3.88}$$

$$\frac{\delta M}{M_0} = \frac{1}{15} \left(\frac{H_r}{H}\right)^2 \quad at \quad H >> H_{ex} \tag{3.89}$$

Most of the experiments were done in the range of the field $H < H_{ex}$. Fig. 3.11 shows the $1/\sqrt{H}$ fit of the magnetization curve for $Gd-Fe-Ga$ amorphous alloy [18]. To this date the $1/\sqrt{H}$ magnetization law has also been observed in $Gd - Fe - Co$ [12], $Ho - Fe$ [9], $Fe - Ni - Pb - B$ [22], $Mn - B$ [22,23], $Dy - Fe - B$ [24,25], $Gd - Fe - B$ [24,25], $Dy - Y - Al$ [26], $Dy - Gd - Ni$ [27,28], etc. This turned out to be the most distinctive feature of the CSG.

For most of random anisotropy systems the crossover from the $1/\sqrt{H}$ regime in low field to $1/H^2$ regime in high field is not easy to observe. First, in amorphous materials with large exchange and R_a of atomic scale, the crossover field H_{ex} may be of the order of the exchange field in crystalline ferromagnets, (i.e. very large). In $Dy_{0.1}Gd_{0.9}Ni$, for instance, the $1/\sqrt{H}$ law is a good approximation up to $10\,T$. The estimate of the crossover field for this material gives [28] $H_{ex} \sim 22\,T$. Second, in large fields δM becomes very small due to the random anisotropy. Consequently δM may become comparable to the contribution of the conducting electrons which is linear in H. Thus, to study this crossover one should select a material which does not have a very large an exchange, and/or does not have extended correlations in the short-range structural order (not too large R_a). Note that the parameter λ must still be small for such a material, otherwise it will not display the $1/\sqrt{H}$ regime at all.

The crossover from $1/\sqrt{H}$ to $1/H^2$ law in high fields has been recently reported for $R_xFe_{80-x}B$ amorphous alloys. Analysis of the data shows that the parameter R_a in these materials should be of the order of 100 Å. Given that the X-ray study did not reveal any clusters of crystalline order beyond the 10 Å scale, this large value of R_a may seem surprising. A possible explanation lies within the model of orientational order discussed in Section 1 of this chapter [29]. Based upon general theoretical arguments,

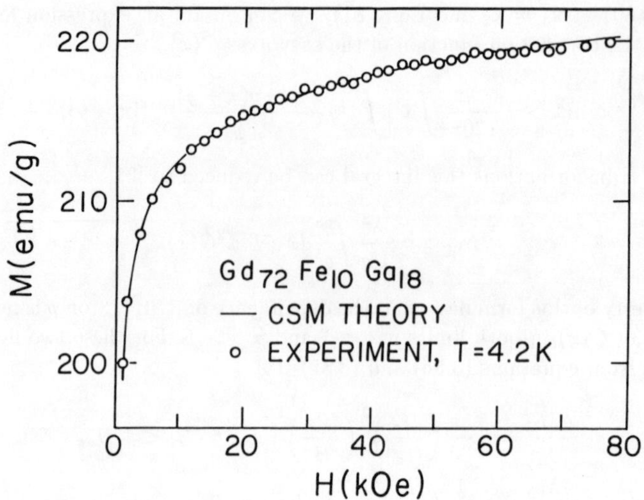

Figure 3.11: Magnetization law of the amorphous ferromagnet $Gd_{22}Fe_{10}Ga_{18}$ and comparison with the CSG=CSM (correlated speromagnet) theory. (From Sellmyer and Nafis, Ref.18)

it seems quite reasonable that the correlation length of the orientational order may in some materials significantly exceed the length of the structural order. While X-ray measurements elucidate the structural correlation length, the magnetic properties are sensitive to the distribution of the easy axes (i.e. sensitive to the orientational order in a solid). Of course, much more work has to be done before this explanation can be fully justified. If proved, this explanation would have an impact in the physics of the amorphous state. It might explain, for example, why many amorphous solids, while appearing completely disordered, have rather low entropies.

It was noticed long ago that the analysis of the magnetization curve on approach to saturation may elucidate structural properties of an inhomogeneous material. For example, the magnetocrystalline anisotropy of randomly arranged crystallites gives a H^{-2} term in approaching saturation [30]. Interactions of the magnetization with point, linear, and layered sources of spin pinning were shown [31] to give $H^{-1/2}$, H^{-1}, and H^{-2} contributions respectively. It has also been demonstrated [32] that the $H^{-1/2}$ law for localized pinning sources must change to H^{-2} in high fields. The idea of extracting information on structural disorder from the magnetization law has been recently revived [21] in relation to amorphous ferromagnets. Such a possibility is based upon equations (3.85) and (3.87) which can be rewritten as

$$\frac{\delta M}{M_0} = \frac{1}{30}\left(\frac{H_r}{H_{ex}}\right)^2 \left(\frac{H_{ex}}{H}\right)^{1/2} \int_0^\infty dx\, C(x) x^2 \exp\left[-x\left(\frac{H}{H_{ex}}\right)^{1/2}\right] \ . \tag{3.90}$$

This equation allows one to obtain an explicit form of $C(x)$, performing an inverse Laplace transformation of the magnetization curve. As shown above, the form of $\delta M(H)$ is not sensitive to $C(x)$ at $H \ll H_{ex}$ and $H \gg H_{ex}$, when δM goes as $H^{-1/2}$ and H^{-2} respectively. To extract from Eq.(3.90) the information about $C(x)$, one, therefore, needs a careful measurement of the magnetization curve at $H \sim H_{ex}$.

Let $S_{1/2}$ and S_2 be the slopes of M versus $H^{-1/2}$ and M versus H^{-2} for the low field and high field regimes respectively. With the help of equations (3.88) and (3.89), Eq.(3.90) can be presented as [33]

$$M(H) = M_0 - \frac{S_{1/2}}{\omega\sqrt{H}} \int_0^\infty dx\, C(x)x^2 \exp\left[-x\left(\frac{2S_{1/2}}{\omega S_2}\right)^{1/3}\sqrt{H}\right] . \qquad (3.91)$$

This formula has been used to extract $C(x)$ from the magnetization law in $Dy_6Fe_{74}B_{20}$ amorphous alloy [33]. The parameters $S_{1/2}, S_2$, and M_0 were fixed by measurements at $H \ll H_{ex}$ and $H \gg H_{ex}$. In this case $M(H)$ depends only on the explicit form of $C(x)$, with no other fitting parameters. Note that this dependence is also hidden in the parameter $\omega = \int_0^\infty dx\, x^2 C(x)$. The important property of Eq.(3.91) is its invariance under the scale transformation $x \to kx$. Consequently, the fit gives the functional form of $C(x)$ and is not sensitive to the value of R_a. Fig. 3.12 shows the fit of the magnetization curve to $C(x) = \exp(-\frac{1}{2}x^2)$. For comparison, also the fit to $C(x) = \exp(-x)$ is shown (dotted line). This analysis demonstrates that, in principle, magnetic measurements allow one to distinguish between different models of amorphous disorder.

3.4. Dimensional crossover

The law $\delta M \propto 1/H^2$ in high fields does not depend on the dimensionality of the amorphous ferromagnet. However, in low field ($H \ll H_{ex}$) the magnetization law does depend on the dimensionality. In three dimensions it is $\delta M \propto 1/\sqrt{H}$. By repeating the arguments of Section 3.3 for $d = 2$, it is easy to verify that in two dimensions the magnetization law at $H \ll H_{ex}$ is $\delta M \propto 1/H$. Thus, decreasing the thickness (h) of the amorphous film, one may expect to observe the crossover from $1/\sqrt{H}$ to $1/H$ dependence. The crossover must occur in the range of the thickness $R_f^{(2)} < h < R_f^{(3)}$, where $R_f^{(2)} \sim R_a/\lambda$ and $R_f^{(3)} \sim R_a/\lambda^2$ are the ferromagnetic correlation lengths in two and three dimensions respectively. Typically it will be in the range from a few tens to a few hundred angströms. The crossover from a $3d$ to a $2d$ behavior has been recently reported in $SmCo_4$ and $TbFe_3$ amorphous films obtained by the electron beam evaporation [34]. It has been observed that as the thickness of the film is reduced from $1000\,\text{Å}$ to $100\,\text{Å}$ the law of approach to saturation in low field changes from $1/\sqrt{H}$ to $1/H$, in accordance with the theoretical prediction. The $2d$ formula for the entire region of approach to saturation, analogous to Eq.(3.90), is

$$\frac{\delta M}{M_0} = \frac{1}{32}\left(\frac{H_r}{H_{ex}}\right)^2\left(\frac{H_{ex}}{H}\right)^{1/2} \int_0^\infty dx\, C(x)x^2 K_1\left[\left(\frac{H}{H_{ex}}\right)^{1/2}x\right] , \qquad (3.92)$$

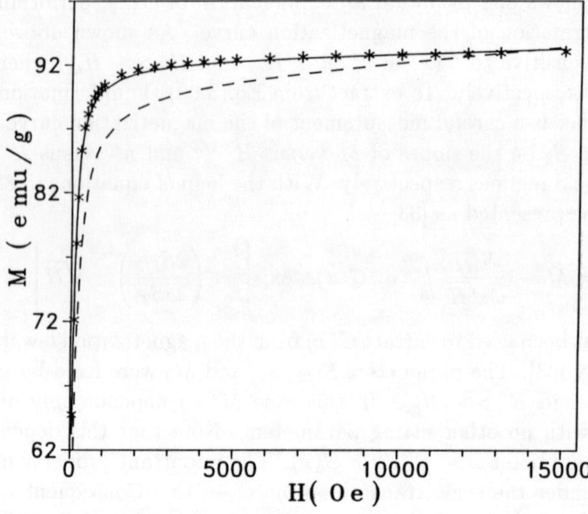

Figure 3.12: Magnetization law in $Dy_6Fe_{74}B_{20}$ at $4.2\,K$. The solid line corresponds to the fit to $C(x) = \exp(-\frac{1}{2}x^2)$. The dotted line shows the fit to $C(x) = \exp(-x)$. (From Tejada et al., Ref.33)

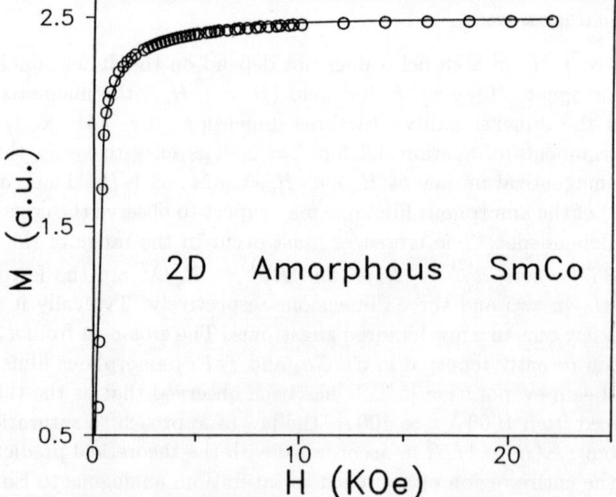

Figure 3.13: Magnetization law in a $2d$ $Sm - Co$ amorphous film, fitted to the theoretical curve (from Ruiz, Ref.34).

where K_1 is the modified Hankel function. This formula gives $\delta M \propto 1/H$ at $H \ll H_{ex}$ and $\delta M \propto 1/H^2$ at $H \gg H_{ex}$. At $H \sim H_{ex}$ the magnetization law depends on the explicit form of $C(x)$. Fig. 3.13 shows the fit of the experimental data to Eq.(3.92) with $C(x) = \exp(-x)$. The agreement with the theory is quite remarkable. The crossover from $1/H$ to $1/H^2$ occurs at $H \sim 10\,kOe$. The values of the parameters obtained from the low and high field regimes are: $H_r \sim 8.7\,kOe$, $H_{ex} \sim 16.7\,kOe$, $\lambda \sim 0.52$, $R_a \sim 10\,\text{Å}$.

Acknowledgements

The work on this Chapter has been supported by NSF Grant No. DMR-9024250.

References

(1) E. M. Chudnovsky and R. A. Serota, Phys. Rev. **B26** (1982) 269.

(2) B. I. Halperin and D. R. Nelson, Phys. Rev. Lett. **41** (1978) 121; D. R. Nelson and B. I. Halperin, Phys. Rev. **B19** (1979) 2457; A. P. Young, Phys. Rev. **B19** (1979) 1855.

(3) E. M. Chudnovsky, Phys. Rev. **B33** (1986) 245; Phys. Rev. **B43** (1991) 7831; Phys. Rev. Lett. **67** (1991) 1809.

(4) C. A. Murray et al., Phys. Rev. Lett. **64** (1990) 2312; R. Seshardi and M. Westervelt, Phys. Rev. Lett. **66** (1991) 2774; H. Dai, H. Chen, and C. M. Lieber, Phys. Rev. Lett. **66** (1991) 3183.

(5) D. R. Nelson, J. Non-Crystal. Solids **61-62** (1984) 475.

(6) Y. Imry and S. Ma, Phys. Rev. Lett. **35** (1975) 1399.

(7) R. Dickman and E. M. Chudnovsky, Phys. Rev. **B44** (1991) 4357.

(8) R. Harris, M. Plischke, and M. J. Zuckerman, Phys. Rev. Lett. **31** (1973) 160.

(9) J. J. Rhyne, IEEE Trans. Magn. **MAG-21** (1985) 1990.

(10) G. Aeppli, S. M. Shapiro, R. J. Birgeneau, and H. S. Chen, Phys. Rev. **B29** (1984) 2589.

(11) E. M. Chudnovsky, W. M. Saslow, and R. A. Serota, Phys. Rev. **B33** (1986) 251.

(12) See, e.g., D. J. Sellmyer, S. Nafis, and M. J. O'Shea, J. Appl. Phys. **68** (1988) 3743.

(13) D. J. Webb and S. M. Bhagat, J. Magn. Magn. Mat. **42** (1984) 109.

(14) E. M. Chudnovsky and R. A.Serota, J. Phys. **C16** (1983) 4181.

(15) E. Callen, Y. I. Liu, and J. R. Cullen, Phys. Rev. **B16** (1977) 263.

(16) A. Aharony and E. Pytte, Phys. Rev. **B27** (1983) 5872; V. S. Dotsenko and M. V. Feigelman, J. Phys. **C16** (1983) L803.

(17) B. Barbara and B. Dieny, Physica **B130** (1985) 245.

(18) D. J. Sellmyer and S. Nafis, J. Appl. Phys. **57** (1985) 3584.

(19) E. M. Chudnovsky and R. A. Serota, IEEE Trans. Magn. **MAG-20** (1984) 1400.

(20) E. M. Chudnovsky, J. Magn. Magn. Mat. **40** (1983) 21.

(21) E. M. Chudnovsky, J. Magn. Magn. Mat. **79** (1989) 127.

(22) M. J. Park, S. M. Bhagat, M. A. Manheimer, and K. Moorjani, Phys. Rev. **B33** (1986) 2070.

(23) W. A. Bryden, J. S. Morgan, T. J. Kistenmacher, and K. Moorjani, J. Appl. Phys. **61** (1987) 3661.

(24) V. A. Ignatchenko, R. S. Iskhakov, and G. V. Popov, Zh. Eksp. Teor. Fiz. **82** (1982) 1518 [Sov. Phys. JETP **55** (1982) 878].

(25) J. Tejada et al., Phys. Rev. **B42** (1990) 898.

(26) P. M. Gehring, M. B. Salamon, A. del Moral, and J. I. Arnaudas, Phys. Rev. **B41** (1990) 9134.

(27) J. Filippi, V. S. Amaral, and B. Barbara, Phys. Rev. **B44** (1991) 2842.

(28) B. Barbara et al., J. Phy. (Paris) **I2** (1992) 101.

(29) E. M. Chudnovsky and J. Tejada, Europhysics Lett. **23** (1993) 517.

(30) R. Becker and W. Döring, Ferromagnetismus (Springer, Berlin, 1939).

(31) W. F. Brown, Jr., Phys. Rev. **58** (1940) 736.

(32) S. Chikazumi, Physics of Magnetism (Wiley, New York, 1964).

(33) J. Tejada, B. Martinez, A. Labarta, and E. M. Chudnovsky, Phys. Rev. **B44** (1991) 7698.

(34) J. M. Ruiz et al., Phys. Rev. B **47** (1993) 11848.

CHAPTER FOUR

MAGNETIC PROPERTIES OF VAPOR-QUENCHED AMORPHOUS AND METASTABLE CRYSTALLINE ALLOYS

C. L. CHIEN

Department of Physics and Astronomy, The Johns Hopkins University
Baltimore, Maryland 21218 USA

1. Introduction

A crystalline solid is characterized by a periodic lattice upon which the unit cells are repeated. In the vicinity of each atomic site there are a definite number of neighbors located at some fixed distances away. A crystalline solid is not isotropic; its properties are different along various crystalline directions. A perfect crystalline solid can be theoretically defined and experimentally approached, but all real materials have many imperfections due to impurities, defects, and dislocations. In polycrystalline specimens, the ones most commonly encountered, there are also many grain boundaries.

The underlying lattice of crystalline solids imposes stoichiometric restrictions on the compositions that are permissible. One generally cannot demand crystalline solids of arbitrary compositions. For example, the phase diagram of Fe-B allows only solids of two stoichiometries (FeB and Fe_2B) [1,2]. The phase diagram does not permit a crystalline solid with an arbitrary composition such as $Fe_{37}B_{63}$ or $Fe_{80}B_{20}$. In immiscible systems (e.g. Fe-Cu and Fe-Ag), the phase diagrams allow no alloy composition. Traditionally, those compositions which are absent from the phase diagrams are not accessible to experimental studies. However, it should be recognized that all tabulated phase diagrams are *equilibrium* phase diagrams for *stable* crystalline solids, in which amorphous and metastable crystalline alloys, the subjects of this book, are excluded.

Amorphous metals [3-7], a new state of matter, have no repetitive lattice nor long range atomic order. They are diametrically different from crystalline solids. A good operational model of amorphous metals is the random packing of atoms [3-7]. Each atom in an amorphous metal has roughly 12 neighbors within the first-neighbor shell, similar to that of fcc solids. In amorphous solids all atomic sites are inequivalent, but all directions are equivalent. Amorphous solids are therefore approximately *isotropic* and *without grain boundaries*, which account for many of their unusual properties such as their soft magnetic properties, ductile mechanical properties, and phenomenal corrosion resistant properties [3-7].

In a sense, an amorphous solid may be viewed as the most disordered form of a crystalline solid. The extremely disordered nature of amorphous metals gives rise to the high electrical resistivities and their weak temperature dependences, as well as high tolerance of radiation damage, for example. While crystal momentum is a good quantum number in crystalline solids, it is no longer so in amorphous alloys, but excitations with long wavelengths, such as sound waves and spin waves, can still be defined and measured.

The amorphous state is absent from the equilibrium phase diagrams because it is a *metastable* state, separated from its thermodynamic ground state by an energy barrier ΔE. One must employ a *non-equilibrium process* to reach the metastable state. Indeed, the advent of various rapid quenching processes are instrumental in the discovery and the development of amorphous metals. The size of the energy barrier ΔE gives a measure of the stability of the metastable state, since the rate of crystallization and the life-time of the metastable state are proportional to $\exp(\Delta E/k_B T)$, where T is the temperature and k_B the Boltzmann constant [8]. When a metastable state is heated to a sufficiently high temperature, the metastable state inevitably and irreversibly transforms to a crystalline solid of one or more phases. However, if ΔE is large, the integrity of an amorphous alloy near ambient temperature is never in doubt. The values of ΔE of many amorphous alloys are in excess of 2 eV. Even if the lifetime at 700 K is only one second, the Arrhenius behavior of $\exp(\Delta E/k_B T)$, assumed to be valid at all temperatures, would assure a lifetime of 3 $\times 10^4$ years at 400 K, let alone its lifetime at room temperature.

Because amorphous solids have no repetitive lattice, the stoichiometry restriction it imposes on crystalline solids no longer holds. It is readily possible to alter the composition of amorphous alloys literally at will. Consequently, the composition and hence the properties of the amorphous alloys can be varied *continuously*. This is a key feature of amorphous alloys, providing the crucial degree of freedom in composition for studying the evolution of fundamental properties and tailoring properties for technological applications. The glassy structure of amorphous alloys also facilitates the inclusion of additional elements. For instance, some of the amorphous alloys in commercial use contain as many as six elements in order to fulfill a variety of material requirements [3-7].

Metastable crystalline alloys are not amorphous, but nevertheless metastable, hence absent from the phase diagrams. Their presence and properties were unknown until the advent of rapid quenching processes. In some cases, the metastable crystalline alloys greatly extend the solubility range allowed by the equilibrium phase diagram. In other cases, most notably the immiscible alloy systems, hitherto unexplored new alloy compositions have been revealed.

A number of review articles on the magnetic properties of liquid-quenched amorphous transition-metal (TM) alloys have already appeared [3-7]. We instead concentrate primarily on the *vapor-quenched* alloy systems, and emphasize the systems and compositions unattainable by liquid quenching, and the new features revealed. For any discussion of fundamental magnetic properties, it is highly desirable to reduce the complexity of the alloy system by keeping the number of constituent elements to a minimum. Since elementally pure amorphous metals (e.g. amorphous pure Fe) cannot be made and studied, we describe for the most part, the magnetic properties of *binary* transition metal systems.

In the crystalline state, the TM elements Fe, Co, and Ni, are ferromagnetic, whereas Mn and Cr are itinerant antiferromagnets. In binary metastable TM systems, only those containing Fe and Co acquire strongly ferromagnetic properties, and are focused in our discussions.

In terms of their constituents, vapor-quenched TM alloys may be classified as :

TM-Metalloid (M) : A substantial amount of investigations have been made in amorphous TM-Metalloid (M) systems, as we discuss in this Chapter.

TM-Early Transition Metal (ET): This is another class of binary amorphous alloys in which a large body of research exists, as described in this Chapter.

TM-Noble Metal : These binary alloys are not amorphous, but metastable crystalline alloys revealing a number of interesting magnetic properties, which are mentioned in this Chapter.

TM-Rare Earth (RE): Amorphous TM-RE alloys are an important class of amorphous alloys, and are extensively discussed in Chapter 8.

TM-Actinide : The TM-Actinide systems, with their unusual magnetic properties and heavy fermion characteristics, are an interesting class of materials. However, there are few reports of TM-actinides [9] largely because all actinides are radioactive and some are highly toxic. These are obstacles for most researchers, except those who are specially equipped to explore these systems.

TM-Alkali and Alkaline Earths : Since the alkali metals are vacuum poisoners, few researchers are interested in venturing into deposition of alkali metals. However, some work in TM-alkaline earths (e.g.Fe-Mg) has been made and the results are similar to those of TM-ET systems [10].

This Chapter includes, primarily, binary amorphous alloys of a transition metal and a metalloid (e.g. Fe-B), a transition metal and an early transition metal (e.g. Fe-Zr), and binary metastable crystalline alloys of a transition metal and a noble metal (e.g. Fe-Cu). Through these examples, we explore the manner with which the magnetic properties depend on the structure of the materials and the alloying elements. From the behaviors established by these alloy systems over wide composition ranges, we also elucidate the properties of elementally pure metals, which defy direct observation. Since the magnetic properties of a large number of binary metastable systems are mentioned, it is necessarily at the expense of the details of each system.

The organization of this Chapter is arranged in the following order. Various rapid quenching methods and the merits of vapor quenching are discussed first. After some brief discussions of the magnetic properties of elemental metals and alloys, we describe the composition ranges of amorphous alloys that can be achieved by vapor quenching. They provide the constraints within which one can vary the compositions. These are followed by the magnetic properties of amorphous transition metal-metalloid systems and amorphous metal-metal systems. We then contemplate the properties of amorphous pure metals. In the last section, we discuss the properties of the metastable crystalline alloys.

2. Rapid Quenching Processes and Vapor Quenching Methods

Prerequisite to achieving amorphous and metastable crystalline alloys are the rapid quenching processes. Various rapid quenching processes have been demonstrated for producing metastable alloys [11]. These include, but are not limited to, liquid-quenching [12], vapor-quenching [13-15], electrodeposition [16], and solid-state reaction [17]. Of those, liquid-quenching (LQ) methods and vapor quenching (VQ) have been the most widely used. In rapid quenching processes, one operates in conditions diametrically opposite to those of growing crystals. As it is well known, even under optimum conditions, a large single crystal generally requires a long growth time. One may therefore argue that the quality of a single crystal specimen may be estimated as $\log(t)$, where t is the crystal growth time. In analogy, the prowess of a rapid quenching process may be measured by $\log(1/\tau)$, where τ is its effective quenching time. In the LQ methods, the random atomic arrangement in the liquid state is frozen into a solid with a high cooling rate (10^4 - 10^6 K/s). In the VQ methods, the random arrival of atoms is preserved on a suitably cooled substrate with an even higher rate (10^8 - 10^{14} K/s). These two types of rapid quenching processes have both advantages and shortcomings. Generally speaking, LQ methods can produce large quantities of amorphous alloys of limited compositions, suitable for large-scale applications such as transformer cores [3]. VQ methods can fabricate alloys with much wider composition ranges, but in the form of thin films, useful for fundamental studies and thin film applications.

The major advantages of LQ methods [3,5,6] are high material production rates (measured in grams/sec) and continuous casting of ribbons or sheets (from 1 mm to as much as 1 m in width), typically 25 μm's in thicknesses. However, despite the rather high quenching rates (up to 10^6 K/s), only compositions near the eutectic points in the phase diagram can be made into amorphous alloys. This is a limitation intrinsic to liquid quenching. For example, in the Fe-B system, only compositions near the eutectic compositions of $Fe_{80}B_{20}$ can be made amorphous. Indeed, hundreds of compositions of liquid-quenched amorphous transition metal (TM)-metalloid (M) alloys have been studied during the last two decades, and *all* of them are near the canonical $TM_{80}M_{20}$ composition, the eutectic composition. In the Fe-Zr system, an example of the metal-metal systems, there are two eutectic points and only compositions near the two eutectic points at Fe-Zr and Fe-Zr can be made amorphous by liquid quenching. LQ methods also rely on homogeneous melts with reasonably low melting points, and thus are incapable of exploring immiscible alloy systems (e.g. Fe-Cu) and alloys with very high melting points (e.g. Fe-W).

Vapor-quench (VQ) methods capture the random arrival of atoms into an amorphous solid. High quenching rates are achieved by low substrate temperature and high deposition rates. The term "vapor-quench" is used, because the substrate is maintained at a sufficiently low temperature, and the arriving atoms in the vapor state are rapidly quenched into a solid. Various estimates have placed the quenching rates from 10^8 K/sec to 10^{14} K/sec [18]. For example, one can estimate the surface diffusion time as $\tau_D = d^2/D_S$, where d is the interatomic distance and D_S is the surface diffusivity. A low surface temperature assures a long τ_D. The other quantity of importance is τ_R, inversely proportional to the adatom arrival rate (deposition rate). Then, in order to "freeze" a metastable state, the condition of $\tau_R < \tau_D$ should be met, accomplished by a low substrate temperature and a high deposition rate. One can question the accuracy of these estimates of the quenching rates. But these quenching rates are certainly much higher than those of liquid quenching, simply because metastable alloys of so many different compositions can be readily made by VQ but are unattainable by LQ methods. The "low" substrate temperature is relative to the crystallization temperature of the metastable alloy. If the crystallization temperature is 700 K, room temperature is sufficiently low, whereas a crystallization temperature of 400 K might require a much lower substrate temperature (e.g. 77 K). However, the crucial temperature is the substrate *surface* temperature which can be significantly higher than the temperature at which the substrate is maintained. Its value is also strongly dependent on the deposition method, the energy of the deposit atoms, and the deposition rate. Unfortunately, surface temperature is extremely difficult to measure with confidence. Operationally, one can place an upper limit for the surface temperature by examining the resulting metastable alloys. For example, if an as-deposited metastable alloy has a crystallization temperature of 425 K, then it is certain that the surface temperature during deposition has never exceeded that value.

Vapor-quenching methods include thermal evaporation, electron-beam (e-beam) deposition, and a variety of sputter depositions [19]. In thermal evaporation, the deposition rates are usually quite low. The evaporated films are often only a few thousand Å thick. The early applications of VQ, involving thermal evaporation onto substrates held at cryogenic temperatures, in fact predated that of LQ. For example, in the 1950's, Buckel and Hilsch [15] discovered the first amorphous superconductors of bismuth and gallium made by vapor deposition onto substrates cooled to cryogenic temperatures. Many of these early efforts were devoted to the study of conductivity and superconductivity of very thin films of highly metastable alloys. Some of these thin metastable films, due to their low crystallization temperatures, could not even withstand room temperature. They were all but inaccessible to experiments except *in situ* measurements. These early practices of vapor-quenching, however, were not followed by others. Even within the field of amorphous superconductors, systematic studies of amorphous transition-metal superconductors by Collver and Hammond [20] occurred belatedly in the mid-1970's.

Since the 1970's, with the development of advanced deposition methods (principally magnetically assisted sputtering and e-beam deposition), VQ have become more widely used. Because the deposition rates of these methods are very much higher than those of thermal evaporation, film thicknesses of 1 - 10 μm or more are common. In e-beam deposition, as in the case of evaporation, one devotes one e-beam source for each element [19]. A multi-element alloy would require as many e-beam sources, and a tight control of the deposition rates of all sources to assure a uniform alloy composition.

Most researchers resort to sputtering [19], which enjoys a number of advantages over other deposition methods. The deposition mechanism is the bombardment of the sputter gas ions (most commonly Ar^+) towards the cathode, which is the "target" of a certain composition. Taking advantage of the equilibrium sputtering yield, an alloy film can be readily fabricated from a multi-element target. Since the sputter gas ions with energies in the hundreds of eV range are capable of overcoming the binding energy of all atoms, sputtering can be administered to any material, no matter how high the melting temperature. Operating in the rf mode, all insulating materials can also be accommodated in sputtering. Consequently, materials such as W, C, BN, Al_2O_3, etc., which are not amenable to evaporation or e-beam deposition, can be readily sputtered. The early successful fabrication of amorphous alloys [21] was made by diode sputtering, which suffers from low deposition rates, high sputtering voltage, inert gas inclusion, and the tendency for columnar structure. These shortcomings are largely eliminated or alleviated in magnetically assisted sputtering (e.g. magnetron) and triode sputtering. The high deposition rates (0.01 - 0.5 μm/min) achieved in these modern versions of sputtering provide films with substantial thicknesses and low impurity contents.

3. Characterizations and Measurements of Magnetic Properties

As in all materials research, characterization is prerequisite to studying vapor-quench samples. Of the various analyses, compositional and structural characterizations are the most essential. The composition of a specimen, and the impurity contents, if any, should be determined by using electron microprobe, energy or wavelength dispersive x-ray analyses, XPS, Auger, or other suitable methods. Since most physical properties are very sensitive to the structure of the specimen, it is imperative to determine, by x-ray diffraction or other relevant techniques, whether the sample is amorphous, crystalline, or a mixture of several phases.

Most magnetic properties can be obtained by performing magnetometry measurements using a variety of instruments, such as vibrating sample magnetometers (VSM), SQUID magnetometers, or B-H loop tracers, from which the relevant magnetic quantities are determined. Magnetic properties can also be profitably supplemented by other techniques, among them, neutron diffraction and methods involving nuclear hyperfine interactions. Neutron diffraction is indispensable for the measurements of spin excitations and magnetic correlation in a variety of amorphous ferromagnets and spin glasses.

Various methods of nuclear hyperfine interactions [22], such as nuclear magnetic resonance (NMR) and Mössbauer effect (ME), provide a microscopic probe of the structural and magnetic properties of a material. Usually, NMR requires a much larger quantity of specimen (~ 1 gram) than ME, a demand which may not be readily accommodated, especially by vapor-quench thin films. Consequently, there are many more studies of vapor-quench alloys using ME than NMR. Among the transition metal elements, ME can in principle be applied using the isotopes of Ni^{61}, Co^{57} and Fe^{57}. However, the Ni^{61} resonance is too broad to be of any use. The isotope of Co^{57} is radioactive and ME measurements can be performed using specimens containing radioactive Co^{57}, a prospect few experimentalists cherish. Fortunately, Fe^{57} is the best of all Mössbauer isotopes, and Fe is a common constituent in many metastable alloys. Consequently, extensive ME studies have been reported providing much important information concerning metastable alloy systems containing Fe.

A wealth of information of metastable alloys can be obtained from ME using Fe^{57} [23]. For simplicity, we mention only a few most common parameters here. At temperatures below the magnetic ordering temperature of the alloy, one obtains a six-line magnetic hyperfine spectrum, from which the isomer shift (the centroid of the spectrum) and the magnetic hyperfine field (proportional to the spectrum splitting) can be obtained. The magnetic hyperfine field (H), to a good approximation, scales with the Fe moment (μ_{Fe}). This is especially useful in systems (e.g. spin glasses) in which the magnetic moments cannot be aligned by an external magnetic field. In such systems, the Fe moments cannot be determined from magnetometry, but can be inferred from ME measurements. In a metastable alloy, a distribution of magnetic moments and magnetic hyperfine fields [P(H)], resulting from a large number of inequivalent sites, are present [24]. Consequently, the ME spectral lines are considerably broadened. The P(H) deduced from the broadened spectrum provides information of the distribution of the Fe moments, which also cannot be obtained from magnetometry.

At temperatures above the magnetic ordering temperature, one measures a quadrupole-split doublet spectrum in a metastable alloy. The centroid of the doublet spectrum defines the isomer shift (IS), which measures the atomic state of the atom. The separation of the doublet determines the quadrupole splitting (QS), which is proportional to the electrical field gradient (EFG) at the nuclear site, since the site symmetry in a metastable alloy is generally non-cubic. The magnetic ordering temperature is the temperature at which the spectrum first transforms from a doublet spectrum to a magnetic hyperfine spectrum. The onset of magnetic hyperfine interaction can be profitably used to determine the magnetic ordering temperature of the sample.

4. Magnetic Properties of Crystalline Fe and Co

Some of the most fascinating aspects of magnetism are found in transition metal elements and alloys, including the prominent role of structure, the strong effects due to the alloying elements, and the diversity of the magnetic orderings. The magnetism of the transition metal elements even plays a crucial role in its thermodynamic ground state. At ambient conditions, most metallic elements crystallize into one of three structures : body-centered cubic (bcc), face-centered cubic (fcc), and hexagonal close-packed (hcp). Not surprisingly, the metallic elements in the same column of the Periodic Table usually share the same crystal structure. However, the magnetic elements of Fe and Co appear to have acquired the "wrong" structures. Unlike its companions in the column with an hcp structure, Fe has a bcc structure. Likewise, Co is hcp instead of fcc as are the other elements in the column. It turns out that the total energies of a transition metal element with different structures are rather close in values. The "wrong" structures of Fe and Co are stabilized by the magnetic energies [25].

Even more interestingly, the type of magnetism that crystalline Fe acquires depends very sensitively on the structure and the nearest-neighbor distance (or lattice parameter) [26-28]. Several states of Fe are known to exist. There is one stable bcc state, the ordinary bcc α-Fe, which is strongly ferromagnetic with a magnetic moment of 2.2 μ_B and T_C = 1045 K. There are at least two fcc γ–Fe states depending on the nearest neighbor distance (d) being smaller or larger than the critical distance of about 2.55 Å. The γ_1–Fe state, with a low atomic volume (d < 2.55 Å), is weakly antiferromagnetic with a low moment (~ 0.5 μ_B). On the other hand, the γ_2–Fe state, with a high atomic volume (d > 2.55 Å), has a high moment (~ 2 μ_B) and it is strongly ferromagnetic. Experimentally, the metastable fcc γ–Fe states can be achieved, among other means, by thin film epitaxy [29]. Most unusually, the interatomic distances of the epitaxial fcc Fe films are accidently very close to the critical value dividing the γ_1–Fe and the γ_2–Fe states. Hence the magnetic properties of the fcc Fe films can be diametrically different depending on very small variations in the lattice parameter. Under high pressure Fe converts to still another state, the hcp state, which turns out to be non-magnetic.

The various states of crystalline Fe are preludes to an even richer variety of magnetic properties found in alloys, where one may alter the atomic species, the composition, the underlying structure, and the strength of the magnetic interactions. Vapor-quench metastable alloys, amorphous or crystalline, provide two crucial ingredients for such endeavors: alloy systems with new structures and extended composition ranges. Using suitable systems one may study topics of fundamental interest, such as the formation of magnetic moment, magnetic dilution, spin glass, magnetic percolation, reentrant behavior, etc.

The magnetic element of Co is known to exist in at least three different structures. At ambient conditions, Co has a hcp structure with a moment of 1.7 μ_B and it is strongly ferromagnetic. At about 700 K, hcp Co transforms to fcc Co, which is still strongly ferromagnetic with a Curie temperature of about 1400 K, the highest of all elements. The bcc Co, which can be achieved by epitaxy onto suitable substrates, is again strongly ferromagnetic [30]. Hence, even though there are several states of Co, their magnetic properties are not strongly dependent on structure. This is a key difference between Fe and Co, a fact also reflected in their metastable alloys.

5. Composition Ranges for Amorphous Alloys

The formation, the stability, and the composition range of the amorphous state are of fundamental and technological importance. The composition range of an amorphous alloy system specifies the latitude within which one can alter the composition and study the resultant properties. The early success of the liquid-quenching methods in producing a large number of amorphous alloys of the eutectic compositions (e.g. $TM_{80}M_{20})$ has prompted theoretical accounts for their apparent stability. The roles of the valence electron concentration [31] and the manner with which the smaller metalloid atoms filling up the Bernal holes in various structure models have been suggested [32,33].

However, we now know that the ability to produce amorphous $TM_{80}M_{20}$ alloys is not a reflection of the exceptional stability of the $TM_{80}M_{20}$ composition, but a limitation of the liquid quenching method, which relies on the depressed melting point near the eutectic composition. The eutectic compositions are *not* the most stable of an amorphous alloy system and can even be the *least* stable. For example, in amorphous Fe_xB_{100-x}, the most stable composition, as judged by the highest crystallization temperature, is not the eutectic point at x = 80 but at x = 65 [34]. In the Fe_xZr_{100-x} system, the eutectic compositions of x = 24 and 91 are both near the limits of the composition range for amorphous alloys. They are essentially the least stable amorphous alloys having the lowest crystallization temperatures. The amorphous alloys produced by solid-state reaction usually do not have eutectic compositions [17]. Therefore, the early models which attempt to account for the stability of the eutectic compositions (e.g. $TM_{80}M_{20}$) cannot be viable models in addressing the questions of composition ranges of amorphous alloys and their stability.

The relevant question of the composition range of the amorphous state at a finite temperature (e.g. 400 K) can best be explored experimentally by using vapor quenching methods. However, since the vapor quenching methods (e.g. sputtering, evaporation) used by various experimentalists differ considerably in details (e.g. deposition rates and substrate temperatures), it is best to examine samples made under very similar conditions. The composition ranges ($x_{min} \leq x \leq x_{max}$) for metastable state of a large number of binary amorphous systems have been investigated by us using samples fabricated under similar conditions [35,36]. In particular, the substrates during deposition were all held at 77 K. The composition ranges for amorphous alloys of these systems having crystallization temperatures higher than 400 K are shown in Table I.

First of all, in each alloy system, the amorphous state exists in only *one* continuous composition range ($x_{min} \leq x \leq x_{max}$), regardless of whether the equilibrium phase diagram has one or two eutectic points. Secondly, in many cases, the boundary separating the amorphous and the crystalline states is very sharp. For example, in the Fe_xTi_{1-x} system, the sample with x = 0.80 is amorphous, x = 0.82 is crystalline having a bcc structure, and x = 0.81 is a mixture of both states [35]. Thirdly, beyond the composition range for the amorphous state (i.e. x < x_{min}, and x > x_{max}), many of the samples are new metastable crystalline alloys, also worthy of exploration.

Also listed in Table I are the constituent atomic radii. It is noted that the width of the composition range (x_{max} - x_{min}) of the amorphous state of each alloy system is closely correlated with the *size difference* of the constituent atomic species -- the larger the size

Table I : Constituent atomic radii (r_A, r_B), ratio, and composition range ($x_{min} \leq x \leq x_{max}$) of vapor-quench binary amorphous alloys. The theoretical predictions [$(x_{min})_t \leq x \leq (x_{max})_t$] are shown in the last column.

A_xB_{1-x}	r_A,	r_B	r_B/r_A	Experimental		Theoretical	
	(Å)			$(x_{min} \leq x \leq x_{max})$		[$(x_{min})_t \leq x \leq (x_{max})_t$]	
Fe-Ti	1.28	1.46	1.14	0.30	0.80	0.28	0.85
Fe-Zr	1.28	1.58	1.234	0.20	0.93	0.19	0.92
Fe-Hf	1.28	1.67	1.305	0.20	0.94	0.16	0.94
Fe-Ta	1.28	1.49	1.164	0.30	0.90	0.24	0.88
Fe-Mo	1.28	1.39	1.086	0.40	0.80	0.41	0.75
Fe-B	1.28	0.78	0.609	0.0	0.90	0.03	0.91
Fe-Cu	1.28	1.27	1.007	----	----	----	----
Fe-Nb	1.28	1.46	1.14	0.25	0.85	0.27	0.85
Ni-Nb	1.28	1.46	1.14	0.20	0.80	0.27	0.85
Cu-Nb	1.27	1.46	1.14	0.20	0.80	0.26	0.86
Fe-W	1.28	1.37	1.07	0.40	0.70	0.38	0.69
Fe-Al	1.28	1.43	1.117	0.10	0.45	0.25	0.82
Fe-Sb	1.28	1.45	1.23	0.03	0.80	0.27	0.84
Co-Cu	1.28	1.27	1.007	----	----	----	----

difference, the wider the composition range for the amorphous state. Egami and Waseda [37] have developed an atomic scale elasticity theory which led to the stress criteria for the topological instability of a solid solution. The amorphous alloys are stabilized because the solid solutions of the corresponding compositions are topologically unstable. They showed that the composition range ($x_{min} \leq x \leq x_{max}$) of amorphous A_xB_{1-x} can be accounted for by the simple expressions of

$$(x_{max})_t \approx 1 - 2\,\lambda_1 \frac{V_A}{|\Delta V|} \tag{4.1}$$

$$(x_{min})_t \approx 2\,\lambda_1 \frac{V_B}{|\Delta V|} \tag{4.2}$$

where V_A and V_B are the atomic volumes, $|\Delta V| = |V_A - V_B|$, and $2\lambda_1$ is a parameter. The two expressions depend only on *geometrical* factors of the atomic species. Once a value of $2\lambda_1$ has been selected, one can calculate $(x_{min})_t$ and $(x_{max})_t$ from Eq.(4.1) and Eq.(4.2) using the known values of the atomic radii. Very remarkably, in most cases, the experimental composition range [$x_{min} \leq x \leq x_{max}$] and the theoretical composition range [$(x_{min})_t \leq x \leq (x_{max})_t$] using $2\lambda_1 = 0.07$ agree exceptionally well as shown in Table I. We know of no other theoretical models which achieve comparable successes.

Although only atomic sizes appear in Eq.(4.1) and Eq.(4.2), effects far beyond geometrical differences are indirectly incorporated. This is because the atomic size difference is intimately related to the heat of formation [38], the electronic structure, and competing equilibrium crystal structures [39], all of which are known to be important in the stability of the amorphous state.

There are, however, some exceptions despite the many successes of this model. One notable one is Fe-Al, for which experimentally one finds that $0.1 \leq x \leq 0.45$ is the composition range for amorphous Fe-Al [40], whereas the model predicts $0.25 \leq x \leq 0.82$. This is probably due to the fact that the Al atoms are highly ionized when alloying with Fe. As a result, the atomic radius of Al is significantly smaller than the metallic radius of 1.43 Å used in the calculations.

From this model, one can also examine the *minimum* size difference below which no amorphous alloy exists. From Eq.(4.1) and Eq.(4.2), one obtains $|\Delta V| \geq 2\lambda_1 (V_A + V_B)$. Using $2\lambda_1 = 0.07$, one has $(V_A/V_B)_{min} \approx 1.15$, or $(r_A/r_B)_{min} \approx 1.05$, i.e. the size difference between the constituent atoms should at least be 5% for the existence of the amorphous state. This result also naturally accounts for the experimental facts that no amorphous state has been found in the vapor-quench binary alloys of Fe, Ni, Co, Cu, and Cr, such as Fe-Cu, Fe-Ni, and Co-Cu because the constituent atoms have nearly identical sizes. However, it should be emphasized that while the vapor-quench alloys, such as Fe-Cu and Co-Cu, are not amorphous, they are nevertheless new metastable crystalline alloys exhibiting hitherto unexplored and new physical properties as described below.

6. Magnetic Properties of TM-Metalloid Systems

Since the discovery of the first amorphous transition metal (TM)-metalloid (M) ferromagnet [41], the magnetic properties of hundreds of liquid-quench amorphous TM-M systems have been studied [3-7]. However, since the eutectic composition (near $TM_{80}M_{20}$) imposes such a strong restriction in the LQ methods, one can merely alter the relative metal composition and the relative metalloid composition, such as $(Fe_{1-y}Co_y)_{80}B_{20}$, and $Fe_{80}(B_{1-y}C_y)_{20}$, without breaking the lock on $TM_{80}M_{20}$. Vapor quenching methods permit the fabrication of binary amorphous TM-M alloys with much wider composition ranges, revealing the compositional dependence of their magnetic properties. A number of binary amorphous TM-M systems with TM = Fe, Co, Ni, Mn, and Cr, and M = B, C, P, Ge, Si, and Sb, have been studied [42-55]; some in very wide composition ranges (e.g. Fe-B) [41,42], while others in only very limited compositions (e.g. $Mn_{50}B_{50}$) [54,55].

Any discussion of the magnetic properties of binary amorphous TM-M alloys must include those of Fe-M and Co-M because they are the only two systems where strongly ferromagnetic properties are realized. In the following, we describe the magnetic properties of amorphous Fe-M and Co-M alloys. Some features of amorphous TM-M, where TM = Ni, Mn, and Cr, will also be mentioned.

186

Fe-M

Vapor-quench amorphous Fe_xB_{100-x} alloys, though not the first vapor-quenched TM-M alloy system studied, provide the widest composition range for investigation. Amorphous Fe_xB_{100-x} ($0 \leq x \leq 90$) was first independently and simultaneously reported by Chien and Unruh using sputtering [42], and Buschow and van Engen using co-evaporation [43]. The magnetic properties of the amorphous Fe-B system reveals a number of features that are typical of other binary Fe-M systems. Because of the exceptional wide composition range, amorphous Fe-B exhibits some features which are unattainable in other Fe-M systems. The general features of the magnetic properties of amorphous Fe-B and other Fe-M systems are summarized as follows:

(1) The magnetic ordering temperature (T_C) of vapor-quenched amorphous Fe_xB_{100-x} as a function of Fe content [42] is shown in Fig.4.1. The values of T_C for the liquid-quench samples ($72 \leq x \leq 86$), shown in Fig.4.2, are in good agreement with those of vapor-quenched alloys [50]. As long as the amorphous state has been achieved, there are no discernable differences between the vapor-quenched and the liquid-quenched samples. The Fe-rich samples of amorphous Fe-B and other amorphous Fe-M systems (e.g. Fe-Si [46] as shown in Fig.4.3) are strongly ferromagnetic; the values of T_C are much higher than room temperature. Only amorphous Co-M systems [49] show comparably high values of T_C. This is the basic reason that virtually all amorphous TM-M alloys for magnetic applications contain Fe and Co as major constituents.

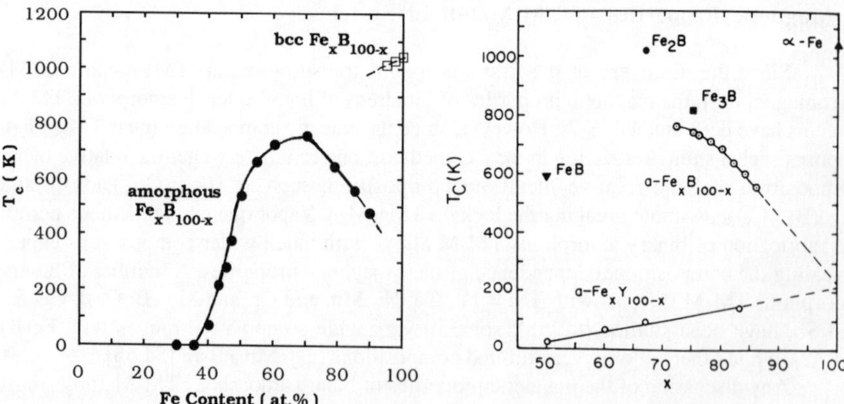

Fig.4.1 : Magnetic ordering temperatures of amorphous Fe_xB_{100-x} (solid circles) and crystalline bcc Fe_xB_{100-x} (open squares) alloys as a function of Fe content (Chien and Unruh, ref.42).

Fig.4.2 : Magnetic ordering temperatures of amorphous Fe_xB_{100-x}, amorphous Fe_xY_{100-x}, crystalline α-Fe, FeB, Fe_2B and Fe_3B (Chien et al. ref.50).

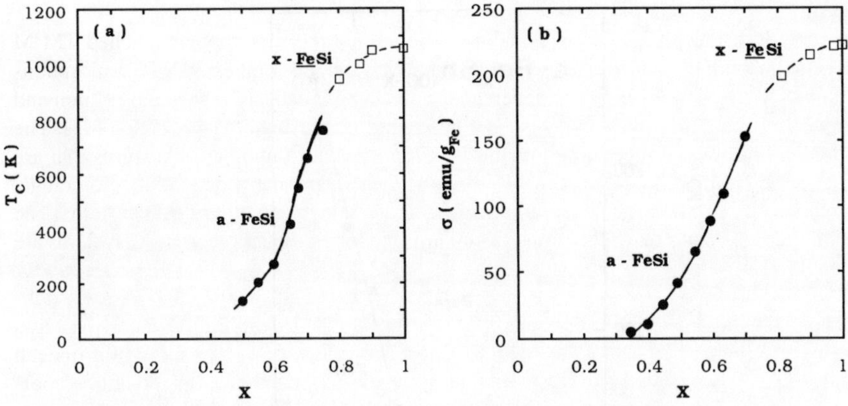

Fig.4.3 : Magnetic ordering temperature (a) and magnetization (b) of amorphous Fe_xSi_{100-x} (solid circles) and crystalline FeSi (open squares) (from ref.46 by Mangin and Marchal).

(2) As shown in Fig.4.1-4.3, the values of T_C change *continuously* as the Fe content is varied. One can tailor the value of T_C by selecting a specific composition. This degree of freedom, which exists in all amorphous solids, is generally deprived from Fe-M compounds. For instance, there are three stable (Fe, FeB, and Fe_2B) and one metastable (Fe_3B) crystalline compounds of Fe and B. Their values of T_C, as shown in Fig.4.2, do not show any systematic variation [50].

(3) In amorphous Fe-B, there exists a critical composition (x_c), below which the samples with low Fe contents are non-magnetic, as shown in Fig.4.1. The critical concentration for amorphous Fe-B has been found to be $x_c \approx 38\%$. Above x_c, the value of T_C of Fe-B rises sharply; every percent increment in x causes a change in T_C of about 40 K [42]. Similar behavior has been observed in Fe-Si (Fig.4.3(a)), Fe-Sb (Fig.4.4) and other systems. In fact, a critical concentration for magnetic ordering has been observed in every amorphous Fe-M system with values of x_c ranging from 40 to 50 [42-47,52].

It should be recognized that although x_c resembles that of, but in fact is not, a magnetic percolation threshold [56,57], which for structures with high coordination numbers should have values less than 20%. Instead, x_c is the critical composition at which the Fe moment first develops, as discussed below.

Near x_c, the magnetic properties are often more complex than what has been indicated in Fig.4.1 and Fig.4.3. These features may be illustrated by amorphous Fe-Sb [52]. Detailed low-field measurements near x_c have been studied in amorphous Fe-Sb. The samples with low Fe contents (x ≤ 54) show a cusp in the zero-field-cooled susceptibility,

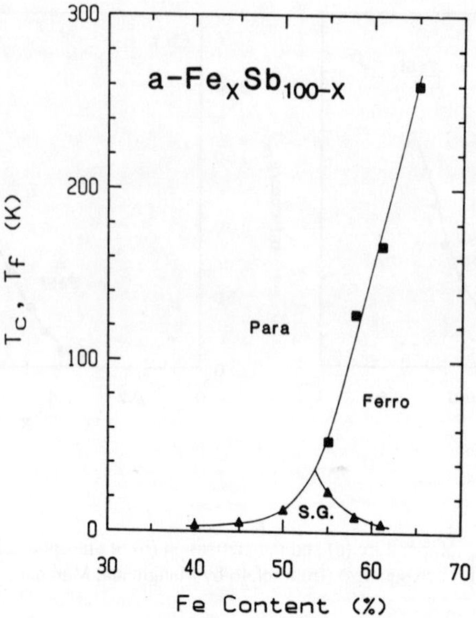

Fig.4.4 : Magnetic phase diagram of amorphous Fe$_x$Sb$_{100-x}$ alloys near the critical concentration for magnetic ordering (Xiao and Chien, ref.53).

indicative of a spin glass behavior. In the composition range of $55 \leq x \leq 61$, there is a double transition in the zero-field-cooled susceptibility; a Curie temperature (T_C) and a spin freezing temperature (T_f) at a lower temperature. The samples with $x > 61$ are ferromagnetic, and whose T_C increases rapidly with increasing Fe content. These results lead to a magnetic phase diagram as shown in Fig.4.4. This reentrant behavior has also been observed in other amorphous magnetic systems, such as Fe-Sn [58], and (Fe-Ni)$_{75}$P$_{16}$B$_6$Al$_3$ [59], and discussed by theoretical models [60].

(4) Most remarkably, in Fe-B, a maximum in T_C occurs near $x = 70\%$, beyond which T_C *decreases* for samples with higher Fe contents as shown in Fig.4.1. This fact has been well established; the liquid-quenched samples of Fe$_x$B$_{100-x}$ ($72 \leq x \leq 86$) [50] show the same behavior as shown in Fig.4.2. Following the trend of the amorphous Fe-B alloys, a reasonable extrapolation to amorphous pure Fe would give a value of $T_C \approx 220$ K, which is much lower than the value of $T_C = 1045$ K for crystalline bcc Fe. Extrapolating the results of other amorphous alloys of iron with other early transition metal elements (as discussed later), and those of amorphous rare earth-iron alloys, gives the same results for amorphous pure Fe. For instance, extrapolating the results of amorphous Fe-Y [61], as shown in Fig.4.2, also gives a value of $T_C \approx 220$ K for amorphous pure Fe.

It may be noted that, beyond the composition range of for the amorphous state ($x \leq 90$), crystalline bcc Fe-B alloys ($95 \leq x \leq 100$) with small amounts of B can also be made by quenching. The values of T_C of the crystalline bcc Fe-B alloys are very high and progressively decrease from the value of $T_C = 1045$ K for pure α-Fe as shown in Fig.4.1. Thus, toward the pure Fe limit ($x = 100$), amorphous and crystalline Fe-B alloys have completely different magnetic ordering temperatures, a direct consequence of the structural differences. Further discussions of the magnetic properties of amorphous pure Fe will be made later.

In amorphous Fe_xB_{100-x} alloys, the net magnetic exchange interaction, which T_C measures, is apparently decreasing with x for $x > 70$. This can be due to a decrease of the ferromagnetic interaction (J). In the Fe-rich end, the Fe-Fe distance decreases with increasing Fe content. It has been suggested that should Fe lie in the region of $\partial J(r)/\partial r > 0$ on the Bethe-Slater curve, then a lower T_C would be resulted [62,63]. Another possibility is the emergence of antiferromagnetic interaction. The Invar characteristics observed in the Fe-rich samples of Fe-B and the complex magnetic properties in the Fe-rich samples of Fe-ET seem to be consistent with the latter assertion [63,64].

The composition range of Fe-B for the amorphous state is exceptionally large. The unusual compositional dependence of T_C would also occur in other amorphous Fe-M systems provided that the Fe content can be extended high enough. However, if the alloys could only be made amorphous, say, up to $x = 70$, such as that in Fe-Si [46], then one would have been led to conclude, mistakenly, that the values of T_C of the amorphous state gradually merge into those of the crystalline alloys, as shown in Fig.4.3(a).

(5) The Fe moment and Fe hyperfine field (shown in Fig.4.5) of amorphous Fe-B are not constant, but increase monotonically with Fe content, despite the fact that there is a maximum in T_C [42]. The information of Fe moment can be obtained from either magnetometry or hyperfine field measurements, since empirically, as established by a large number of amorphous alloys and crystalline compounds of Fe-M, the Fe hyperfine field scales with the Fe moment with a ratio of about $130 - 150$ kOe/μ_B [24]. It should be noted that the Fe-poor samples exhibit no magnetic moment. The Fe moment first appears when the Fe concentration reaches a critical composition [42]. More importantly, for amorphous Fe-B, the critical composition for the formation of Fe moment *coincides* with the critical concentration (x_c) for magnetic ordering with a finite T_C. Therefore, the first appearance of magnetic ordering is dictated by the formation of the Fe moment. These are general features of amorphous Fe-M, and have been observed in other systems such as Fe-Si (Fig.4.3(b)) and Fe-Sb (Fig.4.5). The critical concentration (x_c) for the appearance of the Fe moment and magnetic ordering are quite similar ($x_c \approx 38\text{-}40$) for Fe-B, Fe-Si and Fe-Ge, and it is slightly higher ($x_c \approx 50$) for Fe-Sb.

Two oversimplified models have been extensively used in the literature to account for the existence of x_c for the formation of Fe moment, and both have been open to criticisms. In the rigid-band model, the disappearance of the Fe moment is a consequence of the filling of the $3d$ band by the valence electrons of the sp elements [66]. A straightforward calculation gives critical concentrations of $x_c = 43$ for Fe-Si, Fe-Ge, and

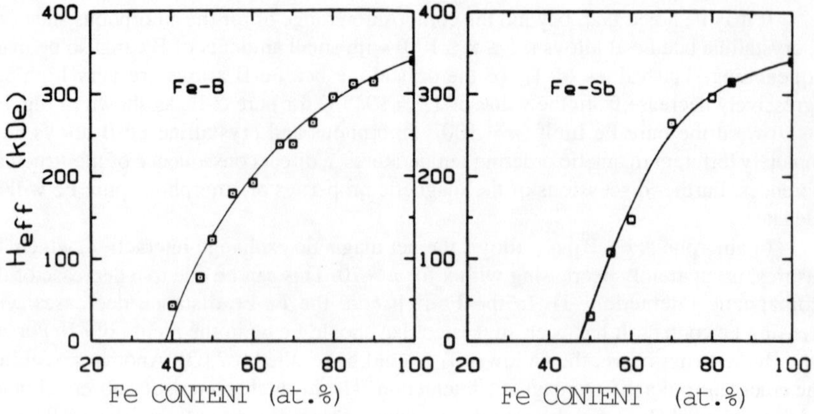

Fig.4.5 : Average magnetic hyperfine field (H_{eff}) of Fe, which scales with the Fe moment, of amorphous Fe-B (open squares) and amorphous (open squares). The solid squares are those of crystalline solids (Xiao and Chien, ref.65).

Fe-Sn, and $x_c = 53$ for Fe-Sb systems. Hence this simple model can reasonably account for the values of x_c, although the actual experimental values are all less than the predicted ones. Another model known as the nearest-neighbor (NN) coordination model [67,68] has been widely quoted even though it is rather *ad hoc*. The NN model asserts that the Fe moment and hyperfine field are determined by the characteristics of the NN shell surrounding a central Fe atom. The physical origin of the NN model is the hybridization of the Fe *3d* band with the *s, p, d* bands of the other component. The degree of hybridization depends sensitively on the coordination number and the radius of the NN shell. It has been suggested that an Fe atom must acquire at least 6 Fe neighbors before a local moment can materialize [68]. Despite the qualitative agreement of these models, their shortcomings are obvious, and only full-fledged theoretical calculations can accurately address the question of the critical concentration for magnetic moment.

(6) Across the boundary separating the amorphous and the crystalline state at $x = 90$, while there is a huge difference in T_C in Fe-B, there is practically *no* difference in the average magnetic moment and average hyperfine field [65], the latter is shown in Fig.4.5. Apparently the value of the average magnetic moment is primarily determined by the Fe concentration. The disordered structure in the amorphous state certainly causes a wide distribution of moments and hyperfine fields, but the averaged values are roughly structure independent. Similar conclusions have been drawn from amorphous Fe-Si (Fig.4.3(b)) and Fe-Sb (Fig.4.5).

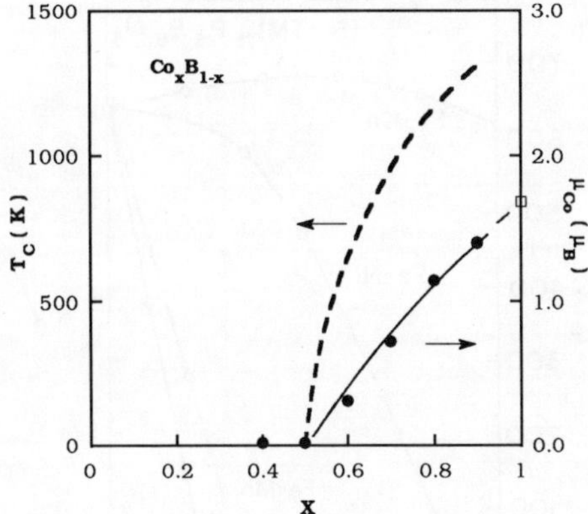

Fig.4.6 : Magnetic ordering temperatures and magnetic moments of amorphous Co$_x$B$_{100-x}$, the latter are from ref.49 by Buschow and van Engen.

The magnetic moment of Fe, and indeed the magnetism of Fe, is a subject of long-standing interest and immense complexity. The accumulated experimental results in crystalline solids show that the value of the Fe moment depends, in a very complex manner, on the structure, the constituents, the bonding, the interatomic distances, the band structure, etc. The magnetic moment of Fe and other aspects of magnetic properties in amorphous Fe-M alloys are no less complex, except that they are much more systematic. As shown in Fig.4.1-5, the values of various magnetic quantities of Fe-M vary smoothly and continuously as the composition is changed.

Co-M

The magnetic properties of amorphous Co-M alloys are less complex than those of amorphous Fe-M alloys. Both the Co moment and the magnetic ordering temperature show monotonic dependences on the Co content. Studies of vapor-quenched Co-M alloys [49] together with those of liquid samples of the eutectic compositions [48], form the basis for the magnetic properties of amorphous Co-M alloys. First of all, the Co-rich alloys are strongly ferromagnetic; their T_C's are so high that they are at times *higher* than their crystallization temperatures. For example, the T_C of amorphous Co$_{80}$B$_{20}$ is about 1000 K, whereas the crystallization temperature is only about 650 K [48]. One can therefore only estimate the compositional dependence of T_C of the Co-rich amorphous alloys. The compositional dependence of T_C of amorphous Co-B alloys is schematically illustrated in Fig.4.6, which shows that the value of T_C of amorphous Co-B alloy increases monotonically with the Co content and approaches 1300 - 1400 K in the limit of amorphous pure Co. The extrapolated value of T_C for amorphous Co is very close to the

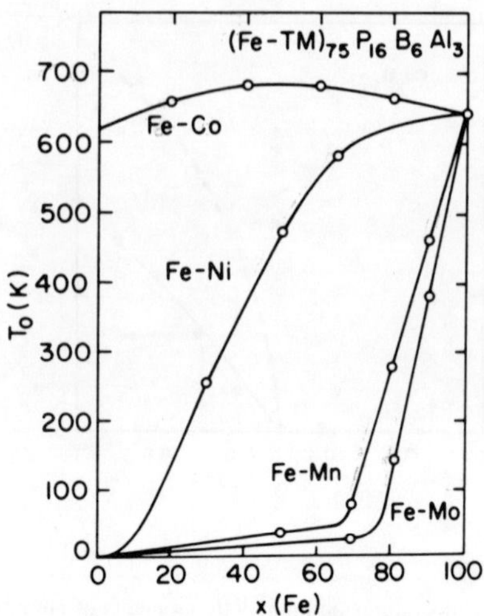

Fig.4.7 : Concentration dependence of magnetic ordering temperatures of $(Fe-TM)_{75}P_{16}B_6Al_3$ for TM = Co, Ni, Mn, and Mo (Chien, ref.69).

value of $T_C = 1400$ K for crystalline fcc Co. On the other hand, the Co-poor samples are non-magnetic, and, just like amorphous Fe-M alloys, there exists a critical concentration (x_c) for the appearance of magnetic ordering in amorphous Co_xB_{100-x}. For a number of binary Co-M alloys (e.g. Co-B and Co-Si), the values of x_c are found to be about 50.

The Co moment of amorphous Co-M is not constant but increases monotonically the Co content. The compositional dependence of the Co moment in amorphous Co-B alloys, observed by Buschow and van Engen [49], is shown in Fig.4.6. The Co moment first appears at x_c, above which magnetic ordering becomes possible. Thus the Co moment, as the Fe moment discussed earlier, develops only if the Co concentration is sufficiently high. This turns out to be a recurring theme in *all* amorphous TM-M as well as TM-ET alloys.

In the limit of amorphous pure Co, the extrapolated Co moment of about 1.6 - 1.7 μ_B is virtually indistinguishable from that of crystalline Co. Hence, the extrapolated values of both the T_C and the Co moment of amorphous pure Co are very close to those of crystalline fcc Co. Evidently, the magnetism of Co, unlike that of Fe, is insensitive to structural differences.

Other TM-M Alloys

Only binary amorphous Fe-M and Co-M alloys exhibit strongly ferromagnetic properties. This fact can be concluded from the results of the liquid-quenched samples of (Fe-TM)-M alloys, when TM = Co, Ni, Cr, or Mn replaces Fe. As illustrated in Fig.4.7, upon

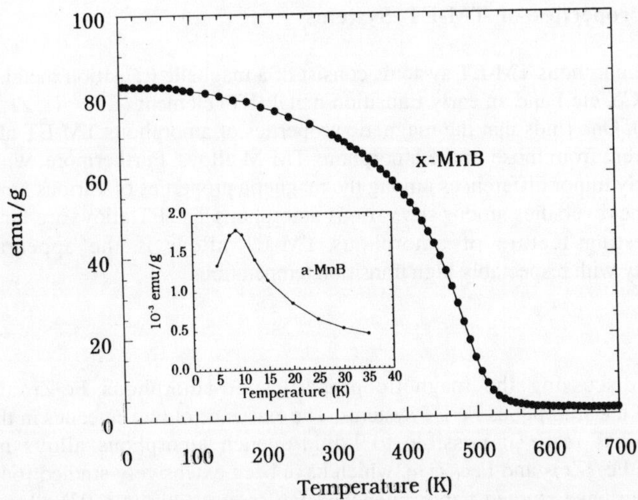

Fig.4.8 : Magnetization curve of crystalline MnB. In the inset, the magnetization curve of amorphous MnB at a low field of 10 G is shown (Sullivan and Chien, ref.55).

substitution for Fe, only Co sustains strongly ferromagnetic properties [69]. This is because in the (Co-Fe)-M alloys, both Fe and Co carry magnetic moments. In the (Fe-Ni)-M alloys, the Fe moment never vanishes, regardless of how low the Fe concentration may be. This is quite different from that of binary Fe-M alloys. The role of Ni in (Fe-Ni)-M alloys approximates that of a non-magnetic diluent. Indeed, the T_C dependence of the (Fe-Ni)-M alloys shows a magnetic percolation threshold of about 20%. In the case of (Fe-Cr)-M and (Fe-Mn)-M alloys, however, both the values of T_C and the Fe moment decrease precipitously as shown in Fig.4.7, most likely due to a rapid emergence of antiferromagnetic interactions.

From the results of amorphous (Fe-TM)-M, one can anticipate the properties of binary Ni-M, Cr-M, and Mn-M alloys. The magnetic moment of Ni never quite develops in binary Ni-M alloys to allow appreciable magnetic ordering. Amorphous Cr-M and Mn-M alloys exhibit weak spin-glass-like ordering at best, and none of them is ferromagnetic [70]. For example, both amorphous MnB and MnB_2 show spin glass ordering below 20 K [54,55], even though crystalline MnB and MnB_2 are ferromagnetic with rather high values of T_C = 530 K and 157 K respectively. Upon crystallizing amorphous MnB and MnB_2, the ferromagnetic properties of crystalline MnB and MnB_2 are readily recovered [55]. The contrasting behaviors of amorphous and crystalline MnB are shown in Fig.4.8. Amorphous MnB shows a spin-glass ordering below about T_G = 10 K. After crystallization, a strong ferromagnet with T_C = 530 K is obtained in crystalline MnB. Thus the magnetic properties of MnB depend critically on the structure.

194

7. Magnetic Properties of TM-ET Systems

Binary amorphous TM-ET systems consist of a magnetic transition metal element (e.g. TM = Fe, Co, etc.) and an early transition metal (ET) element (ET = Ti, Zr, Hf, Nb, Ta, Mo, and W). One finds that the magnetic properties of amorphous TM-ET alloys are altogether different from those of the amorphous TM-M alloys. Furthermore, while there are only relatively minor differences among the magnetic properties of various amorphous TM-M alloys, the diversities among the various amorphous TM-ET alloys are far greater. Another interesting feature of amorphous TM-ET alloys is the appearance of superconductivity with respectably high transition temperatures.

Fe-ET

We begin by discussing the magnetic properties of amorphous Fe-Zr alloys, a representative of the amorphous Fe-ET systems. The existence of two eutectics in the Fe-Zr phase diagram has made it possible to liquid-quench amorphous alloys near the compositions of $Fe_{25}Zr_{75}$ and $Fe_{90}Zr_{10}$, which have been extensively studied [64,71,72]. Unruh and Chien have studied vapor-quenched Fe_xZr_{100-x} ($20 \leq x \leq 93$) alloys over a much wider composition range, thus revealing the evolution of various properties [73]. The key features of amorphous Fe-Zr and other Fe-ET alloys are as follows:

(1) The magnetic ordering temperatures (T_C) of Fe-Zr are shown in the right side of Fig.4.9, exhibiting a number of features shared by other Fe-ET systems. The values of T_C

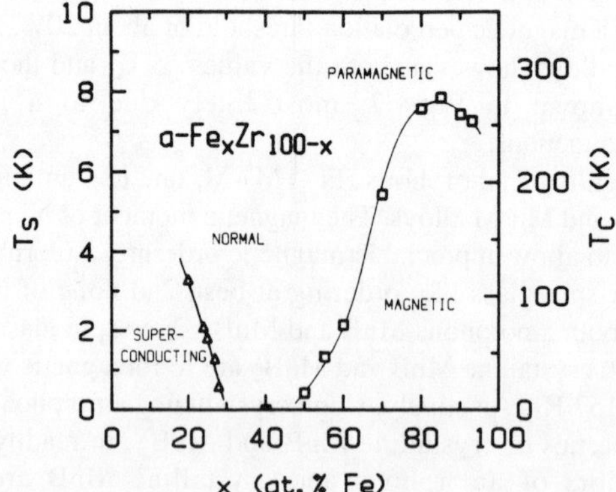

Fig.4.9 : Phase diagram of amorphous Fe_xZr_{100-x}. The magnetic ordering temperatures (T_C) and the superconducting transition temperatures (T_s) are denoted as squares and triangles respectively (Unruh and Chien, ref.73)

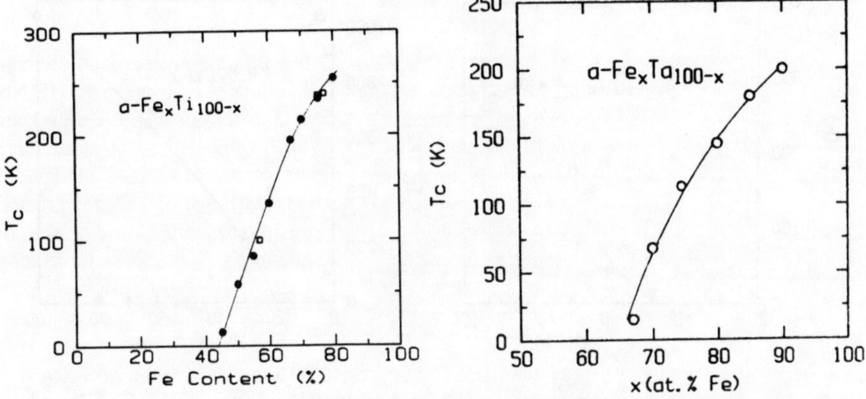

Fig.4.10 : Magnetic ordering temperatures of amorphous Fe_xTi_{100-x} alloys (Liou and Chien, ref.77) and amorphous Fe_xTa_{100-x} alloys (Chien et al., ref.78)

for all compositions of Fe-Zr are below 300 K [73-76]. As a matter of fact, the values of *all* compositions of *all* binary amorphous Fe-ET, where ET = Zr, Hf, Ti, Ta, Nb, Mo, and W, are below 300 K, as illustrated in Fig.4.10 (Fe-Ti and Fe-Ta) and Fig.4.11 (Fe-Hf and Fe-Mo) [77-81].

(2) As in the cases of amorphous Fe-M alloys, in each of the amorphous Fe-ET systems, there exists a critical concentration x_c for magnetic ordering. The values of $x_c \approx 40$ (Fe-Zr), $x_c \approx 42$ (Fe-Ti), $x_c \approx 45$ (Fe-Hf), $x_c \approx 55$ (Fe-Nb), $x_c \approx 63$ (Fe-Ta), have been observed in these amorphous Fe-ET systems [77-80]. Even higher values of x_c have been observed in Fe-Mo and Fe-W as discussed below. Again, this critical concentration (x_c) has nothing to do with magnetic percolation, but merely the critical composition for the formation of the Fe moment.

(3) The samples with $x < x_c$ exhibit Pauli paramagnetic behavior without a localized magnetic moment.

(4) The Fe moment first develops at x_c, above which the Fe moment increases monotonically. This is shown in Fig.4.12 for amorphous Fe-ET (ET = Ti, Ta, Hf, and Mo).

(5) The magnetic moments and the hyperfine fields in all amorphous Fe-ET alloys are much reduced; their values are much lower than those of amorphous Fe-M alloys of comparable compositions.

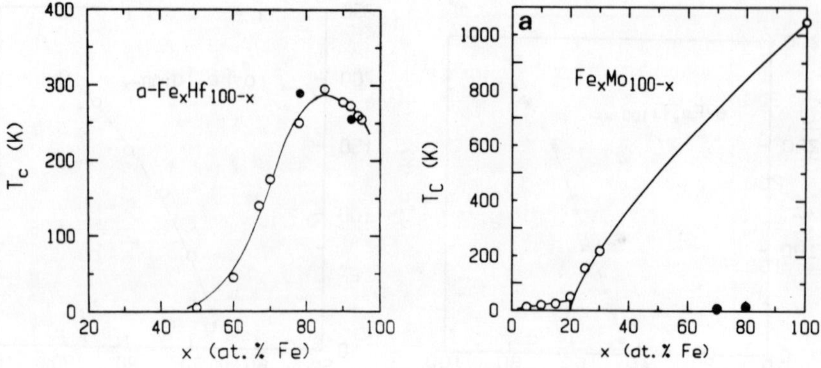

Fig.4.11 : Magnetic ordering temperatures of amorphous Fe_xHf_{100-x} alloys (Liou et al., ref.79), metastable crystalline and amorphous Fe_xMo_{100-x} alloys (Chien et al., ref.80)

(6) One of the unusual features of amorphous Fe-Zr is the realization of superconductivity [72,73] with sizable superconducting temperatures (T_S) in samples of low Fe contents, as shown on the left side of Fig.4.9. In amorphous Fe-Zr, one can, therefore, observe the evolution of both superconductivity and magnetic ordering, the disappearance of one and the emergence of the other. The existence of superconductivity is a definitive proof that there are no Fe moments in these alloys with low Fe contents. It is also interesting to note that for pure crystalline Zr, the value of T_S is only 0.6 K, whereas much higher values of T_S were found in amorphous Fe-Zr. The decrease of T_S with increasing Fe content has been attributed to spin fluctuation, whose strength increases with Fe content, resulting in a rapid increase in the susceptibility [72]. This precursor to magnetic order eventually destroys the superconductivity. Superconductivity exists in many amorphous TM-ET alloys, particularly those with ET = Zr and Nb. Substantial transition temperatures (> 4.2 K) have been observed in Fe-Zr, Cu-Zr, Ni-Zr, Co-Zr and Ni-Nb alloys with low TM concentrations [20,82].

(7) In Fe-Zr, there is a maximum in T_C, at about x = 85, above which T_C decreases with increasing Fe concentration, as shown in Fig.4.9. A similar maximum has also been observed in amorphous Fe-Hf (Fig.4.11). The maxima in T_C again reinforce the notion that amorphous pure Fe should have $T_C \approx 220$ K, a conclusion also made in amorphous Fe-M alloys. Such maxima in T_C are not seen in other amorphous Fe-ET systems (e.g. Fe-Ta shown in Fig.4.10), because their T_C's are already so depressed that they never exceed 220 K within the composition range for the amorphous alloys.

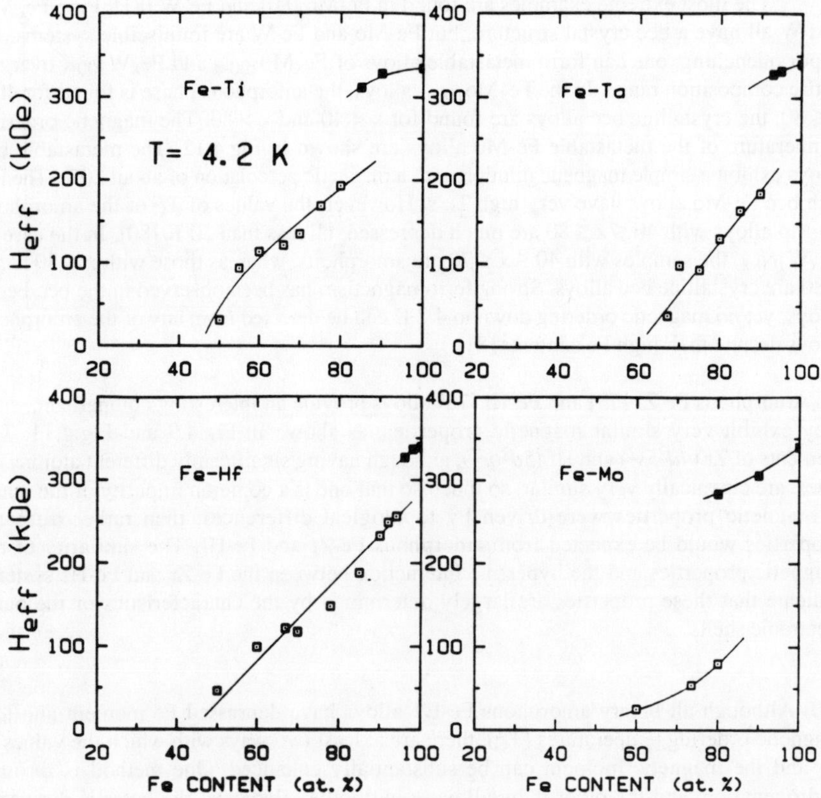

Fig.4.12 : Concentration dependencies of effective magnetic hyperfine field at 4.2 K of amorphous
Fe-early transition metal systems of Fe-Ti, Fe-Ta, Fe-Hf, and Fe-Mo (Xiao and Chien, ref.65).

(8) The boundaries separating the amorphous state and the crystalline state of the Fe-ET
systems are rather sharp; often within a few atomic percent. Across the boundary, there is a
dramatic difference in the magnetic properties and hyperfine interactions [65]. The magnetic
hyperfine fields of Fe-Ti, Fe-Ta, Fe-Hf, and Fe-Mo are shown in Fig.4.12. In each case,
there is an abrupt change in the hyperfine field at the boundary, illustrating that the
magnetic moments of the Fe-ET systems are very sensitive to structure. Furthermore, there
is a sudden loss of the magnetic moment, when the structure changes from crystalline to
amorphous.

The most extreme examples are found in Fe-Mo [80] and Fe-W [81]. Pure Fe, Mo, and W all have a bcc crystal structure, but Fe-Mo and Fe-W are immiscible systems. By vapor quenching, one can form metastable alloys of Fe_xMo_{100-x} and Fe_xW_{100-x} over the entire composition range. In the Fe_xMo_{100-x} alloys, the amorphous phase is found for $40 \leq x \leq 80$, the crystalline bcc alloys are found for $x < 40$ and $x > 80$. The magnetic ordering temperature of the metastable Fe-Mo alloys are shown in Fig.4.12. The metastable bcc alloys exhibit a simple magnetic dilution with a magnetic percolation of about 20%. The Fe-rich bcc Fe-Mo alloys have very high T_C's. However, the values of T_C of the amorphous Fe-Mo alloys with $40 \leq x \leq 80$ are much depressed; all less than 20 K [80]. In the case of Fe_xW_{100-x}, the samples with $40 \leq x \leq 70$ are amorphous, whereas those with $x \leq 30$ and $x \geq 80$ are crystalline bcc alloys. Strong ferromagnetism has been observed in the bcc Fe-W alloys, yet *no* magnetic ordering down to 4.2 K can be detected from any of the amorphous alloys despite their high Fe contents [81].

(9) Amorphous Fe-Zr [73] and Fe-Hf [78] alloys provide an interesting comparison, since they exhibit very similar magnetic properties, as shown in Fig.4.9 and Fig.4.11. The elements of Zr ($4d^2 5s^2$) and Hf ($5d^2 6s^2$), although having significantly different atomic sizes, are chemically very similar, so much so that one is a common impurity of the other. If magnetic properties were driven by topological differences, then rather different properties would be expected from amorphous Fe-Zr and Fe-Hf. The similarity of the magnetic properties and the hyperfine interaction between the Fe-Zr and Fe-Hf systems indicate that these properties are largely determined by the characteristics of the outer electronic shells.

(10) Although all binary amorphous Fe-ET alloys have depressed Fe moment and low magnetic ordering temperatures (T_C), there are at least two ways with which the values of T_C and the magnetic moment can be substantially enhanced. One method is through hydrogenation, and the other is by alloying with other elements; the latter is discussed below. Hydrogenation in amorphous Fe-ET systems was earlier performed in liquid-quenched samples with compositions near $Fe_{90}Zr_{10}$; both the Fe moment and T_C were substantially increased upon hydrogenation [83,84]. Fries et al. provided the first systematic study of hydrogenation of Fe-Zr across a wide composition range using vapor-quenched samples [85]. As shown in Fig.4.13, the values of T_C are increased for all compositions, some by as much as 120 K. The magnetic moment and hyperfine field are also substantially increased, while the critical concentration (x_c) for magnetic magnetic ordering is reduced from 40 to about 30.

(11) Another effective way of boosting the depressed magnetism of binary amorphous Fe-ET solids is through alloying, in which Fe is partially replaced by other TM's. The results for the case of amorphous $Fe_{70}Ti_{30}$ are shown in Fig.4.14. Both the spontaneous magnetization and the values of T_C are dramatically increased when the Fe content of $Fe_{70}Ti_{30}$ is partially replaced by Co, Ni, and Cu [86]. The most prominent increase is

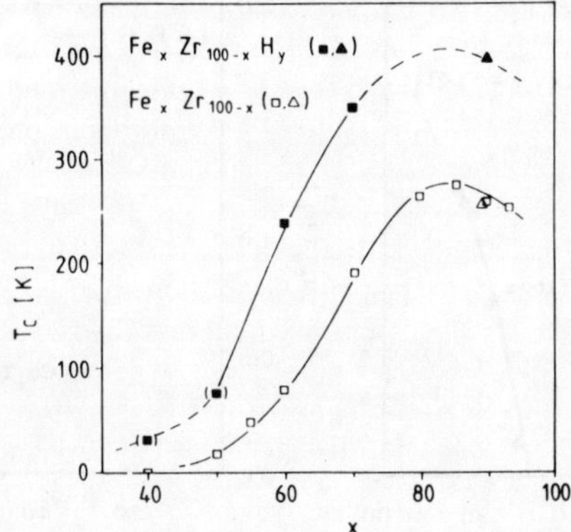

Fig.4.13 : Magnetic phase diagrams of amorphous Fe$_x$Zr$_{100-x}$ and Fe$_x$Zr$_{100-x}$H$_y$ alloy systems. The magnetic transition temperatures T$_C$ of the vapor-quenched and the liquid-quenched alloys are denoted as squares and triangles respectively (Fries et al., Ref.85).

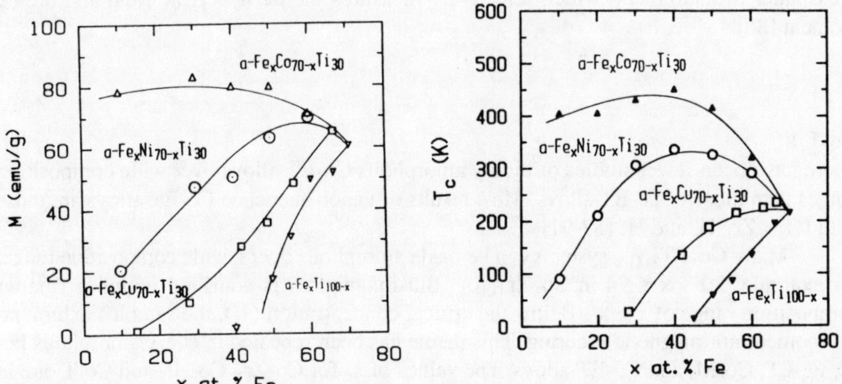

Fig.4.14 : Spontaneous magnetization measured at 6 K (left) and magnetic ordering temperatures (T$_C$) (right) of amorphous Fe$_x$Co$_{70-x}$Ti$_{30}$, Fe$_x$Ni$_{70-x}$Ti$_{30}$, Fe$_x$Cu$_{70-x}$Ti$_{30}$, and Fe$_x$Ti$_{100-x}$ as a function of Fe content (Liou et al., ref.86).

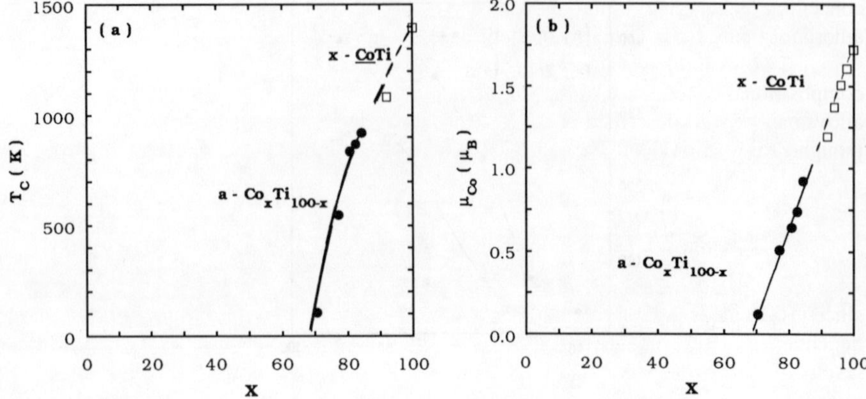

Fig.4.15 : (a) Magnetic ordering temperature (T_C) and (b) Co moment of amorphous alloys (solid circles) and crystalline alloys (open squares) of Co-Ti (Suran et al., ref.91).

achieved, not surprisingly, by substituting Co for Fe. Intermediate increase is realized when Fe is partially substituted by Ni. Even non-magnetic Cu has a sizable effect in increasing T_C and the Fe moment. These enhanced magnetic properties are caused primarily by the increased Fe moment upon alloying. This is concluded from the Mössbauer measurement, which selectively measures the Fe hyperfine field and the Fe moment [86].

Co-ET

There have been fewer studies of binary amorphous Co-ET alloys over wide composition ranges than those of Fe-ET alloys. Most results of vapor-quenched Co-ET alloys are those with ET = Zr, Ti, and Hf [87-91].

Many Co_xET_{100-x} systems can be made amorphous over a wide composition range; for example, $20 \leq x \leq 94$ in Co_xZr_{100-x}. But magnetism is confined within a smaller composition range of $x > x_c$. Below the critical concentration (x_c), the samples exhibit no Co moment nor magnetic ordering. This theme has been repeated in every amorphous Fe-M, Fe-ET, Co-M, and Co-ET alloys. The values of x_c for Co-Zr , Co-Hf, and Co-Ti are in the range of $x_c \approx 58 - 68$. These values are much higher than those of amorphous Fe-Zr and Co-M alloys [87-91].

The magnetization and the Co moment of Co-ET alloys increase monotonically beginning from x_c toward the values of crystalline Co. Some results (e.g. Co-Ti) indicate a quasi linear increase [91] as shown in Fig.4.15(b), while others (e.g. Co-Zr) suggest a slight curvature [87]. In amorphous Co-Ti, the value of the Co moment increases with Co concentration, and merges into those of crystalline Co-Ti alloys. The extrapolated value for amorphous pure Co is virtually the same as that for crystalline Co.

Two variations of the charge transfer model have been proposed to account for the compositional dependence of the Co moment (μ_{Co}). In one version, the data for the Co-Zr alloys can be well described if 2.2 electrons are assumed to transfer from Zr to Co, even though Zr is quadrivalent. Then the Co moment, expressed in Bohr Magneton is simply

$$\mu_{Co} = 1.72 - 2.2 \, \frac{100-x}{x}, \tag{4.3}$$

since $(100-x)/x$ is the relative number of Zr alloying with Co. According to Eq.(4.3), the dependence of μ_{Co} on x is not linear but contains a slight curvature as indicated by the results of amorphous Co-Zr [87]. By the choice of the amount of charge transfer Eq.(4.3) gives the observed critical concentration of $x_c \approx 60$. Since the values of x_c for Co-Ti and Co-Hf are similar to that of Co-Zr, Ti, and Hf would transfer similar number of electrons upon alloying.

In the other version of the charge transfer model, the Co moment in Co-Ti can be expressed as

$$\mu_{Co} = 1.72 - 5 \, \frac{100-x}{100}, \tag{4.4}$$

where 5 is the excess charge, and $(100-x)/100$ is the composition of Ti. Eq.(4.4) is linear in x and gives a critical concentration of $x_c \approx 66$, in agreement with the results in Co-Ti [91]. The magnetic ordering temperature of Co-Ti has been measured over a wide composition range. Its compositional dependence is believed to be representative of other amorphous Co-ET alloys [91]. As shown in Fig.4.15(a), T_C increases monotonically as the Co content is increased, and equally important, the value merges into those of crystalline Co-Ti alloys.

From the extrapolations to the amorphous Co limit, one concludes that amorphous pure Co would be strongly ferromagnetic with a moment of about 1.7 μ_B and a T_C of about 1400 K. These are precisely the characteristics of crystalline fcc Co. These extrapolated results, together with those of Co-M alloys, indicate that the properties of amorphous pure Co are very similar to those of crystalline fcc Co. Different structures have little influence on the magnetic properties of Co.

Other TM-ET Alloys

Just as in TM-M alloys, only Co-ET and Fe-ET show magnetic ordering at reasonably high temperatures. Amorphous TM-ET alloys with ET = Ni, Mn, Cr, and Cu do not show appreciable magnetic ordering. Most of the investigations center on the transport properties and especially superconductivity; the latter is readily observed in many TM-ET alloys with ET = Zr, Nb, Hf, and Ti, due to a complete loss of the TM magnetic moment [82].

8. Toward the Amorphous Pure Metal Limit

Ever since the discovery of amorphous metallic alloys, amorphous pure metals, the simplest possible amorphous systems, have been the obvious targets for theoretical and experimental investigations. Experimentally, many researchers including this author, have attempted amorphous pure metals using deposition at cryogenic temperatures, but without success. Samples deposited at cryogenic temperatures were inevitably found to be crystalline at room temperature. Results obtained from *in-situ* Mössbauer measurements at 4.2 K also indicated crystalline Fe. The prospect of stabilizing amorphous pure Fe at a reasonable temperature appears to be rather bleak. The few claims of amorphous Fe and other TM's have been met with a high degree of skepticism [92]. Furthermore, since the magnetic ordering temperature of amorphous Fe, and especially that of amorphous Co, are likely to be quite high, there is little hope that their values can ever be directly and convincingly established.

An attractive alternative is to utilize various amorphous systems having a stable amorphous phase up to very high Fe or Co concentrations and *extrapolate* the properties toward those of amorphous Fe or Co. The large number of vapor-quench binary amorphous TM-M and TM-ET alloys over wide composition ranges can be used to elucidate the properties of amorphous pure TM. Of course, inherent to all extrapolations uncertainties are unavoidable. But given the futile situations in amorphous pure metals, there are few alternatives.

The characteristics of amorphous pure Co can be readily concluded by extrapolating the results of amorphous Co-M and Co-ET alloys. In both cases, as shown in Fig.4.6 and Fig.4.15, the magnetic properties of amorphous Co are strikingly similar to, and in fact indistinguishable from, those of crystalline fcc Co. Both the Co moment and the Curie temperature of amorphous Co are virtually the same as those of crystalline fcc Co. These are manifestations of the fact that the magnetic properties of Co are insensitive to structure.

An entirely different situation is encountered in amorphous Fe. The results of the magnetic ordering temperature, the Fe moment, and the Fe hyperfine field (as described in sections 6 and 7), and those of the isomer shift and the quadrupole splitting described in the following, all point to the conclusion of *polymorphism* of amorphous pure Fe [65]. The key evidences are the following:

(1) In the Fe-rich end, the magnetic ordering temperatures of the Fe-M alloys are much higher (> 300 K) that those of the Fe-ET alloys (< 300 K).

(2) In all Fe-ET systems (Fig.4.12), there is an abrupt change in the Fe moment and hyperfine field at the boundary separating the amorphous and the crystalline states, whereas there are no discontinuities in the Fe-M systems (Fig.4.5).

(3) The isomer shift (IS) relative to α-Fe is another key quantity differentiating Fe-M and Fe-ET systems [65]. The signs of IS of the Fe-M systems are exclusively positive, with an extrapolated value of 0.04 mm/s for amorphous pure Fe, whereas those of the Fe-ET are

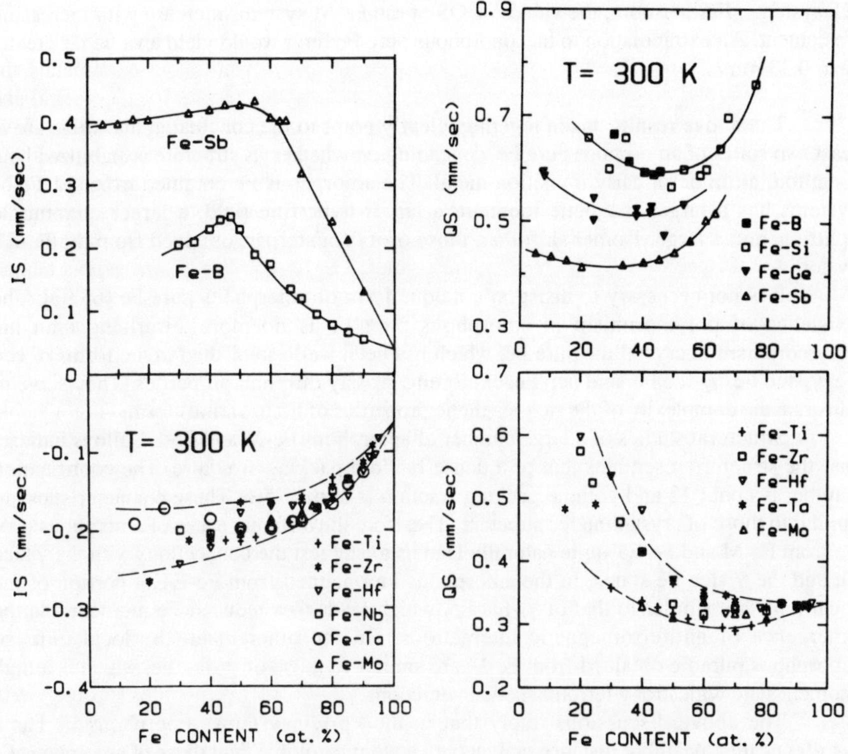

Fig.4.16 : Concentration dependences of isomer shift with respect to bcc a-Fe and quadrupole splitting at 300 K of amorphous Fe-M (M = Sb and B) and amorphous Fe-ET (ET = Ti, Zr, Hf, Nb, Ta, Mo) alloys (Xiao and Chien, ref.65).

always negative, with an extrapolated value of - 0.06 mm/sec for amorphous Fe, as shown in Fig.4.16. This sizable difference of about 0.1 mm/s is comparable to that between crystalline bcc Fe and fcc Fe, where the IS of fcc Fe is 0.08 mm/s less than that of bcc Fe.

(4) The quadrupole splitting (QS), measured in the paramagnetic state, measures the local charge distributions surrounding the probe nucleus Fe. As shown in Fig.4.16, in all the Fe-ET systems, the values of QS are decreasing with Fe concentration, and converge to a value of QS = 0.33 mm/s for amorphous pure Fe [65]. In the Fe-M systems, the values of QS cannot be determined from samples with high Fe contents because of the presence of the magnetic hyperfine splitting. However, as shown in Fig.4.16, from the data up to about x = 80, the QS's of the Fe-M systems have much larger values that those of the Fe-

ET systems. Furthermore, the values of QS of the Fe-M systems increase with increasing Fe content. An extrapolation to the amorphous pure Fe limit would yield a value far greater than 0.33 mm/s.

The above results, taken together, clearly point to the conclusion that there are at least two states of amorphous pure Fe, depending on whether its structure is stabilized by a metalloid atom or an early transition metal. The amorphous Fe obtained from the Fe-M systems has a larger magnetic moment, a larger hyperfine field, a larger quadrupole splitting, and a larger isomer shift than those of its counterpart obtained from the Fe-ET systems.

It is not necessary to insist on a unique form of amorphous pure Fe [65,93]. The existence of polymorphism in amorphous pure Fe is no more surprising than the polymorphism in crystalline pure Fe, which has been well established in the forms of bcc Fe, γ_1-fcc Fe, γ_2-fcc Fe, and hcp Fe, exhibiting grossly different properties. They serve to illustrate the complexity of the rich magnetic properties of Fe in various forms.

Structural studies of a large number of amorphous Fe-M and Fe-ET alloys indicate that the structure resembles that of a dense random packing structure. The coordination number is about 12 and volume packing fraction is about 75%. These characteristics are similar to those of crystalline fcc structure. Therefore the different states of amorphous pure Fe from Fe-M and Fe-ET quite naturally lead us to suggest the connections with the γ_1-fcc Fe and the γ_2-fcc Fe states. In the amorphous Fe obtained from Fe-ET, a portion of the local units are similar to that of γ_1-fcc Fe, which leads to a reduced Fe moment and the emergence of antiferromagnetic interactions. On the other hand, the local units of amorphous pure Fe obtained from Fe-M are similar to those of γ_2-fcc Fe, which is a high moment state with strong ferromagnetic interactions.

The above discussions imply that in the Fe-ET systems, amorphous Fe has a *smaller* nearest-neighbor distance or average Fe atomic volume than those of amorphous Fe in the Fe-M systems. Structural studies support this assertion. Values of the nearest-neighbor distance and atomic Fe volume obtained structural data, extrapolated from Fe-B, are about $d_{Fe} = 2.60$ Å and $V_{Fe} = 12.3$ Å^3 respectively [94-96]. These values are above the critical values for the $\gamma_1 - \gamma_2$ transition. On the other hand, in the Fe-ET systems, the opposite situation is realized. Both EXAFS [97] and neutron diffraction [98] of $Fe_{90}Zr_{10}$ indicate $d_{Fe} \approx 2.415$ Å and 2.45 Å respectively. Density study [99] of the Fe-Zr system gives an average Fe atomic volume of $V_{Fe} = 10.7$ Å^3 for amorphous Fe.

9. Metastable Crystalline Systems

In the equilibrium state, a magnetic 3d element (e.g. Fe, Co, and Ni) generally forms alloys or intermetallic compounds with almost all metallic elements in the Periodic Table. The physical properties in general, and magnetic properties in particular, of the alloy systems have been extensively studied. Excluded from such studies are the *immiscible* systems in which no alloy or compound exists. Many transition metal-noble metal systems are examples of immiscible systems. For example, in the Fe-Cu system, the equilibrium phase diagram allows practically no alloys despite very similar atomic radii [1]. Only at elevated temperatures above 900° C can the solubility at the extreme ends of the composition range be extended to a mere few percent. In other systems (e.g. Fe-Ag, Fe-Pb), the elements are immiscible even in the liquid state [1]. In the past, these immiscible systems could not be investigated except in the dilute limit via impurity studies. The advent of vapor quenching processes has opened up hitherto unattainable alloy compositions of immiscible systems, permitting the exploration of their properties for the first time.

In the 1960's, Kneller first fabricated Fe_xCu_{100-x} ($20 \leq x \leq 100$) [100] and Co_xCu_{100-x} ($15 \leq x \leq 60$) [14] and by co-evaporation of thin films (~1000 Å) onto quartz substrates held at room temperature. From the magnetic measurements, it was concluded that a simple dilution of the magnetic moment occurs, and that the Co-rich and the Fe-rich samples are strongly ferromagnetic, but no systematic determinations of the magnetic ordering temperatures were made. Unfortunately, these early successes were not noted nor followed up. Since the early 1980's, renewed interest in these and other metastable crystalline alloy systems occurred when sputtering as a more effective vapor-quenching technique was recognized [101-106]. The thick samples (5 - 15 μm) permit detailed measurements of the structural and magnetic properties. In this section, we will describe a few representative metastable crystalline systems.

Fe-Cu

Despite the immiscibility indicated by the phase diagram, metastable Fe_xCu_{100-x} alloys over the *entire* composition range can be readily achieved by sputtering. Sumiyama et al., with substrates held at room-temperature, found that the samples with $x > 70$ were bcc, those with $x < 42$ were fcc, and those with $42 \leq x \leq 70$ were mixed-phase [101-103]. Chien et al., using liquid-nitrogen-cooled substrates, found that the Fe-rich samples with $x > 75$ are in the bcc structure, those with $x < 60$ are in the fcc structure, and the samples with the intermediate compositions between $x = 60$ and 75 contain both bcc and fcc phases [104,105]. In other words, with a higher substrate temperature, the mixed-phase composition range became larger, and at the expense of the single-phase bcc and fcc states.

The magnetic ordering temperatures (T_C) of the Fe-Cu samples are shown in Fig.4.17. For the Fe-rich ($x > 60$) samples, because the values of T_C are so high, partial or complete phase transformation occurred after the T_C measurements. These values are expressed by open circles. It is clear from Fig.4.17 that the values of T_C are structure-dependent; those for bcc-Fe_xCu_{100-x} are much higher than those of fcc-Fe_xCu_{100-x}.

A large composition range for the fcc Fe-Cu alloys is available for investigation. The behavior of T_C of the fcc samples follows closely that of a simple magnetic dilution, which leads to a percolation threshold of $x_p = 18.5$. The samples with $x < x_p$ are not ferromagnetic, but spin glasses, as shown in Fig.4.17. The values of the ferromagnetic samples with $x > x_p$ follow a power law $(x - x_p)^\gamma$ of

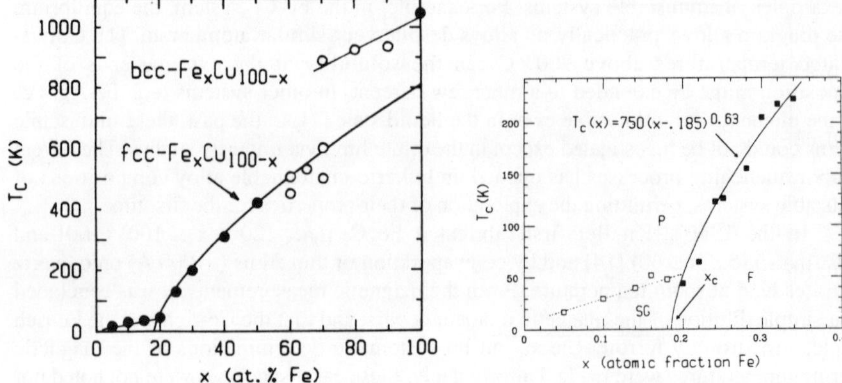

Fig.4.17 : Magnetic ordering temperatures of metastable crystalline fcc and bcc Fe_xCu_{100-x} alloys (Chien et al., ref.104). On the right, the magnetic phase diagram near the percolation threshold is displayed, where P = paramagnetic, F = ferromagnetic, and SG = spin glass (Chien et al., ref.105).

$$T_C(K) = 41 \ (x - 18.5)^{0.63} \qquad \text{for } Fe_xCu_{100-x}$$

or

$$T_C(K) = 750 \ (x - 0.185)^{0.63} \qquad \text{for } Fe_xCu_{1-x}$$

as shown by the solid curve in Fig.4.17. The values of $x_p = 18.5\%$ and $\gamma = 0.63$ are in good agreement with theoretical calculations made for nearest-neighbor interactions on an fcc lattice [57]. In the vicinity of, and just above, the magnetic percolation threshold x_p, some evidences of reentrant behavior have been observed, although no detailed systematic measurements have been made.

In Fig.4.17, if one extrapolates the T_C values of the fcc-Fe_xCu_{100-x} to the limit of pure fcc Fe, a value of 800 K would be obtained for γ_2-fcc Fe. This is significantly lower than the value of $T_C = 1045$ K for bcc Fe, again reflecting the sensitivity of the magnetic properties of Fe on structure.

Mössbauer spectra of the Fe-Cu samples at 4.2 K show that all the samples are magnetically ordered at 4.2 K. As the Fe concentration is reduced from pure Fe, the saturation hyperfine field (H_{eff}) decreases only slightly, but it is finite for all compositions. The same conclusion has been made from magnetometry measurements [104]. This is shown in Fig.4.18 where we plot the normalized hyperfine field $H_{eff}(x)/H_{eff}(100)$ (open circles) or $\mu_{Fe}(x)/\mu_{Fe}(100)$ (solid circles) per Fe atom as a function of x. The data nearly coincide with the straight line, which would be the result for a perfect dilution. This is different from those of *all* amorphous Fe-M and Fe-TM alloys, where the μ_{Fe} and H_{eff} vanish at a finite concentration, varying from 40 to 60 at.%.

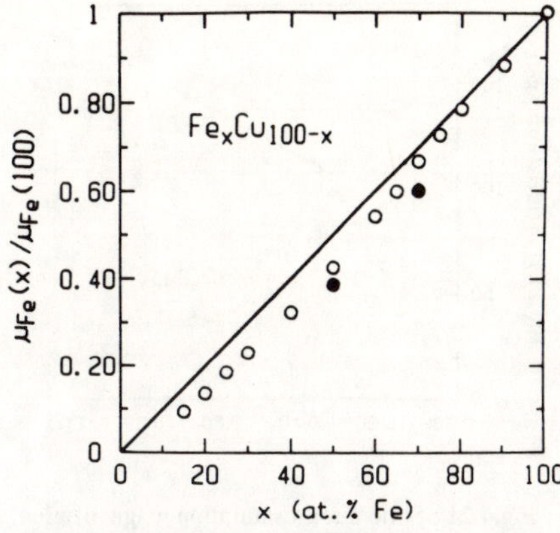

Fig.4.18 : Normalized hyperfine field $H_{eff}(x)/=H_{eff}(100)$ (open circles) and moments $\mu_{Fe}(x)/\mu_{Fe}(100)$ (solid circles) per transition metal atom versus Fe concentration. The straight line is the result for a simple magnetic dilution (Chien et al., ref.104).

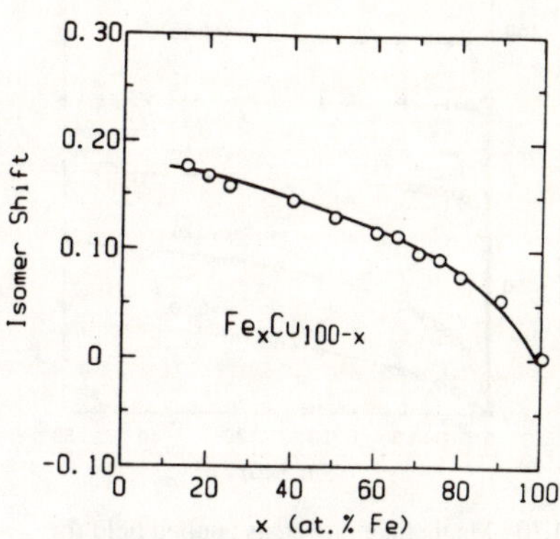

Fig.4.19 : Isomer shift relative to α-Fe of metastable crystalline fcc (x < 70) and bcc (x > 70) Fe_xCu_{100-x} alloys as a function of Fe component (Chien et al., ref.104).

The isomer shift (IS) results of Fe-Cu are shown in Fig.4.19. Alloying with Cu causes the Fe isomer shift to increase. Extrapolating to the limit of dilute Fe in Cu, one obtains a value of IS = 0.18 mm/sec relative to α-Fe, close to the value observed in dilute FeCu alloys. The behaviors of H_{eff}, μ_{Fe}, and IS across the whole composition range bring up another interesting point. Across x $\approx$ 75, where the structure changes from fcc to bcc, there is a sudden change in the magnetic ordering temperature as shown in Fig.4.17. Quite unexpectedly, across this phase boundary, there is *no* discernable change in the H_{eff}, μ_{Fe}, nor IS. Apparently in Fe-Cu, the Fe moment, the Fe hyperfine field and the isomer shift are dominated by the Fe concentration and not structure. This is quite different from amorphous Fe-ET alloys, where at the phase boundaries pronounced changes occur.

Co-Cu

In the case of the Co_xCu_{1-x} system [106], another immiscible system, the vapor-quench samples reveal that all the samples with $0 \leq x \leq 0.7$ are fcc, and the hcp phase appears for samples with $x \geq 0.8$. The large composition range available in the fcc state allows a detailed study of the magnetic properties of Co in the fcc state, while bulk Co is in the hcp state at ambient temperatures.

The magnetization curves for various Co-Cu samples at 5 K are shown in Fig.4.20. For the samples with x > 0.3, the magnetization at 5 K can be easily saturated using fields of only a few kG, indicative of ferromagnetic order. For the samples with $x \leq 0.3$, the magnetization does not saturate, even in fields up to 50 kG, indicating a breakdown of

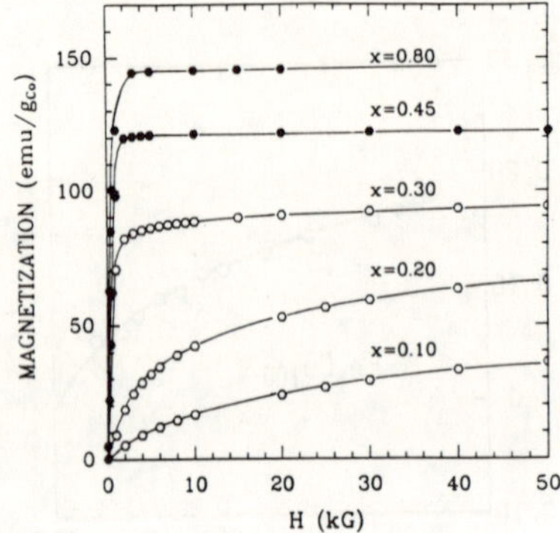

Fig.4.20 : Magnetization versus applied field for metastable crystalline fcc Co_xCu_{1-x} alloys in units of emu per gram of Co at T = 5 K. Full saturation is not reached at H = 50 kOe for x ≤ 0.30 (open circles) (Childress and Chien, ref.106).

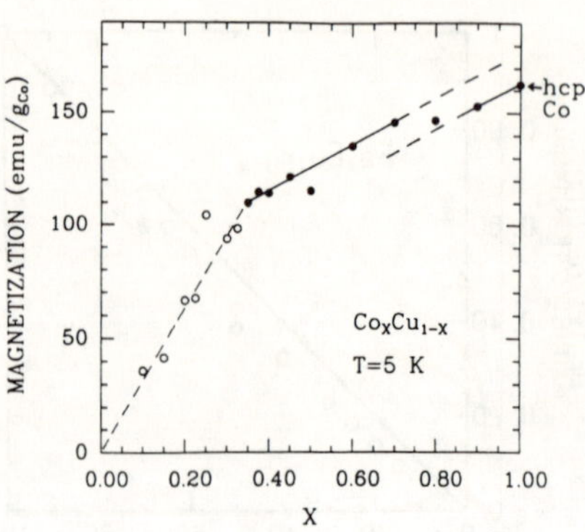

Fig.4.21 : Solid circles:saturation magnetization (in units of emu/g of Co) of metastable crystalline Co_xCu_{1-x} alloys as a function of x. Open circles : magnetization under an applied field of 50 kOe (Childress and Chien, ref.106).

long-range ferromagnetic order. The saturation magnetization (for x > 0.3) and the magnetization at 50 kG (for x ≤ 0.3) at 5 K in units of emu/gram of Co, are shown in Fig.4.21 as a function of Co content. For the ferromagnetic samples with fcc structure (0.3 < x < 0.8), the saturation magnetization is not constant but decreases linearly with x. But the Co moment does not vanish at any composition. An extrapolation of this trend to x = 0 yields a magnetization of 80 emu/gCo, roughly half of the bulk hcp value of 162 emu/gCo (or 1.7 μ_B per Co atom). At low Co concentrations (x ≤ 0.3), since the magnetization does not saturate because of the spin glass ordering, the data (open circles) do not reliably yield information on the Co moment.

The magnetization in the fcc phase extrapolated to x = 100 from the data below x = 80 is approximately 175 emu/gCo, slightly higher than the bulk hcp. This is consistent with the enhancement of magnetization seen in bulk Co upon transformation to the fcc state at 700 K. Thus the ramification of different structures on the Co moment, at least between fcc and hcp, is much less than those of Fe.

Fig.4.22 shows the zero-field-cooled (ZFC) magnetization curves of samples with x = 0.23, 0.25, and 0.28 under a small field of 10 G. The results of samples with x up to 0.23, with a sharp cusp-like feature, are typical of spin glasses. The samples with x = 0.25 and 0.28 show a difference behavior. Two distinct transitions are now present; one is characteristic of a ferromagnetic transition at a high temperature, while the other is a

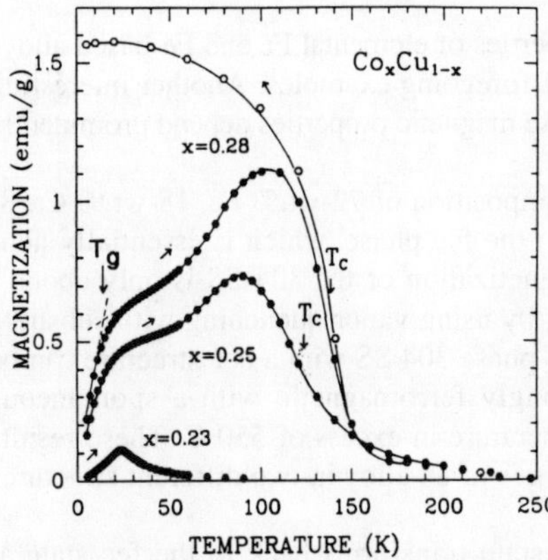

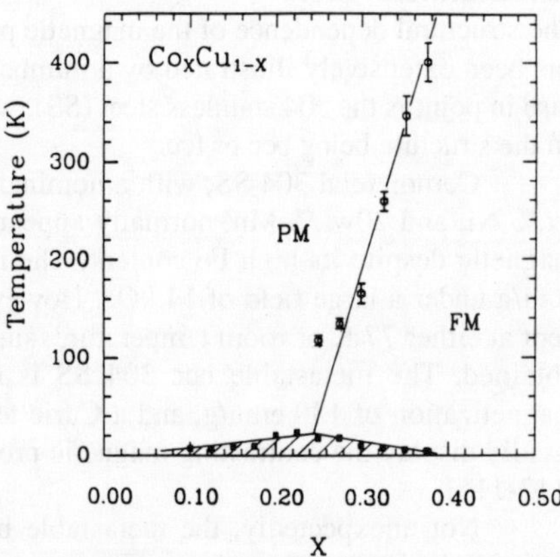

Fig.4.22 : Zero-field-cooled magnetization curves (solid circles) for metastable fcc Co_xCu_{1-x} alloys in the reentrant regime with H_{ext} = 10 G. The field-cooled data are shown for x = 0.28 (open circles) (Childress and Chien, ref.106).

Fig.4.23 : Magnetic phase diagram of metastable fcc Co_xCu_{1-x} alloys showing paramagnetic (PM), ferromagnetic (FM) and spin-glass (shaded area) phases (Childress and Chien, ref.106).

reentrant spin glass transition at a lower temperature. Thus, evidences for both ferromagnetic and spin-glass transitions are simultaneously present. All samples with 0.24 $\leq$ x $\leq$ 0.40 exhibit a double-transition reentrant behavior. For samples with x > 0.40, only a single ferromagnetic transition can be observed. These results lead to a magnetic phase diagram shown in Fig.4.23. At low Co concentrations (x $\leq$ 0.23), only the spin-glass phase is present at low temperatures. In that region, T_g scales with the Co content, reaching a maximum of 24 K. Ferromagnetic order appears above a critical transition of about x $\approx$ 0.24, with the value of T_C increasing sharply with Co concentration at a rate of approximately 23 K per atomic %. For samples up to x = 0.40, there is also a spin glass transition at low temperatures. In this case, T_g decreases with increasing Co content, and becomes zero at about x = 0.40, above which only ferromagnetic transition may be observed.

It may be mentioned that annealing a metastable immiscible alloy, such as Fe-Cu, and Co-Cu, causes the formation of a *granular* magnetic system consisting of single-domain ferromagnetic particles embedded in a metallic medium [107,108]. Most remarkably, in such magnetically inhomogeneous systems, giant magnetoresistance (GMR), which previously have only been observed exclusively in multilayers [109] has been uncovered [110,111]. Subsequently, GMR as large as 80% has been observed in a variety of granular magnetic systems [112].

304 Stainless Steel

The structural dependence of the magnetic properties of elemental Fe and Fe-based alloys has been extensively illustrated by a number of foregoing examples. Another interesting case in point is the 304 stainless steel (SS), whose magnetic properties depend prominently on the structure being bcc or fcc.

Commercial 304 SS, with a nominal composition of 72-wt.% Fe, 18-wt.% Cr, 8-wt.% Ni, and 20wt.% Mn, normally appears in the fcc phase, which is essentially non-magnetic despite its high Fe content. The magnetization of fcc 304 SS is only about 1 emu/g under a large field of 14 kOe. However, by using vapor-quenching with substrate kept at either 77 K or room temperature, single-phase 304 SS with a bcc structure can be obtained. The metastable bcc 304 SS is strongly ferromagnetic with a spontaneous magnetization of 130 emu/g, and a Curie temperature in excess of 550 C. These results vividly display the contrasting magnetic properties of an alloy in two different structures [113-115].

Not unexpectedly, the metastable bcc state transforms back to the fcc state at sufficiently high temperatures. This transformation can be followed by both x-ray diffraction and magnetometry [114]. The bcc state is stable up to 500 C, above which diffraction lines of fcc begin to appear. At 500 C, the yield of the fcc state is about 5%, and the yield quickly reaches 50% at 550 C, and then reaches completion near 800 C. As a result of the transformation, the spontaneous magnetization decreases precipitously beginning at 500 C. When the 304 SS transforms completely into the fcc state, it loses its strong ferromagnetic characteristics, and exhibits a weak magnetization of less than 1 emu/g at room temperature.

Quasicrystals

As a final example, we will discuss the various states of a quasicrystal. Quasicrystalline order with icosahedral symmetry in metastable form was first discovered by Shechtman et al.[116], prompting an intense interest worldwide. Subsequently, stable icosahedral quasicrystals, most notably Al-Li-Cu and Al-Cu-Fe, have been discovered [117], dispelling the notion that the quasicrystals are inherently metastable. Recently, $Al_{65}Cu_{20}Fe_{15}$ in *three* different structural phases have been achieved [118]. This rather rare occurrence allows the comparison of properties in three difference structures.

Vapor quenching has been used to first capture the amorphous state of $Al_{65}Cu_{20}Fe_{15}$, which is not attainable by rapid solidification from the melt. Upon annealing to suitably high temperatures, a metastable crystalline phase with a cubic structure is first formed. At still higher temperatures, the formation of the quasicrystalline phase follows. A DSC scan of $Al_{65}Cu_{20}Fe_{15}$ exhibits *two* exothermic peaks at $T_1 = 310°C$ and $T_2 = 515°C$, indicating the two phase transformations.

The as-sputtered sample of $Al_{65}Cu_{20}Fe_{15}$ is amorphous with a diffraction pattern (Fig.4.24-a) typical of those of amorphous metallic solids. In the diffraction pattern (Fig.4.24-b) of a sample annealed to $T_a = 450°C$, where $T_1 < T_a < T_2$, all the diffraction peaks can be indexed to a cubic CsCl crystal structure with a lattice constant of $a = 2.94$ Å, a structure consisting of two interpenetrating simple cubic sublattices. When the sample is

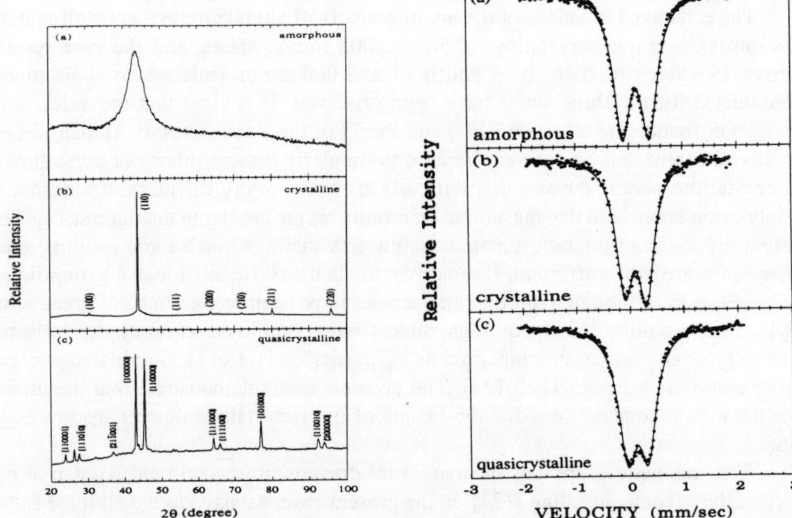

Fig.4.24 : (Left panel) X-ray diffraction patterns of $Al_{65}Cu_{20}Fe_{15}$: (a) as-sputtered, amorphous (b) annealed at 450 C, indexed as a cubic CsCl structure with a lattice parameter of a = 2.94 Å, and (c) annealed at 600 C, indexed using Bancel's scheme of 6 Miller indices with a quasilattice parameter of 4.46 Å.
(Right panel) Mössbauer spectra at room temperature of $Al_{65}Cu_{20}Fe_{15}$: (a) as-sputtered, amorphous (b) annealed at 420 C, crystalline and (c) annealed at 600 C, quasicrystalline (Chien and Lu, ref.118)

annealed at $T_a = 600°C$, where $T_a > T_2 > T_1$, the diffraction pattern changes completely into that of the quasicrystalline phase as shown in Fig.4.24-c. All the diffraction peaks can now be indexed using the Bancel's scheme [119] with six Miller indices, as shown in Fig.4.24-c. A quasilattice parameter of 4.46 Å has been determined for the quasicrystalline phase, in agreement with previous results [117]. Therefore, the amorphous $Al_{65}Cu_{20}Fe_{15}$ alloy is highly unusual in that the transformation proceeds first to yet another metastable but crystalline cubic phase before finally descending into the quasicrystalline state. Furthermore, in all three states, the materials are essentially single phase. Precisely due to this rare occurrence, one can study and compare the physical properties of three distinct phases of $Al_{65}Cu_{20}Fe_{15}$.

Mössbauer spectra of the three structures of $Al_{65}Cu_{20}Fe_{15}$ at room temperature are shown in the right panel of Fig.4.24. It is immediately clear that the Fe site symmetry in all three cases are non-cubic, resulting in quadrupole-split doublet spectra. Non-cubic environment are not surprising for the amorphous and the quasicrystalline states. For the intermediate crystalline state with a cubic CsCl structure, the non-cubic symmetry comes from the disordered atomic occupation on the two interpenetrating simple cubic sublattices.

The effective QS values of the amorphous (0.50 ± 0.008 mm/s), crystalline (0.41 ± 0.008 mm/s), and quasicrystalline (0.38 ± 0.008 mm/s) states, and the corresponding effective IS values of 0.26, 0.23 and 0.24 ± 0.008 mm/s (relative to α-Fe at room temperatures) for the three states have been observed. It is clear that the values of the effective QS (hence the effective EFG) and the IS of the three states of $Al_{65}Cu_{20}Fe_{15}$ are distinctly different, but do not vary greatly, particularly between those of crystalline and quasicrystalline states, despite diametrically different x-ray diffraction patterns. One certainly cannot conclude that the atomic environments are similar in the three states. This is merely a reflection of the unfortunate situation in which the quadrupole splitting and the isomer shift are not sufficiently sensitive to changes in the local surroundings of $Al_{65}Cu_{20}Fe_{15}$, even though hyperfine interactions have been proven to be of great value in studying many solids. Many previous studies have attempted to relate the Mössbauer results to those of various structure models of quasicrystals, from a simple two-site model to more elaborate schemes [120-123]. The present results demonstrate that the utility of hyperfine interactions measured at the Fe site of quasicrystals for such purposes is quite limited.

The magnetic properties of many quasicrystals in general, and $Al_{65}Cu_{20}Fe_{15}$ in particular, have been unsettling [124]. In the present case, we have found all three states of $Al_{65}Cu_{20}Fe_{15}$ are paramagnetic with no sign of magnetic ordering. From the susceptibility data, we have determined the effective moment of Fe, assumed to be the only species carrying the local moments (p_{eff}), to be 0.45, 0.24 and 0.28 ± 0.02 μ_B for the amorphous, crystalline, and quasicrystalline states respectively. This variation of moments results solely from the differences in the structures of the three states.

The room temperature resistivity $\rho(300\ K)$ has a value of 250, 3200, and 2000 $\mu\Omega$-cm for the amorphous, crystalline, and quasicrystalline states respectively. The value for the amorphous state is similar to those observed in other amorphous and disordered metals [125]. However, both the intermediate crystalline and the quasicrystalline states have values about an order of magnitude higher. While many quasicrystals of $Al_{65}Cu_{20}Fe_{15}$ show resistivities in the few tens to few hundred $\mu\Omega$-cm range [126], very large resistivities of 4500 $\mu\Omega$-cm [127,128] have been observed for highly perfect quasicrystalline specimens of $Al_{65}Cu_{20}Fe_{15}$. The resistivities of both our quasicrystalline and intermediate crystalline samples have comparable values to these values.

10. Conclusions

Before the advent of rapid quenching processes, it would not have been feasible to fabricate, let alone study, any of the metastable alloy systems described in this Chapter. Among the various rapid quenching methods, the prowess of the vapor quenching technique is amply demonstrated through many examples of metastable binary alloy systems. The wide composition ranges afforded by the vapor-quenched amorphous and metastable crystalline alloy systems provide unique opportunity for studying their magnetic properties.

The first question we have addressed concerns the composition range of vapor-quenched amorphous alloys; in particular, the composition is very wide in Fe_xB_{100-x} ($0 \leq x \leq 90$), less wide in Fe_xMo_{100-x} ($40 \leq x \leq 80$), and none at all in Fe-Cu and Co-Cu. We show experimentally that the size difference of the constituent atoms is of overriding importance, the larger the size difference the larger the composition range for the amorphous alloys. When the size difference is less than 5%, no amorphous alloy can be formed. The predictions by the theory of Egami and Waseda give quantitative agreement with the experimental results.

When the magnetic properties of binary metastable binary alloy systems of transition metal elements are examined, it is evident that all magnetic quantities (whether the values of the ordering temperature, the magnetic moment, and the exchange interaction discussed here, or magnetostriction, anisotropy, and other magnetic quantities discussed elsewhere) vary continuously as the composition is varied. This crucial degree of freedom in composition, unique to metastable alloys, enables the magnetic properties to continuously evolve.

Of the five transition metal elements (TM) of Fe, Co, Ni, Cr, and Mn, which show magnetic orderings in their crystalline states, only binary amorphous alloys containing Fe and Co show strongly ferromagnetic properties in the TM-rich compositions. However, in every amorphous binary Fe-M, Fe-ET, Co-M, and Co-ET alloy system, neither the Fe nor the Co magnetic moment is constant; they both increase with increasing Fe or Co concentration. Equally important, the Fe and Co moment first develop at a critical concentration (x_c), below which there is no magnetic moment nor magnetic ordering. Generally speaking, the values of x_c for the Co alloys are higher than those of the Fe alloys, even though both are also dependent on the alloying element. The Ni moment is apparently more precarious; none of the binary Ni-M and Ni-ET alloys show ferromagnetic characteristics. It is probable that the value of x_c for the amorphous Ni alloys is beyond the composition range for the formation of the amorphous state. Finally, none of the binary amorphous (Cr or Mn)-M and (Cr or Mn)-ET alloys show ferromagnetic characteristics -- at most only spin-glass ordering at very low temperatures. This is perhaps not too surprising since Cr and Mn are itinerant antiferromagnets in their crystalline forms.

The magnetic properties of binary amorphous Co-M and Co-ET alloys, are far simpler than those of their Fe counterparts. Both the Co moment and the magnetic ordering temperature first emerge at the critical composition (x_c), increase monotonically with the Co composition, and eventually merge into those of crystalline Co alloys. Extrapolations to the amorphous pure Co limit show that amorphous pure Co would have a Co moment of about 1.7 μ_B and a T_C of about 1400 K, which are indistinguishable from those of crystalline Co (hcp Co, fcc Co, and bcc Co), all of which have similar magnetic properties. Hence, the magnetism of Co is insensitive to structural differences.

Entirely different situations are encountered in amorphous Fe-M and Fe-ET alloys, which display a much richer variety of magnetic properties. The amorphous Fe-M and amorphous Fe-ET alloys exhibit completely different magnetic properties. Most of amorphous Fe-M alloys show high values of T_C (> 300 K), whereas none of the amorphous Fe-ET alloys show T_C higher than 280 K. While the Fe moment in the amorphous Fe-M alloys eventually merges into those of crystalline Fe alloys, the Fe moment of the amorphous Fe-ET alloys is much depressed. These differences, together with those in isomer shift and quadrupole splitting, indicate that there are at least two different states of amorphous pure Fe, stabilized by either a metalloid or an early transition metal element. These two states for amorphous pure Fe are reminiscent of the crystalline γ_1-fcc Fe and γ_2-fcc Fe states.

The amorphous state of pure Fe is of considerable scientific importance. To date, there are only theoretical studies and no direct experimental measurements because of its extreme instability. Using amorphous Fe-M and Fe-ET alloys with wide composition ranges and extending to very high Fe concentration, we show that amorphous pure Fe would have a T_C of about 220 K, which is much lower than the value of $T_C = 1045$ K for crystalline bcc Fe. Unlike that of Co, the magnetic properties of pure Fe are extremely sensitive to the structural details. We now know there are at least 6 different states of pure Fe. In addition to the two amorphous pure Fe states stabilized by the metalloid (M) and the early transition metal (ET) elements, there are also four crystalline states of bcc Fe, γ_1-fcc Fe, γ_2-fcc Fe, and hcp Fe. These different states vividly illustrate the complexity of the magnetism of pure Fe.

Very different magnetic properties are realized in the metastable crystalline alloys of Fe-Cu and Co-Cu. Unlike those in the amorphous alloys, the Fe and Co moments never vanish at any composition. Consequently, a true magnetic percolation with a percolation threshold of about 20% has been observed in both cases.

Through various examples, we also show the sensitivity with which the magnetic properties of the TM alloys depend on structure. At the boundary separating the amorphous state and the crystalline state in various Fe-M and Fe-ET alloys, a large change in the magnetic ordering temperature and the Fe moment are observed. The 304 stainless steel in its ordinary fcc state is hardly magnetic, whereas in the bcc state it is strongly ferromagnetic. Likewise, MnB, strongly ferromagnetic in the crystalline state, reduces to a weak spin glass in the amorphous state.

The example of vapor-quenched $Al_{65}Cu_{20}Fe_{15}$ shows a rare instance where three different states of the same composition can be formed and individually studied.

Acknowledgements - Several associates have contributed to the results in this Chapter. Among them are Dr. K. M. Unruh now at University of Delaware, Dr. J. H. Hsu now at Taiwan University, Dr. S. H. Liou now at the University of Nebraska, Dr. G. Xiao now at Brown University, and Dr. J. R. Childress now at Laboratoire de Physique des Solids, Universite Paris-Sud, Orsay, France. This work has been supported by NSF Grant No.DMR-9200280.

References :

1. See e.g. M. Hanson and K. Anderko, *Constitution of Binary Alloys* (McGraw Hill, New York 1958); R. P. Elliot, *Constitution of Binary Alloys, First Supplement* (McGraw Hill, New York 1965); F. A. Shunk, *Constitution of Binary Alloys, Second Supplement* (McGraw Hill, New York, 1969).
2. O. Kubaschewski, *Iron-Binary Phase Diagrams*, (Springer-Verlag, Berlin, 1982).
3. *Amorphous Metallic Alloys*, edited by F. E. Luborsky (Butterworth, 1983),
4. K. Moorjani and J. M. D. Coey, *Magnetic Glasses*, (Elsevier, 1984).
5. *Metallic Glasses*, edited by J. J. Gilman and H. J. Leamy (American Society of Metals, 1978).
6. H. S. Chen, Rep. Prog. Phys. **43** (1980) 23.
7. See e.g.*Glassy Metals I*, edited by H. J. Güntherodt and H. Beck (Springer, Berlin, 1981); *Glassy Metal II*, edited by and H. Beck and H. J. Güntherodt (Springer, Berlin, 1983); *Glassy Metals, Magnetic, Chemical and Structural Properties*, edited by R. Hasegawa, (CRC, Boca Raton, 1983); *Rapidly Quenched Metals V*, edited by C. N. J. Wagner and W. L. Johnson, (North-Holland, Amsterdam, 1984); *Rapidly Quenched Metals*, edited by S. Steeb and H. Warlimont (Elsevier, Amesterdam, 1985).
8. See e.g. M. G. Scott in ref.3, p.44.
9. P. P. Freitas and T. S. Plaskett, Phys. Rev. Lett. **64** (1990) 2184.
10. K. H. J. Buschow, J. Appl. Phys. **54** (1983) 2578.
11. *Synthesis and Properties of Metastable Phases*, edited by E. S. Machlin and T. J. Rowland, (Metallurgical Society of AIME, 1980).
12. K. Klement, R. H. Willens, and P. Duwez, Nature, **187** (1960) 869.
13. S. Mader and A. S. Nowick, J. Vac. Sci. and Tech. **2** (1965) 23.
14. E. F. Kneller, J. Appl. Phys. **35** (1964) 2210.
15. W. Buckel and R. Hilsch, Z. Phys. **138** (1954) 109; **138** (1954) 118.
16. A. W. Simpson and P. R. Brambley, Phys. Stat. Solidi (b) **43** (1971) 291.
17. R. B. Schwartz and W. L. Johnson, Phys. Rev. Lett. **51** (1983) 415.
18. T. R. Barbee, Jr. and D. L. Keith in ref. 11.
19. See e.g. *Thin Film Processes*, edited by C. J. Vossen and K. Kern, (Academic, New York, 1980).
20. M. M. Collver and R. H. Hammond, Phys. Rev. Lett. **30** (1973) 92.
21. P. Chaudhari, J. J. Cuomo, and R. J. Gambino, IBM J. Res. Dev. **17** (1973) 66.
22. See e.g. *Nuclear and Electron Resonance Spectroscopies Applied to Materials Science*, edited by E. N. Kaufmann and G. Shenoy (Elsevier 1981).
23. See e.g. *Chemical Applications of Mössbauer Spectroscopy*, edited by V. I. Goldanskii and R. H. Herber (Academic, New York, 1968).
24. C. L. Chien, Phys. Rev. **B18** (1978) 1003.
25. V. L. Moruzzi, P. M. Marcus, K. Schwarz, and P. Mohn, Phys. Rev. **B34** (1984) 1784.
26. D. Bagayoko and J. Callaway, Phys. Rev. **B28** (1983) 5419.
27. C. S. Wang, B. M. Klein, and H. Krakaner, Phys. Rev. Lett. **54** (1985) 1852.

28. F. J. Pinski, J. Stunten, B. L. Gyorffy, D. D. Johnson, and G. M. Stocks, Phys. Rev. Lett. **56** (1986) 2096.
29. See e.g. T. Shinjo. Surface Science Rep. **12** (1991) 51, and references therein.
30. G. A. Prinz, Phys. Rev. Lett. **54** (1985) 1051.
31. S. R. Nagel and J. Tauc, Phys. Rev. Lett. **35** (1975) 380.
32. J. D. Bernal, Proc. Roy. Soc. London, Ser. A, **280** (1964) 299.
33. D. E. Polk, Scr. Metall. **4** (1970) 117.
34. F. Kanamaru, S. Miyazaki, M. Shimada, M. Koizumi, K. Oda, and Y. Mimura, J. Solid State Chem. **49** (1983) 1.
35. S. H. Liou and C. L. Chien, Phys. Rev. **B35** (1987) 244.
36. S. H. Liou and C. L. Chien, Mat. Res. Soc. Symp. **80** (1987) 145.
37. T. Egami and Y. Waseda, J. Non-Cryst. Solids, **64** (1984) 113.
38. A. R. Miedema, Z. Metallk. **70** (1979) 345.
39. B. Liu, W. L. Johnson, M. A. Nicolet, and S. S. Lau, Appl. Phys. Lett. **42** (1983) 45.
40. J. H. Hsu and C. L. Chien, Hyper. Int. **69** (1991) 451.
41. P. Duwez and S. C. H. Lin, J. Appl. Phys. **38** (1967) 4096.
42. C. L. Chien and K. M. Unruh, Phys. Rev. **B24** (1981) 1556; **B25** (1982) 5790.
43. K. H. J. Buschow and P. G. van Engen, J. Appl. Phys. **52** (1981) 3557.
44. G. Marchal, Ph. Mangin, and C. Janot, Solid State Commun. 18 (1976) 739.
45. G. Marchal, Ph. Mangin, M. Piecuch, C. Janot, and J. Hubsch, J. Phys. **F7** (1977) L165.
46. Ph. Mangin and G. Marchal, J. Appl. Phys. 49 (1978) 1709.
47. G. Suran, H. Daver, and J. Sztern, Physica **86-88B** (1977) 810.
48. M. Takahashi, C. O. Kim, M. Koshimura, and T. Suzuki, Japan J. Appl. Phys. **17** (1978) 1911.
49. K. H. J. Buschow and P. G. van Engen, Mat. Res. Bull. **16** (1981) 1177.
50. C. L. Chien, D. Musser, E. M. Gyorgy, R. C. Sherwood, H. S. Chen, F. E. Luborsky, and J. L. Walter, Phys. Rev. **B20** (1979) 283.
51. T. Shigematsu, T. Shinjo, Y. Bando, and T. Takada, J. Magn. Magn. Mater. **15-18** (1980) 1367.
52. C. L. Chien, G. Xiao, and K. M. Unruh, Phys. Rev. **B32** (1985) 5582.
53. G. Xiao and C. L. Chien, J. Magn. Magn. Mater. **54-57** (1986) 241.
54. W. A. Bryden, J. S. Morgan, T. J. Kistenmacher, and K. Moorjani, J. Appl. Phys. **61** (1987) 3661.
55. Z. Sullivan and C. L. Chien, to be published.
56. J. W. Essam in *Phase Transition the Critical Phenomena, Vol.2*, edited by C. Domb and M. S. Green (Academic, New York 1972).
57. J. W. Essam, Rep. Prog. Phys. **43** (1980) 833.
58. D. Teirlinck, M. Piecuch, J. F. Geny, G. Marchal, Ph. Mangin, and C. Janot, IEEE Trans. Mag. **MAG-17** (1981) 3079.
59. J. A. Goehegen and S. M. Bhagat, J. Magn. Magn. Mater. **25** (1981) 17.
60. M. Gabey and G. Toulouse, Phys. Rev. Lett. **47** (1981) 201.

61. J. Chappert, T. Abbese-Boggiano, and J. M. D. Coey, J. Magn. Magn. Mater. **7** (1978) 175.

62. H. S. Chen, Phys. Stat. Solidi(a) **17** (1973) 561.

63. R. C. O'Handley, in Ref.3, p.257 and references therein.

64. K. Fukamichi, in Ref.3, p.317 and references therein.

65. G. Xiao and C. L. Chien, J. Appl. Phys. **61** (1987) 3246.

66. J. Friedel, Nuovo Cimento **7 Suppl.** (1957) 287.

67. J. W. Garland and A. Gonis in *Magnetism in Alloys*, edited by P. A. Beck and J. Y. Weber, (American Institute of Mining, Metallurgical and Petroleum Engineering, New York, 1972) 79.

68. G. Marchal, Ph. Mangin, M. Piecuch, and C. Janot, J. Phys. (Paris), Colloq. **37** (1976) C6-763.

69. C. L. Chien in ref.22, p.157.

70. J. J. Hauser, Phys. Rev. **B22** (1980) 2554.

71. H. Hiroyoshi and K. Fukamichi, Phys. Lett. **85A** (1981) 242.

72. Z. Altounian and J. O. Strom-Olsen, Phys. Rev. **B27** (1983) 4149.

73. K. M. Unruh and C. L. Chien, Phys. Rev. **B30** (1984) 4968 .

74. K. H. J. Buschow and P. H. Smit, J. Magn. Magn. Mater. **23** (1981) 85.

75. N. Heiman and N. Kazama, Phys. Rev. **B19** (1978) 1623.

76. K. Fukamichi and R. J. Gambino, IEEE Trans. Mag. **MAG-17** (1981) 3059.

77. S. H. Liou and C. L. Chien, J. Appl. Phys. **55** (1984) 1820 .

78. C. L. Chien, S. H. Liou, B. K. Ha, and K. M. Unruh, J. Appl. Phys. **57** (1985) 3539

79. S. H. Liou, G. Xiao, J. N. Taylor, and C. L. Chien, J. Appl. Phys. **57** (1985) 3536.

80. C. L. Chien, S. H. Ge, S. H. Liou, and G. Xiao, J. Magn. Magn. Mater. **54-57** (1986) 291.

81. M. Lu and C. L. Chien, J. Appl. Phys. **67** (1990) 5787.

82. See e.g. S. J. Poon in Ref.3, p.432.

83. H. Fujimori, K. Nakanishi, H. Hiroyoshi, and N. S. Kazama, J. Appl. Phys. **53** (1982) 7792.

84. Y. Boliang, D. H. Ryan, J. M. D. Coey, Z. Altounian, J. O. Strom-Olsen, and F. Razavi, J. Phys. **F13** (1983) 217.

85. S. M. Fries, C. L. Chien, J. Crummenauer, H. G. Wagner, and U. Gonser, J. Less-Common Metals, **130** (1987) 17 .

86. S. H. Liou, S. H. Ge, J. N. Taylor, and C. L. Chien, J. Appl. Phys. **61** (1987) 3243.

87. N. Heiman and K. Kazama, Phys. Rev. **B17** (1978) 2215.

88. K. H. J. Buschow and N. M. Beckmans, Phys. Rev. **B19** (1979) 3834.

89. R. Krishnan, T. Tarhouni, and M. Tessier, J. Appl. Phys. **53** (1982) 2243.

90. Y. Shimada and H. Kejima, J. Appl. Phys. **53** (1982) 3156.

91. G. Suran, J. Sztern, J. A. Aboaf, and T. R. McGuire, IEEE Trans. Mag, **MAG-17** (1981) 3065 .

92. See e.g. J. G. Wright, IEEE Trans. Mag., **MAG-12** (1976) 95.

93. J. M. D. Coey and D. H. Ryan, IEEE Trans. Mag., **MAG-20** (1984) 1278.

94. N. Cowlam and G. E. Carr, J. Phys. **F15** (1985) 1117.

95. T. Kukunaga, M. Misawa, K. Fukamichi, T. Masumoto, and K. Suzuki in *Rapidly Quenched Metals III, Vol.2*, edited by B, Cantor (Institute of Metals, London, 1978) 325.

96. T. Kemeny, F. J. Litterst, I. Vincze, and R. Wäppling, J. Phys. **F13** (1983) L37.

97. H. Maeda, T. Terauchi, N. Kamijo, M. Hida, and K. Osamura, in *Proc. 4th Int. Conf. Rapidly Quenched Metals*, edited by T. Masumoto and K. Suzuki (Japan Institute of Metals, Sendai, 1982) 397.

98. T. Mizoguchi, S. Yamada, T. Suemasa, J. Nishioka, N. Akutsu, N. Wanatabe, and S. Takayama, in *Proc. 4th Int. Conf. Rapidly Quenched Metals*, edited by T. Masumoto and K. Suzuki (Japan Institute of Metals, Sendai, 1982) 363.

99. H. U. Krebs and H. C. Freyhardt in *Rapidly Quenched Metals*, edited by S. Steeb and H. Warlimont (Elsevier, Amsterdam, 1985) 439.

100. E. F. Kneller, J. Appl. Phys. **33** (1962) 1355.

101. K. Sumiyama and Y. Nakamura, J. Magn. Magn. Mater. **35** (1983) 219.

102. K. Sumiyama and Y. Nakamura, Phys. Stat. Solidi (a) **81** (1984) L109.

103. K. Sumiyama, T. Yoshitake, and Y. Nakamura, J. Phys. Soc.(Japan) **53** (1984) 3160.

104. C. L. Chien, S. H. Liou, D. Kofalt, W. Yu, T. Egami, and T. R. McGuire, Phys. Rev. **B33** (1986) 3247.

105. C. L. Chien, S. H. Liou, G. Xiao, and M. A. Gatzke, MRS Symp. Proc. **80** (1987) 395.

106. J. R. Childress and C. L. Chien, Phys. Rev. **B43**, (1991) 8089.

107. J. R. Childress, C. L. Chien, and M. Nathan, Appl. Phys. Lett. **56** (190) 95.

108. J. R. Childress and C. L. Chien, J. Appl. Phys. **70** (1991) 5885.

109. M. N. Baibich, J. M. Broto, A. Fert, F. Nguyen van Dau, F. Petroff, P. Etienne, G. Creuzet, A. Friederich, and J. Chazeles, Phys. Rev. Lett. **61** (1988) 2472.

110. A. Berkowitz, A. P. Young, J. R. Mitchell, S. Zhang, M. J. Carey, F. E. Spada, F. T. Parker, A. Hutten, and G. Thomas, Phys. Rev. Lett. **68** (1992) 3745.

111. J. Q. Xiao, J. S. Jiang, and C. L. Chien, Phys. Rev. Lett. **68** (1992) 3749.

112. J. Q. Xiao, J. S. Jiang, and C. L. Chien, to appear.

113. T. W. Barbee, Jr., B. E. Jacobson, and D. L. Keith, Thin Solid Films, **63** (1979) 143.

114. J. R. Childress, S. H. Liou, and C. L. Chien, J. Appl. Phys. **64** (1988) 6059.

115. J. R. Childress, S. H. Liou, and C. L. Chien, J. de Phys.(C8) **49** (1988) 113.

116. D. Shechtman, I. Blech, D. Gratias, and J. W. Cahn, Phys. Rev. Lett. **53** (1984) 1951.

117. A. P. Tsai, A. Inoue, and T. Masumoto, Jpn. J. Appl. Phys. **26** (1987) L1505.

118. C. L. Chien and M. Lu, Phys. Rev. **B45** (1992) 12793.

119. P. A. Bancel, P. A. Heiny, P. W. Stephens, A. J. Goldman, and P. M. Horn, Phys. Rev. Lett. **54** (1985) 2422.

120. L. J. Swartzendruber, D. Shechtman, L. Bendersky, and J. W. Cahn, Phys. Rev. **B32** (1984) 1383.

121. M. Eibschütz, H. S. Chen, and J. J. Hauser, Phys. Rev. Lett. **56** (1986) 169.
122. Z. M. Stadnik and G. Stroink, Phys. Rev. **B38** (1988) 10447.
123. N. Kataoka, A. P. Tsai, A. Inoue, T. Masutomo, and Y. Nakamura, Jpn. Appl. Phys. **27** (1988) L1125.
124. Z. M. Stadnik, G. Stroink, H. Ma, and G. Williams, Phys. Rev. **B39** (1989) 9797.
125. See e.g. K. V. Rao in ref.3, p.401.
126. A. P. Tsai, A. Inoue, and T. Masumoto, J. Mat. Sci. Lett. **7** (1988) 322.
127. T. Klein, A. Gozlan, C. Berger, F. Cyrot-Lackmann, Y. Calvayrac, and A. Quivy, Europhys. Lett. **13** (1990) 129.
128. J. C. Phillips and K. M. Rabe, Phys. Rev. Lett. **66** (1991) 923.

CHAPTER FIVE

NEUTRON SCATTERING STUDIES OF THE
SPIN DYNAMICS OF AMORPHOUS ALLOYS

Jeffrey W. LYNN

Reactor Radiation Division,
National Institute of Standards and Technology,
Gaithersburg, MD 20899

Jaime A. FERNANDEZ-BACA

Solid State Division,
Oak Ridge National Laboratory,
Oak Ridge, TN 37831

1. Introduction

A fluid system has a well defined density, which means we know how many atoms are contained in any given volume. This simple fact requires that the material has a well defined average nearest-neighbor distance, well-defined average number of nearest neighbors, and well defined lattice-dynamical excitations (sound waves) at sufficiently long phonon wavelengths. Individual atoms, on the other hand, are mobile and thus are free to diffuse in the system. An amorphous material or glass can then be thought of as such a fluid state that has been rapidly quenched to low temperatures in order to freeze-in the disorder and prohibit the atoms from any further diffusion in the system. In covalently bonded glasses such as Si or Ge there is strong short range order that is maintained in the liquid and amorphous phases in order to accommodate the tetrahedral bonding arrangements that are strongly favored energetically. There are few such materials, however, that are magnetic, and these materials are not the subject of the present review.[1] Metallic systems, on the other hand, can be best described as a dense random packing of hard spheres. The nature of the cooperative magnetic states and their corresponding spin excitations in such metallic magnets has been the subject of many investigations in the last two decades due both to the wide range of interesting physics these systems exhibit and to the broad range of applications for these materials.

In the present chapter we review the experimental work that has been carried out to explore the nature of the magnetic excitations in metallic glasses. We first note that a fluid or glass is by definition *globally* isotropic, by which we mean that there is no preferred direction in the material. This requires that the energies of the excitations, whether they be lattice dynamical, electronic, or magnetic in origin, depend only on the magnitude q of the wave vector $\mathbf{q}$. This also means that since there are no periodicities in the system, Bloch's theorem cannot be applied, and the wave vectors for all excitations must be referred to the origin of reciprocal space. In the conventional notation this means that the reciprocal lattice vector $\boldsymbol{\tau} \equiv (0,0,0)$ and then $Q = q$. The classification of amorphous magnets then falls into two general categories, depending on the nature of the *local* anisotropy. In rare earth systems, for example, the single-site crystal field anisotropy is large compared with the exchange energy and varies randomly in magnitude and direction from site to site. This has the effect of severely disrupting the exchange energies in the material and consequently lowering the effective interactions so the system is below the lower marginal dimensionality, eliminating the possibility of any long range ordering phenomena. The magnetic properties for this class of materials are found to be typical of spin glass systems, and they have proved to be interesting examples of random field systems. However, in the last decade there has been little new work addressing the dynamics of such materials, and thus the reader is referred elsewhere for a review of their properties.[2]

The case of primary interest then is when the anisotropy is weak compared to the exchange interaction, which is typically the case for most transition metal based glasses. In such locally isotropic materials long range ferromagnetic order can develop, and indeed such materials now provide the best known examples of isotropic ferromagnets. For any isotropic ferromagnet the nature of the low temperature excitations at long wavelengths is well established; we define an average magnetization density and apply linear spin wave theory. In this hydrodynamic or continuum limit the small wave vector excitations are the familiar spin waves or magnons, and as we will see spin wave theory works well over a surprisingly wide range of temperatures. Hydrodynamic theory can also be used effectively in the vicinity of the Curie temperature T_C, and we anticipate that the critical dynamics will follow the conventional behavior for the universality class of a three dimensional $(d = 3)$ isotropic $(n = 3)$ ferromagnet. Nevertheless, the weak anisotropy can still play an important role and give rise to phenomena such as "wandering axis" ferromagnetism and transverse freezing of the spin components, and we will review the work along these lines. At larger $\mathbf{Q}$, however, the inherent randomness of the system must dominant the physics. This will change the nature of the excitations, and generally one must think of these in terms of a general density of excitations rather than in terms of well defined excitations which obey a dispersion relation.

One of the noteworthy advantages of amorphous systems is that they can generally be fabricated over a wider range of compositions than the corresponding crystalline systems since thermodynamic equilibrium is not a requisite. This allows one to better control and tailor the magnetic properties, both in the investigation of physical phenomena of interest as well as when developing a material for a specific application. Fabrication techniques such as rapid quenching from the melt tend to minimize clustering and chemical short range order, which often is a problem in crystalline alloy systems, and this can simplify the interpretation of measurements and thereby clarify the underlying physics of interest. Hence amorphous magnets have been chosen for a variety of studies as a function of magnetic concentration, in particular in the low concentration regime where the network of spins is highly ramified and long range order is becoming disrupted. This leads to phenomena such as reentrant ferromagnetic order, frustration, transverse freezing, and spin glass behavior.

In this chapter we will first review the neutron scattering techniques as they apply to measurements of the dynamics of these materials. We will then review the nature of the magnetic excitations in well ordered isotropic amorphous ferromagnets, and finally we will survey the interesting effects that develop when such materials are magnetically diluted.

2. Neutron scattering

Neutron inelastic scattering is the ideal probe of the dynamical behavior of magnetic systems. This technique provides a direct measurement of the imaginary part of the generalized susceptibility $\chi(\mathbf{Q}, \omega)$, which contains all the information of the spin correlations of the system under study. The use of neutron scattering to study the spin excitations of magnetic systems has been reviewed extensively [2–7] and in this chapter we will only summarize the most relevant features of this technique. In section (2.1) we will describe the neutron scattering cross sections that are measured in the experiments that use unpolarized neutrons. In section (2.2) we describe the cross sections for polarized neutron scattering, and in section (2.3) we discuss some issues specific to the scattering of neutrons from amorphous systems.

2.1. Unpolarized neutron scattering cross sections

The partial differential cross section for the scattering of unpolarized neutrons from a system of unpaired electron spins, located at the coordinates $\mathbf{r}_n$, is [4] :

$$\frac{d^2\sigma}{d\Omega d\omega} = \left(\frac{\gamma e^2}{m_e c^2}\right)^2 [\tfrac{1}{2}g\, f(\mathbf{Q})]^2 \frac{k_f}{k_i} \sum_{\alpha,\beta}(\delta_{\alpha\beta} - \hat{Q}_\alpha \hat{Q}_\beta)\, S_{\alpha\beta}(\mathbf{Q}, \omega) \ , \qquad (5.1)$$

where the scattering function $S_{\alpha\beta}(\mathbf{Q}, \omega)$ is the space and time Fourier transform of the time-dependent spin pair correlation function, *i.e.*

$$S_{\alpha\beta}(\mathbf{Q}, \omega) = \frac{1}{2\pi} \int dt \, exp(-i\omega t) \sum_{m,n} exp[i\mathbf{Q} \cdot (\mathbf{r}_m - \mathbf{r}_n)] \langle S_\alpha(\mathbf{r}_m, 0) S_\beta(\mathbf{r}_n, t) \rangle \quad . \quad (5.2)$$

In the above equations k_i and k_f are the initial and final wave vectors of the neutrons, $\left(\frac{\gamma e^2}{m_e c^2}\right)^2 = 0.291$ barns is the coupling constant of the neutron to the unpaired electron spins; g is the Landé factor and $f(\mathbf{Q})$ is the magnetic form factor which is related to the Fourier transform of the radial part of the electronic wave function. The scattering wave vector $\mathbf{Q}$ and the energy transfer $\hbar\omega$ are related by the conservation laws

$$\mathbf{Q} = \mathbf{k}_i - \mathbf{k}_f \, ,$$
$$\hbar\omega = \frac{\hbar^2}{2m_n} (k_i^2 - k_f^2) \, , \qquad\qquad (5.3)$$

and $\hat{\mathbf{Q}} = \mathbf{Q}/|\mathbf{Q}|$. In the above equations the indices α and β denote Cartesian coordinates, thus $\hat{Q}_\alpha$ and S_α are the α components of the unit vector $\hat{\mathbf{Q}}$ and the magnetic moment $\mathbf{S}$.

The essential information contained in Eq. (5.1) is that neutrons scattered with momentum and energy transfer $\mathbf{Q}$ and $\hbar\omega$ directly measure a single Fourier component of the spin-pair correlation function. In this chapter we are mainly concerned with dynamical processes and for this reason we will use $S_{\alpha\beta}(\mathbf{Q}, \omega)$ to denote the inelastic ($\hbar\omega \neq 0$) part of the scattering function, unless specified otherwise. By using the fluctuation-dissipation theorem we can also relate $S_{\alpha\beta}(\mathbf{Q}, \omega)$ to the imaginary part of the generalized susceptibility:

$$S_{\alpha\beta}(\mathbf{Q}, \omega) = [n(\omega) + 1] \, Im \, \chi_{\alpha\beta}(\mathbf{Q}, \omega) \, , \qquad\qquad (5.4)$$

where $n(\omega) = [exp(\hbar\omega/kT) - 1]^{-1}$ is the Bose thermal population factor. Thus, as noted at the beginning of this section, the neutron scattering technique provides a direct measurement of $Im \, \chi_{\alpha\beta}(\mathbf{Q}, \omega)$. Another important feature is that $S_{\alpha\beta}(\mathbf{Q}, \omega)$ can also be expressed in terms of the isothermal susceptibility $\chi_{\alpha\beta}(\mathbf{Q})$ and the spectral weight function $F_{\alpha\beta}(\mathbf{Q}, \omega)$, *i.e.*

$$S_{\alpha\beta}(\mathbf{Q}, \omega) = [n(\omega) + 1] \, \omega \, \chi_{\alpha\beta}(\mathbf{Q}) \, \mathbf{F}_{\alpha\beta}(\mathbf{Q}, \omega) \quad . \qquad\qquad (5.5)$$

The isothermal susceptibility $\chi_{\alpha\beta}(\mathbf{Q})$ is the equilibrium response function of the αth component of the magnetization to an external magnetic field in the direction β with an spatial dependence of the form $H[exp(i\mathbf{Q} \cdot \mathbf{r})]$. The spectral weight function $F_{\alpha\beta}(\mathbf{Q}, \omega)$ is the normalized temporal Fourier transform of the magnetization response when the same external field is suddenly switched off. This function describes the shape of the scattering as a function of energy, and satisfies the normalization condition

$$\int_{-\infty}^{\infty} F_{\alpha\beta}(\mathbf{Q}, \omega) d\omega = 1 \quad . \qquad\qquad (5.6)$$

When the total z-component of the spin is a constant of the motion, as in the case of a Heisenberg ferromagnet, the terms in the cross section with $\alpha \neq \beta$ vanish and Eq. (5.1) simplifies to:

$$\frac{d^2\sigma}{d\Omega d\omega} = \left(\frac{\gamma e^2}{m_e c^2}\right)^2 [\tfrac{1}{2} g \, f(\mathbf{Q})]^2 \frac{k_f}{k_i}$$
$$\times \left\{ (1 + \hat{Q}_z^{\,2}) \, S_{xx}(\mathbf{Q},\omega) + (1 - \hat{Q}_z^{\,2}) \, S_{zz}(\mathbf{Q},\omega) \right\} \ . \tag{5.7}$$

This cross section is the sum of two terms, which we denote as transverse (*i.e.* perpendicular to the z-axis) and longitudinal (*i.e.* along the z-axis). The transverse term corresponds to the creation and annihilation of (transverse) magnons or spin waves, and is the term in which we are mostly interested. In particular, the transverse term of the scattering function is

$$S_{xx}(\mathbf{Q},\omega) = [n(\omega) + 1] \, \omega \, \chi_{\alpha\beta}(\mathbf{Q}) \, F_{xx}(\mathbf{Q},\omega) \ . \tag{5.8}$$

Unfortunately, linear response theory does not predict the shape of the spectral weight function $F_{xx}(\mathbf{Q},\omega)$. For a periodic lattice at low temperatures, where the spin-wave interactions are negligible, the spin waves are approximate eigenstates of the system and we have the familiar result

$$F_{xx}(\mathbf{Q},\omega) = \frac{1}{2}[\delta(\hbar\omega - \hbar\omega_{\mathbf{Q}}) + \delta(\hbar\omega + \hbar\omega_{\mathbf{Q}})] \ , \tag{5.9}$$

where δ is Dirac's delta function and $\hbar\omega_{\mathbf{Q}}$ is the spin-wave energy. The two terms in this spectral weight function correspond to the creation and annihilation of spin-wave excitations, while the longitudinal term of the cross section is expected to lead to elastic or quasielastic scattering only.

If we introduce disorder such as for an amorphous alloy, or elevate the temperature, then intrinsic lifetimes to the excitations will be introduced and this needs to be taken into account in the spectral weight function. $F_{xx}(\mathbf{Q},\omega)$ can often be approximated by a double Lorentzian function, where one Lorentzian represents the creation of a spin wave of energy $E_{\mathbf{Q}} = \hbar\omega_{\mathbf{Q}}$, and the other represents the annihilation of a spin wave excitation at $-E_{\mathbf{Q}}$. $F_{xx}(\mathbf{Q},\omega)$ can then be written as [8]

$$F_{xx}(\mathbf{Q},\omega) = \frac{1}{2\pi} \left\{ \frac{2\Gamma}{4[E - E_{\mathbf{Q}}]^2 + \Gamma^2} + \frac{2\Gamma}{4[E + E_{\mathbf{Q}}]^2 + \Gamma^2} \right\} \ . \tag{5.10}$$

When $\frac{\Gamma}{E_{\mathbf{Q}}} \ll 1$ we have well-defined spin wave excitations, with Γ the full width at half maximum (FWHM) intrinsic linewidth of the spin wave, which is simply related to the inverse of the excitation lifetime. An alternate form for the spectral weight function is the damped harmonic oscillator (DHO) [8]:

$$F_{xx}(\mathbf{Q},\omega) = \frac{1}{\pi} \frac{\Gamma E_{\mathbf{Q}}^{\,2}}{[E^2 - E_{\mathbf{Q}}^2]^2 + (\Gamma E)^2} \ . \tag{5.11}$$

The interpretation of the DHO form is relatively straightforward.[8] For the mechanical oscillator $\omega_\mathbf{Q}$ is the natural (undamped) frequency of oscillation ($\propto \sqrt{\frac{k}{m}}$)and for the magnetic system $\omega_\mathbf{Q}$ simply represents the bare exchange energy between two spins. $E_\mathbf{Q}$ is then just the bare (unrenormalized) spin wave energy. For small damping Γ again is the FWHM linewidth of the spin wave. For the mechanical oscillator the actual frequency of oscillation decreases with increasing damping Γ, and for the magnetic system the spin wave energy, as determined by the real part of the poles of the spectral weight function, is given by

$$E_{sw} = \sqrt{E_\mathbf{Q}^2 - \left(\frac{\Gamma}{2}\right)^2} . \tag{5.12}$$

Typically the damping Γ increases with increasing temperature, causing a renormalization of the spin waves to lower energies along with a broadening of the linewidth. For sufficiently large damping the spin wave energy E_{sw} vanishes leading to spin diffusion, characterized by the scattering having a maximum at $E = 0$. However for the present review we will not need to consider the overdamped case when $\frac{\Gamma}{2E_\mathbf{Q}} \geq 1$; details about the critical dynamics can be found elsewhere.[2]

Finally, the static part of the scattering function (Eq. (5.2)) can be written as:

$$S_{\alpha\beta}(\mathbf{Q}, \omega = 0) = \delta(\omega) S_{\alpha\beta}(\mathbf{Q}) , \tag{5.13}$$

where $S_{\alpha\beta}(\mathbf{Q})$ is the infinite-time (*i.e.* static) spin pair correlation function

$$S_{\alpha\beta}(\mathbf{Q}) = \sum_{m,n} e^{[i\mathbf{Q}\cdot(\mathbf{r}_m - \mathbf{r}_n)]} \langle S_\alpha(\mathbf{r}_m, 0) S_\beta(\mathbf{r}_n, \infty) \rangle . \tag{5.14}$$

Generally this corresponds to the magnetic structure of the system, which for crystalline materials is the magnetic Bragg peaks. For amorphous magnets this is just the conventional magnetic structure factor, analogous to the nuclear structure factor. It is also of interest to note that the elastic part of the scattering function provides, in principle, a direct measurement of the Edwards–Anderson order parameter [9] $q_{EA} = \overline{\langle S \rangle}^2$ of a spin glass, where the bar denotes a configurational or site average.

2.2. *Polarized neutron cross sections*

Most of the instruments that are used in neutron scattering experiments do not have the capability of distinguishing between the polarization states of the incident and scattered neutrons. For this reason the cross sections that are normally quoted in the literature are those for "unpolarized neutrons", like the one of Eq. (5.1), in which a summation has been performed over the initial and final spin states of the neutrons. There is some valuable information, however, that is lost when such a summation is performed. In some particular applications it is advantageous to make use of the spin polarization dependence of the neutron scattering cross section to separate the magnetic from non-magnetic scattering, the transverse from the longitudinal spin excitations, etc. This is possible with the "Polarization Analysis" technique which is fully described in the classic paper by Moon, Riste and Koehler.[10] In this technique the sample is maintained in a magnetic field which defines the direction of the polarization for both incident and scattered neutrons. By the use of polarizing crystals and "spin flipper" devices one can conveniently measure four partial cross sections, two of which involve spin-flip (SF) scattering and two which involve non-spin-flip (NSF) scattering. The SF scattering involves the reversal of the spin direction of the incident neutrons from a $|+\rangle$ state to a $|-\rangle$ state, or from a $|-\rangle$ state to a $|+\rangle$ state, while the NSF scattering preserves the $|+\rangle$ or $|-\rangle$ polarization of the incident neutrons. The SF cross sections are usually denoted as $+-$ and $-+$, while the NSF cross sections are denoted as $++$ and $--$. More general types of spin rotations can also be employed, such as in the spin-echo spectrometer or neutron spin polarimetry, but these techniques have not been used in the investigation of amorphous magnets and thus they will not be discussed in this review.[5]

The general cross section for a scattering process in which the neutron spin changes from state $|s\rangle$ to $|s'\rangle$ while the scattering system goes from state $|q\rangle$ to $|q'\rangle$ is [10, 4, 6]

$$\frac{d^2\sigma^{ss'}}{d\Omega d\omega} = \sum_q P_q \sum_{q'} \frac{k_f}{k_i} |\langle q' | \sum_m e^{i\mathbf{Q}\cdot\mathbf{r}_m} U_m{}^{ss'} |q\rangle|^2 \times \delta(\Delta E_n + \Delta E_q) \ . \tag{5.15}$$

In this equation P_q is the probability that the system is in the initial state q and $U_m{}^{ss'}$ is the effective atomic scattering amplitude that describes the processes in which the neutron goes from the spin state $|s\rangle$ to $|s'\rangle$. If we neglect the effects of the nuclear spin $U_m{}^{ss'}$ can be written as [10] :

$$U_m{}^{ss'} = \langle s' | b_m - p_m \mathbf{S}_{\perp m} \cdot \sigma | s \rangle \ . \tag{5.16}$$

In Eq. (5.16) b_m is the nuclear coherent scattering amplitude, p is the magnetic amplitude

$$p = \left(\frac{\gamma e^2}{m_e c^2}\right) [\tfrac{1}{2} g\, f(\mathbf{Q})] \ , \tag{5.17}$$

$\mathbf{S}_\perp$ is the magnetic interaction operator

$$\mathbf{S}_\perp = \mathbf{S} - (\mathbf{S} \cdot \hat{\mathbf{Q}})\hat{\mathbf{Q}} \ , \tag{5.18}$$

σ is the neutron spin operator and $\mathbf{S} = S_x\hat{\mathbf{x}} + S_y\hat{\mathbf{y}} + S_z\hat{\mathbf{z}}$ is the atomic spin vector. The relevance of $\mathbf{S}_\perp$ in Eq. (5.16) is that only the atomic spin component that is perpendicular to the scattering wave vector $\mathbf{Q}$ is effective in scattering neutrons.

For the four partial cross sections mentioned above, the atomic scattering amplitudes are given by [10]

$$U^{++} = b - pS_{\perp z}$$
$$U^{--} = b + pS_{\perp z}$$
$$U^{+-} = -p(S_{\perp x} + iS_{\perp y})$$
$$U^{-+} = -p(S_{\perp x} - iS_{\perp y}) \ , \tag{5.19}$$

In these expressions, z refers to the direction of the neutron polarization determined by the magnetic field on the sample. In the particular case of a ferromagnetic sample that is magnetized to saturation, the magnetization will also be parallel to the neutron polarization $\mathbf{P}$, and along the z-axis direction. As mentioned in section 2.1 the cross section for the scattering of neutrons by spin waves is given by the correlations between the S_x and S_y (*i.e.* transverse) components of the spins on the different sites, and we need to consider only those components of $\mathbf{S}$ when evaluating the magnetic interaction operator. Thus for (transverse) ferromagnetic spin waves $\mathbf{S}_\perp$ is:

$$\mathbf{S}_\perp = S_x[\hat{\mathbf{x}} - (\hat{\mathbf{x}} \cdot \hat{\mathbf{Q}})\hat{\mathbf{Q}}] + S_y[\hat{\mathbf{y}} - (\hat{\mathbf{y}} \cdot \hat{\mathbf{Q}})\hat{\mathbf{Q}}] \ . \tag{5.20}$$

For arbitrary orientations of $\hat{\mathbf{z}}$ and $\mathbf{Q}$, $\mathbf{S}_\perp$ has a NSF component along the $\hat{\mathbf{z}}$-vector and a SF component transverse to $\hat{\mathbf{z}}$. There are two cases of particular interest,which correspond to experimental configurations in which (1) $\mathbf{Q}$ is parallel to $\hat{\mathbf{z}}$, (*i.e.* $\mathbf{Q} \parallel \mathbf{P}$) and (2) $\mathbf{Q}$ is perpendicular to $\hat{\mathbf{z}}$ (*i.e.* $\mathbf{Q} \perp \mathbf{P}$). When $\mathbf{Q}$ is parallel to $\hat{\mathbf{z}}$ (*i.e.* $\mathbf{Q} \parallel \mathbf{P}$) we get:

$$\mathbf{S}_\perp = S_x\hat{\mathbf{x}} + S_y\hat{\mathbf{y}} \ , \tag{5.21}$$

i.e. $S_{\perp x} = S_x$, $S_{\perp y} = S_y$, $S_{\perp z} = 0$ and the magnetic contributions to the NSF amplitudes of Eqs. (5.19) vanish, while the corresponding SF amplitudes are the familiar spin raising and lowering operators $S^\pm = S_x \pm iS_y$. Replacing these in Eq. (5.15) we obtain the SF cross sections for this experimental configuration:

$$\frac{d^2\sigma^{\pm\mp}}{d\Omega d\omega} = \left(\frac{\gamma e^2}{m_e c^2}\right)^2 [\tfrac{1}{2}g\,f(\mathbf{Q})]^2 \sum_q P_q$$
$$\times \sum_{q'} \frac{k_f}{k_i} |\langle q'| \sum_m e^{i\mathbf{Q}\cdot\mathbf{r}_m} S^\pm_m |q\rangle|^2 \times \delta(\Delta E_n + \Delta E_q) \ . \tag{5.22}$$

These SF cross sections correspond to the creation (S^+) and annihilation (S^-) of spin waves in a Heisenberg ferromagnet. The $\delta(\Delta E_n + \Delta E_q)$ in this expression indicates the energy conservation in the system: the creation of spin waves is achieved with neutron energy loss, while the annihilation of spin waves results in neutron energy gain.

When the neutron polarization is perpendicular to $\mathbf{Q}$ (*i.e.* $\mathbf{Q} \perp \mathbf{P}$), we can select a coordinate system where the y axis is along $\mathbf{Q}$. In this way we get:

$$\mathbf{S}_\perp = S_x\hat{\mathbf{x}} + S_z\hat{\mathbf{z}} \ , \tag{5.23}$$

i.e. $S_{\perp x} = S_x$, $S_{\perp y} = 0$, $S_{\perp z} = S_z$. The SF amplitudes are then $U^{+-} = U^{-+} = -pS_x = (S^+ + S^-)/2$ and the SF cross sections are:

$$\frac{d^2\sigma^{\pm\mp}}{d\Omega d\omega} = \left(\frac{\gamma e^2}{m_e c^2}\right)^2 [\tfrac{1}{2}g\,f(\mathbf{Q})]^2 \sum_q P_q$$

$$\times \sum_{q'} \frac{1}{2}\frac{k_f}{k_i}|\langle q'|\sum_m e^{i\mathbf{Q}\cdot\mathbf{r}_m}(S^+{}_m + S^-{}_m)|q\rangle|^2 \times \delta(\Delta E_n + \Delta E_q) \ . \qquad (5.24)$$

These SF cross sections involve both the creation and annihilation of spin waves. The longitudinal component of the magnetic fluctuation spectrum, on the other hand, is contained in the NSF amplitudes as shown explicitly in Eqs. (5.19), and this will give a NSF cross section equivalent to Eq. (5.24) which contains the S_z operator. Thus using the two experimental configurations mentioned above the polarization analysis technique allows us to separate the transverse spin excitations (SF cross sections) from the longitudinal spin excitations and nonmagnetic excitations (NSF cross sections).

2.3. Neutron scattering from amorphous alloys

In general the scattering wave vector $\mathbf{Q}$ defined in Eq. (5.3) can be expressed as $\mathbf{Q} = \boldsymbol{\tau} + \mathbf{q}$, where $\boldsymbol{\tau}$ is a reciprocal lattice vector and $\mathbf{q}$ is the reduced wave vector. This is a consequence of Bloch's theorem which requires that the energy of any (wave-like) eigenstate in a periodic lattice must be periodic in $\boldsymbol{\tau}$ and thus the relevant excitation wave vector is $\mathbf{q}$. Amorphous systems, however, do not have a reciprocal lattice and the only $\boldsymbol{\tau}$ available is $\boldsymbol{\tau} = 0$, with $\mathbf{Q} = \mathbf{q}$. This affects not only the nature of the excitations in these systems, but also the manner in which the scattering experiments are conducted. In particular, for amorphous systems only the long wavelength (small q) excitations are well defined, and these must be measured at small scattering angles, about the forward direction of the incident neutron beam. The experimental situation is shown in Fig. 5.1. On the left is a typical scattering diagram for measuring an inelastic excitation at a small reduced wave vector $\mathbf{q}$. These measurements can be carried out around any convenient reciprocal lattice vector. For an amorphous material where we must use $\boldsymbol{\tau} = 0$, on the other hand, $\mathbf{Q}$ $(= \mathbf{q})$ and the corresponding scattering angle 2θ are quite small. The energy transfer as given by Eq. (5.3) is directly related to the difference in length of $\mathbf{k_i}$ and $\mathbf{k_f}$, and this difference is quite restricted because $\mathbf{q}$ is small; the best we can do is chose $2\theta = 0$, where $\mathbf{k_i}$, $\mathbf{k_f}$ and $\mathbf{q}$ are collinear and we have the scalar relation $k_i = k_f \pm q$. At these small q's the maximum energy transfer that satisfies the conservation relations [Eq. (5.3)] is given by

$$E_{max} = \pm 2Cqk_i \ , \qquad (5.25)$$

where $C = \hbar^2/2m_n = 2.0719$ meV-Å^2. For a given incident k_i this defines the range of energies accessible to measurement. For example, with a typical value of $k_i = 2.6$ Å^{-1} (corresponding to 14 meV neutrons) and a magnon wave vector $q = 0.1$ Å^{-1} we have

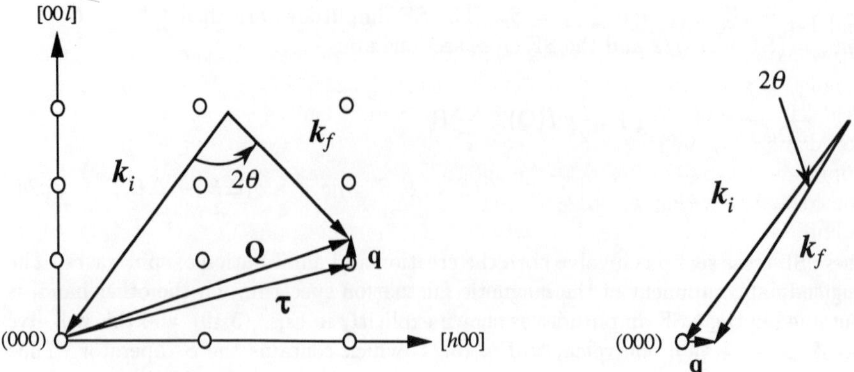

Fig. 5.1: Scattering diagram for measuring a small q excitation in a material. The schematic at the left is for a crystalline system measured around a reciprocal lattice point, while the schematic at the right is for an amorphous system.

$E_{max} = \pm 1.07$ meV. This presents a number of experimental challenges. Firstly, it is clear that the instrumental resolution must be much better than 1 meV in order to make reliable determinations of the spin wave energies and linewidths. The accessible energy range as determined by Eq. (5.25) can be increased by increasing k_i but this usually results in rapidly broadening the instrumental resolution. Secondly the restricted energy range means that determining the appropriate "background" of the inelastic spectra is often difficult. Finally, very tight angular collimation must be employed before and after the sample since the measurements are done very close to the incident beam. There are, however, some advantages that come with these types of measurements; the magnetic form factor is almost unity, so that the magnetic scattering is at its strongest, and phonon scattering ($\propto Q^2$) is negligible.

Because of the restricted range of energies in these types of measurements, the elastic region ($E = 0$) is included in most experiments. The elastic component of the spectrum originates from nuclear incoherent scattering from the sample and environment, and from any magnetic scattering that may develop in the ordered state. The scattering function $S_{\alpha\beta}(\mathbf{q})$ of this magnetic component is given by Eq. (5.13). Thus in general the neutron scattering cross section will consist of two terms:

$$\frac{d^2\sigma}{d\Omega d\omega} = \left(\frac{\gamma e^2}{m_e c^2}\right)^2 [\tfrac{1}{2} g\, f(\mathbf{Q})]^2 \frac{k_f}{k_i} \times \left(\frac{A}{q^2}\,[n(\omega)+1]\,\omega\,F_{xx}(\mathbf{Q},\omega) + B\,\delta(\omega)\right) \quad , \quad (5.26)$$

where A is a parameter proportional to the transverse part of the susceptibility, and B is the strength of the elastic component.

Finally we mention some of the additional considerations that are needed when carrying out polarized beam inelastic scattering measurements at small q in these amorphous systems. The standard technique is to employ the relations in Eq. (5.22) and Eq. (5.24), that is, measure the scattering with $\mathbf{Q} \parallel \mathbf{P}$ and with $\mathbf{Q} \perp \mathbf{P}$. In ferromagnets, however, the neutron polarization $\mathbf{P}$ will generally follow the internal magnetization in the sample, and *not* necessarily the direction of the applied magnetic field. In general then we must apply a field sufficient in magnitude to sweep out the domain boundaries, and we must have a ferromagnetic sample that is of an appropriate geometry so that the internal field is uniform in direction. A second consideration is that at small $\mathbf{Q}$ the direction of $\mathbf{Q}$ changes substantially over the volume of the instrumental resolution, and the conditions $\mathbf{Q} \parallel \mathbf{P}$ or $\mathbf{Q} \perp \mathbf{P}$ are only fulfilled at the center of the resolution ellipsoid. Hence significant additional resolution corrections need to be applied to the data in the polarized beam case.[11]

3. Spin dynamics of isotropic ferromagnets

We start our review of the magnetic scattering in amorphous systems by considering materials which are magnetically concentrated and develop well defined ferromagnetic order below the Curie temperature T_C. At long wavelengths (small q) where we can simply consider an average magnetization density and apply hydrodynamic theory, the excitations in the ordered state are conventional spin waves. Since typically they are sharp excitations it is relatively straightforward to make measurements of the energies and linewidths as a function of temperature, and it is not too surprising then that most of the research has concentrated in this regime. The theory is well developed and the experimental results can be readily compared with results from their crystalline counterparts.

3.1. Small q

The spin dynamics of isotropic ferromagnets at modest temperatures is known to be well described by linear spin wave theory. In the long wavelength (small q) hydrodynamic limit there is a "Goldstone mode" with dispersion relation given by the familiar expression $E_{sw} = D(T)q^2$.[12, 13] The meaning of Goldstone's theorem as it applies to this problem is the following.[14] We assume first that the system is truly isotropic in the paramagnetic state. As the temperature is lowered through the Curie temperature this symmetry is broken, and a unique direction is spontaneously defined, the direction of the magnetization. However, this direction is chosen simply by chance, and if we raise the temperature above T_C and lower it again the system will choose another direction, spontaneously breaking the (continuous) rotational symmetry. Another way to view this situation is that it takes no energy whatsoever to rotate *all* the spins together, defining a new magnetization direction. This infinite wavelength spin wave mode is then gapless by definition, and consequently we must have an excitation whose dispersion relation obeys $E \to 0$ as $q \to 0$, which is Goldstone's theorem.

232

Most of the transition-metal amorphous magnets are in fact excellent approximations of isotropic ferromagnets. More generally we can write the dispersion relation
in the small wave vector regime as

$$E_{sw} = \Delta + D(T)q^2 + E(T)q^4 + \cdots \tag{5.27}$$

where Δ is a (small) gap due to anisotropy and/or pseudodipolar interactions, D is
the spin wave "stiffness constant", and the rest of the terms are higher-order terms
in a power law expansion. The quantitative value of the stiffness constant D depends
on the details of the interactions and the nature of the magnetism, such as whether
the magnetic electrons are localized or itinerant, or the structure is amorphous or
crystalline, but the general form of the spin wave dispersion relation, and hence the
spin wave density of states, is invariant. At small wave vectors the dispersion is
dominated by the quadratic term, and in fact for amorphous systems there is no
definitive evidence for the need of any higher-order terms from neutron scattering
experiments. Knowing the magnetic density of states we can then calculate the bulk
magnetization. The leading order temperature dependence to the magnetization is
given in spin wave theory by

$$M(T) = M(0)[1 - BT^{3/2} + \cdots] \tag{5.28}$$

where the coefficient B is related to the spin wave dispersion relation by

$$B = \frac{2.612 g\mu_B}{M(0)} \left(\frac{k}{4\pi D}\right)^{\frac{3}{2}} . \tag{5.29}$$

This relationship assumes that the spin waves are the only excitations in the system, and of course it is obeyed very well for a wide variety of both crystalline and
amorphous magnets (with the exception of Invar materials as discussed below). A
measurement of the spin wave dispersion relation can then be directly related to the
bulk magnetization, and vice versa.

An example of measurements of the spin waves are shown in Fig. 5.2 for the
$Fe_{86}B_{14}$ amorphous alloy.[15, 16] This is obviously a high magnetic concentration
system, with only 14% boron glass former. The three scans are for three different
values of the momentum transfer, 0.08, 0.10, and 0.12 Å^{-1}. There is a peak at the
elastic scattering position, $E = 0$ (Eq. 5.13), which originates from nuclear incoherent
scattering and static magnetic scattering, while the peaks on either side are the spin
wave excitations. The spin waves are found both on the energy gain ($E < 0$) side,
where the neutron gains energy as it destroys a (quantum) spin wave excitation, and
on the energy loss ($E > 0$) side where the neutron loses energy as it creates a spin
wave in the material. Note that the probability for creation or annihilation of a spin
wave is given by $n(\omega)$ for energy gain, and $1 + n(\omega)$ for energy loss. Thus if these data
were taken at $T = 0$ then there would be no peak on the left and a much weaker peak
on the right; at this high temperature we obviously have $kT \gg \hbar\omega$ (recall that 1 meV

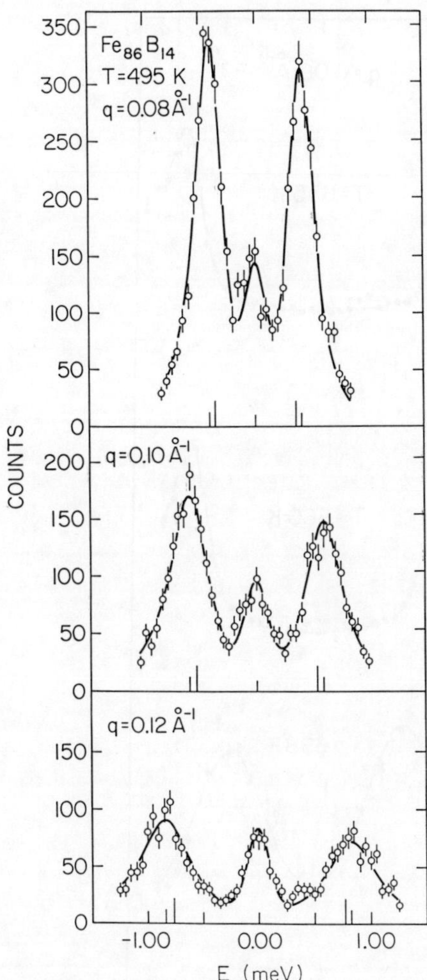

Fig. 5.2: Spin wave excitations observed at 495 K in Fe$_{86}$B$_{14}$ in energy gain ($E < 0$) and energy loss ($E > 0$) for three different values of q. The solid curves at least-squares fits of the instrumental resolution convoluted with spin wave cross sections (after ref. [15])

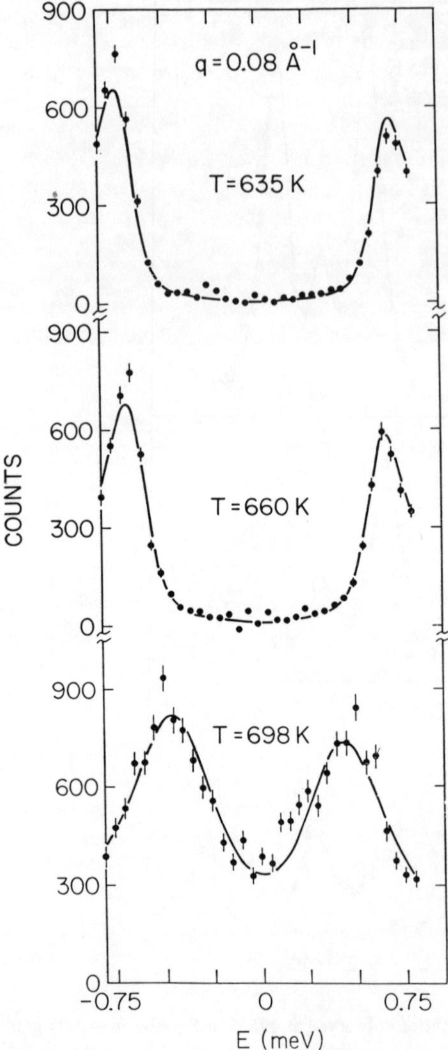

Fig. 5.3: Observed spin wave spectrum at 0.08Å^{-1} for three different temperatures in $Fe_{78}B_{13}Si_9$, which has a Curie temperature of 710 K.[17] The solid curves are least-squares fits to spin wave theory. Note that for this material there is no observable elastic peak.

$\rightarrow$ 11.605 K) and the probabilities are essentially equal. Note also that the scans do not extend far enough in energy to get down to background, and this is due to the kinematic restrictions (Eq. 5.25). Finally we note that with increasing q the peaks shift to higher energy as expected on the basis of Eq. (5.27), broaden, and decrease in intensity due to the decrease in $n(\omega)$. The solid curves are the fits to spin wave theory assuming a Lorentzian linewidth as given in Eq. (5.10).

The spin wave energies and linewidths are also temperature dependent, especially as one approaches the ordering temperature from below, as shown in Fig. 5.3 for the $Fe_{78}B_{13}Si_9$ material at a momentum transfer of 0.08 Å^{-1}.[17] At 635 K the kinematic range of the instrument is barely sufficient to observe the spin wave peaks, and measurements at lower T became increasingly more difficult to obtain. The solid curves are again least-squares fits to the spin wave theory convoluted with the instrumental resolution. Note that for this material the elastic scattering is too weak to be observable, in contrast to the case of most amorphous as well as crystalline alloys. At 660 K the spin waves are measurably shifted to lower energy, while at 698 K, which is close to the Curie temperature of $T_C = 710K$, the energy is clearly much lower and the widths have obviously broadened due to spin-wave – spin-wave interactions.

Data such as shown in the previous two figures can be quantitatively fit to obtain the intrinsic spin wave energies at any given temperature. The results of such fits for the $Fe_{85}Ni_5Zr_{10}$ system are shown in Fig. 5.4.[18] The excellent fits to the straight lines indicate that Eq. (5.27) is obeyed very well, and these fits yield values of the spin stiffness constant D and the spin wave gap. The gap may arise from (small) anisotropy, or could originate from the pseudodipolar interactions. In all amorphous systems studied to date, there has been no need to include higher order terms in the dispersion relation, and thus we set $E(T) = 0$ in Eq. (5.27).

The value of the spin wave stiffness constant $D(T)$ decreases with increasing temperature, as indicated in Fig. 5.4. We can understand this behavior qualitatively by considering mean-field theory. The magnetization is given by $M(T) = N \langle S^z \rangle$ where N is the number of spins in the system, and the average energy of a spin in the system is then $M(T) \langle S^z \rangle$. Hence the energy to make an excitation by reversing a spin against this magnetization will decrease with increasing temperature. The behavior is shown Fig. 5.5, which gives the spin stiffness as a function of T for $Fe_{78}B_{13}Si_9$.[17] The solid curve is just a fit to mean field theory, and hence looks just like the behavior of the magnetization. Note that there are no data below about 400 K, and this is due to the kinematic restrictions [Eq. (5.25)] as discussed for Fig. 5.3. Of course, one can go beyond mean field theory in these comparisons. The leading order temperature dependence is given in spin wave theory by the Dyson result[19]

$$D(T) = D(0)\left[1 - a\left(\frac{T}{T_C}\right)^{5/2} + \ldots\right] \tag{5.30}$$

where $D(0)$ is the low temperature saturation value and a is a constant. Note that the leading-order temperature dependence has an exponent of 5/2, while the leading-

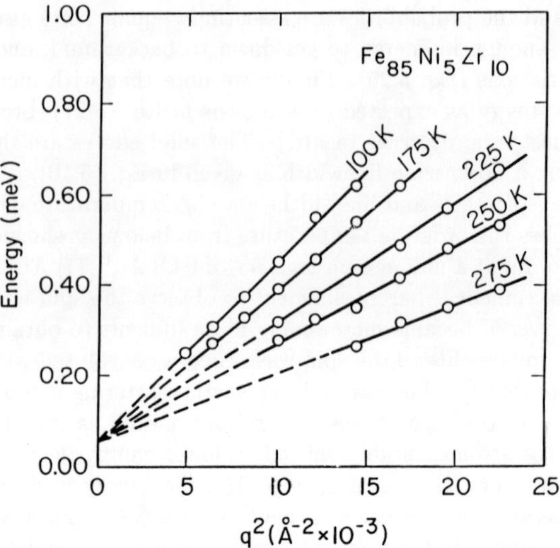

Fig. 5.4: Spin wave energies as a function of q^2 measured on Fe$_{85}$Ni$_5$Zr$_{10}$ for a series of temperatures.[18] The straight lines are fits to Eq. (5.27), and yield values of the gap and the spin stiffness D. There is no need for higher order terms in the dispersion relation for any amorphous system.

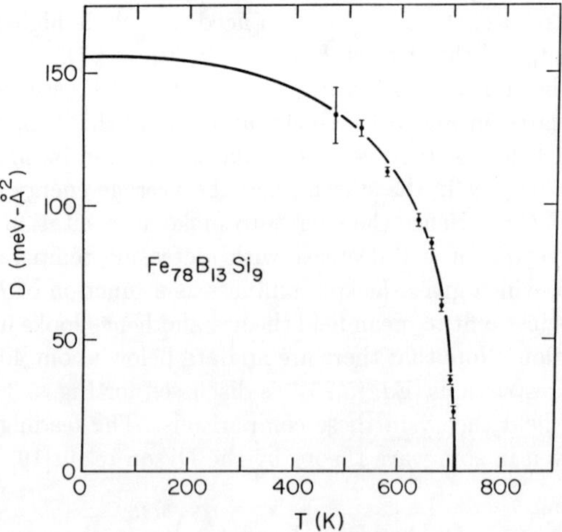

Fig. 5.5: Temperature dependence of the spin wave stiffness constant for Fe$_{78}$B$_{13}$Si$_9$. The solid curve is a fit to a Brillouin function (after [17]).

order temperature dependence of the magnetization has an exponent of 3/2 [Eq. (5.28)]. In the critical regime just below T_C, on the other hand, $D(T)$ should follow a power law given by[13]

$$D(T) \propto \left[1 - \frac{T}{T_C}\right]^{\nu' - \beta} \tag{5.31}$$

where ν' is the critical exponent for the correlation length below T_C and β is the critical exponent for the temperature dependence of the magnetization. There is ample evidence that in the appropriate regime these relations are obeyed for all amorphous ferromagnets studied to date.

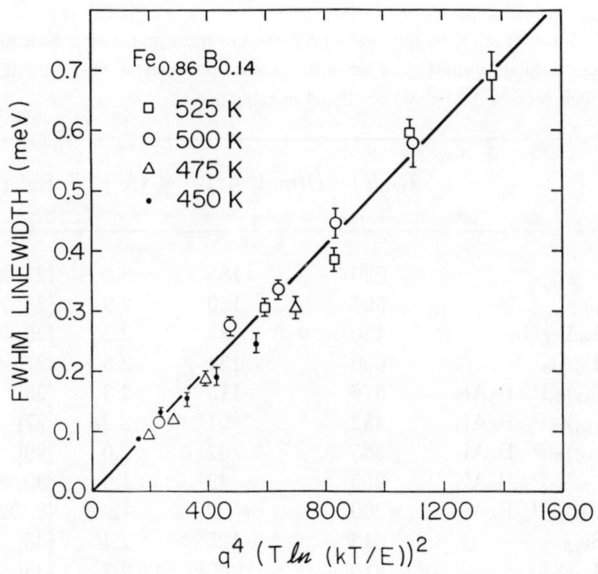

Fig. 5.6: Widths obtained from fits using the Lorentzian spectral weight function vs. the expected (q, T) dependence [Eq. (5.32)] for Fe$_{86}$B$_{14}$ (after ref. [15]).

Finally we turn to the determination of the intrinsic linewidths of the spin wave excitations in the long wavelength regime. The linewidths Γ can be extracted by use of a spectral weight function such as the Lorentzian as given in Eq. (5.10) or for a damped harmonic oscillator as given in Eq. (5.11), suitably convoluted with the instrumental resolution and fitted to the data. The linewidths are expected to follow the spin wave theoretical relation [20]

$$\Gamma(q, T) \propto q^4 \left[T \ln\left(\frac{kT}{E_{sw}}\right)\right]^2 \quad . \tag{5.32}$$

Figure 5.6 shows the experimentally observed spin wave linewidths for the $Fe_{86}B_{14}$ material, which has been studied extensively.[15] Note that the data collapse onto a single universal curve when plotted according to Eq. (5.32). Similarly good agreement is observed for all the (concentrated) amorphous magnetic systems when properly taking into account the instrumental resolution, demonstrating that conventional spin wave theory works remarkably well for all these amorphous ferromagnets. To date a wide variety of concentrated amorphous ferromagnets have been investigated in some detail with neutron scattering as well as with a variety of other experimental (and theoretical) techniques. We close this section by providing a table listing the materials that have been studied along with some of the materials properties and references to the literature.

Table 5.1: Long-wavelength spin wave properties of some amorphous isotropic ferromagnets. Similar properties of amorphous isotropic systems that exhibit Invar behavior (see section 3.2 below) are listed in table 5.2.

Material	$T_C(K)$	$D(meV\,\text{Å}^2)$	$\frac{D}{kT_C}(\text{Å}^2)$	References
Co_4P	620	185	3.5	[21, 22]
$Fe_{75}P_{15}C_{10}$	597	150	2.9	[23, 24, 25]
$(Fe_{93}Mo_7)_{80}B_{10}C_{10}$	450	85	2.2	[26, 25]
$(Fe)_{75}P_{16}B_6Al_3$	630	134	2.5	[27]
$(Fe_{0.65}Ni_{0.35})_{75}P_{16}B_6Al_3$	576	115	2.3	[28]
$(Fe_{0.50}Ni_{0.50})_{75}P_{16}B_6Al_3$	482	91	2.2	[27]
$(Fe_{0.40}Ni_{0.60})_{75}P_{16}B_6Al_3$	365	62	2.0	[29]
$(Fe_{0.30}Ni_{0.70})_{75}P_{16}B_6Al_3$	255	42	1.9	[30, 31]
$(Fe_{0.25}Ni_{0.75})_{75}P_{16}B_6Al_3$	200	32	1.9	[8, 32]
$Fe_{83}B_{16.5}Si_{0.5}$	613	125	2.4	[33]
$Fe_{20}Ni_{60}B_{19}P_1$	410	80	2.3	[33]
$Fe_{70}Cr_{10}P_{13}C_7$	360	60	1.9	[34]
$Fe_{40}Ni_{40}P_{14}B_6$	513	110	2.5	[35]
$(Fe_{0.77}Cr_{0.23})_{75}P_{16}B_6Al_3$	184	26	1.6	[36]
$(Fe_{0.70}Mn_{0.30})_{75}P_{16}B_6Al_3$	143	9	0.7	[37]
$(Fe_{0.75}Mn_{0.25})_{75}P_{16}B_6Al_3$	221	21	1.1	[37]
$(Fe_{0.80}Mn_{0.20})_{75}P_{16}B_6Al_3$	340	47	1.6	[37]

3.2. *Invar systems*

As we saw in the previous section, the spin dynamics of amorphous isotropic ferromagnets appears to be well accounted for by conventional spin wave theory. The theoretical predictions discussed, as well as many others provided by spin wave theory, have been found to be in excellent accord with experimental observations for the vast majority of isotropic ferromagnetic materials, whether they be amorphous or crystalline. The singular exception to this good agreement is for Invar systems. The physical characteristic that distinguishes Invar systems is that they have a very large magnetoexpansion, so large that below T_C it compensates for the ordinary thermal contraction such that the density is approximately independent of temperature.[38] From the point of view of the spin dynamics, for all the Invar materials, whether they be amorphous or crystalline, Eq. (5.29) is found to fail in a major way, with the observed stiffness constant obtained from inelastic neutron scattering measurements as much as a factor of two larger than the stiffness constant that is needed to explain the temperature dependence of the magnetization. This means that the measured magnetization decreases much more rapidly with T than can be accounted for based on the measured dispersion relations. Table 5.2 lists the amorphous Invar-type systems whose dynamics has been investigated to date.

Table 5.2: Amorphous Invar Systems

Material	$T_C(K)$	$D(meV\,Å^2)$	$\frac{D}{kT_C}(Å^2)$	$\frac{D_N}{D_M}$	Reference
$Fe_{86}B_{14}$	556	131	2.7	2.0	[15, 39, 40]
$Fe_{82}B_{18}$	617	165	3.1	2.3	[15]
$Fe_{80}B_{20}$	647	170	3.0	1.9	[41]
$Fe_{76}B_{24}$	723	>175	>2.8	>1.8	[39]
$Fe_{89}Ni_1Zr_{10}$	250	30	1.4		[42]
$Fe_{85}Ni_5Zr_{10}$	306	47	1.8	1.1	[18, 43]
$Fe_{80}Ni_{10}Zr_{10}$	359	78	2.5	1.6	[18, 43]
$Fe_{70}Ni_{20}Zr_{10}$	455	113	2.9	1.6	[44]
$Fe_{72}B_{10}Si_{18}$	705	>230	>3.8	>1.8	[45]
$Fe_{75}B_{10}Si_{15}$	710	220	3.6	1.7	[45]
$Fe_{78}B_{13}Si_9$	708	157	2.6	1.3	[17]
$Fe_{81}B_{10}Si_9$	662	192	3.4	1.7	[45]

The conventional explanation for this observed behavior is that there are additional excitations, "hidden" in the neutron measurements, which participate in reducing the magnetization. The combination of magnetization and neutron measurements already puts stringent conditions on the form that such excitations might take, since there is no freedom to change the *form* of the theory, viz the $T^{3/2}$ behavior for the magnetization, the $T^{5/2}$ behavior for $D(T)$, etc. Hence we must have a density of "hidden" excitations that has precisely the same *form* as the conventional spin wave excitations themselves. One possibility which has been suggested [46] is that the (transverse) spin wave excitations couple to the longitudinal fluctuations, yielding propagating longitudinal excitations which peak at or very near the transverse spin wave energies. The idea experimentally is that when one makes measurements such as shown in Figs. 5.2 and 5.3, the observed excitations are *assumed* to be spin waves, that is, that they correspond to the (transverse) raising and lowering operators $S^{\pm}$ in the theory. In an unpolarized neutron beam experiment the conventional spin wave excitations cannot be distinguished from the proposed longitudinal excitations. Such longitudinal excitations would of course have precisely the same form for the density of states as the ordinary spin waves since they lie on top of each other, and would provide a simple and natural explanation for the "Invar anomaly". Such longitudinal spin excitations have in fact been observed in amorphous systems. However, they have been observed in both Invar *and* non-Invar materials, whereas such longitudinal excitations have not been observed in crystalline Invar systems. Thus they do not appear to provide an explanation for the Invar anomaly, but rather may be an interesting type of new excitation that is found in amorphous materials.

The polarization analysis technique as applied to this problem is in principle straightforward. All the transverse spin wave scattering, represented in the Hamiltonian by the raising and lowering operators $S^{\pm}$, causes a reversal of the neutron spin. These spin-flip (SF) cross sections are denoted by $(+-)$ and $(-+)$. If the neutron polarization $\mathbf{P}$ is parallel to the momentum transfer $(i.e.\,\mathbf{Q} \parallel \mathbf{P})$, then we may create a spin wave in the $(-+)$ configuration, or destroy a spin wave in the $(+-)$ configuration (See Eq. (5.22)).[†] Longitudinal fluctuations, on the other hand, are invisible in this situation. Figure 5.7 shows a measurement of the spin flip scattering in the amorphous Invar alloy $Fe_{86}B_{14}$.[47, 48]

In the $(+-)$ configuration (open circles) we have a strong peak on the energy gain side (E < 0), corresponding to the annihilation of spin wave excitations, and no peak on the energy loss side. For the $(-+)$ configuration (filled circles), on the other hand, we have the opposite behavior of finding the peak on the energy loss side only. These data demonstrate that not only is the instrument working properly and the sample

---

[†]Although in section 2.2 the z direction was defined to be the direction of the neutron polarization, there are different conventions in use for the designation $(+)$ or $(-)$ when using polarized neutrons. These can be confusing not only because the spin and moment are antiparallel for both the electron and the neutron, but also because different polarization devices yield different senses for the polarization direction. Thus a Heusler alloy monochromator, such as used in the measurements of Figs. 5.7 and 5.8, reflects a polarization opposite to that of a Fe^{57} monochromator or supermirror. Experimenters often use $(+)$ for when the spin-flipper is off, and $(-)$ when it is on, and in the present case this produces a notation opposite to Eq (5.19), even though the meaning is clear.

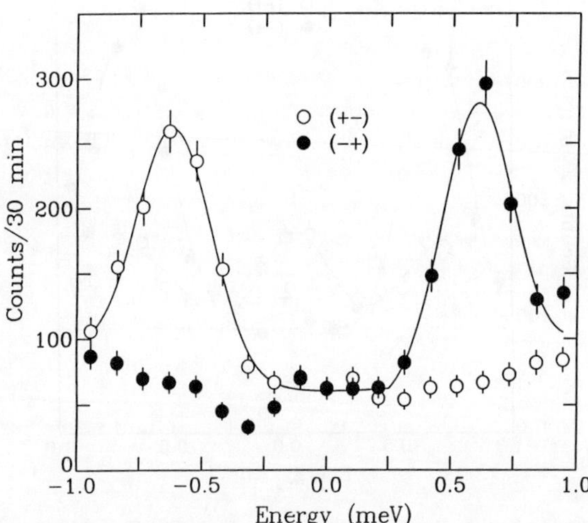

Fig. 5.7: Spin-flip scattering observed for the amorphous Invar $Fe_{86}B_{14}$ system in the $\mathbf{Q} \parallel \mathbf{P}$ configuration. Spin waves are observed for neutron energy gain (E < 0) in the (+−) cross section (open circles) and for neutron energy loss (E>0) in the (−+) configuration (filled circles) [after ref. [48]].

is saturated magnetically, but also that the basic polarization cross section derived from the Heisenberg Hamiltonian is appropriate to describe the system to a good approximation.

In the configuration where $\mathbf{Q} \perp \mathbf{P}$, the spin wave scattering still causes a neutron spin-flip, but it shows up with equal intensity in the energy gain and energy loss cross section, with $\frac{1}{4}$ the intensity of the $\mathbf{Q} \parallel \mathbf{P}$ configuration. The non-spin-flip (+ +) or (− −) scattering, on the other hand, is directly related to the longitudinal (S^z) scattering. Figure 5.8 shows a measurement in this vertical field configuration. The spin-flip scattering clearly shows spin waves in energy gain and energy loss, as expected. The non-spin-flip data, on the other hand, also display peaks very close in energy to the spin waves. There is also a peak at $E = 0$, which originates from nuclear incoherent and magnetic elastic scattering. The scattering at the spin wave positions is $\sim \frac{1}{3}$ the strength of the spin-flip scattering, while the flipping ratio is ~ 10. There are several important features for this non-spin-flip scattering: 1) The peak in energy obeys a q^2 dependence. 2) The ratio of the intensity of the spin-flip to non-spin-flip scattering did not change significantly when experimental improvements doubled the flipping ratio. 3) The ratio did not change significantly as a function of q, while the resolution effects [11] change substantially.

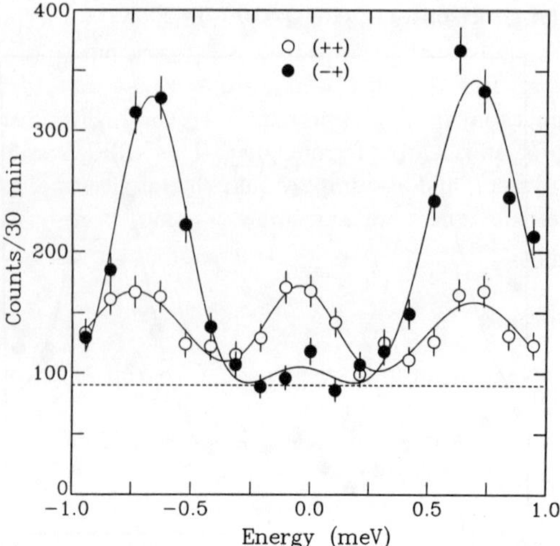

Fig. 5.8: Observed scattering for the amorphous Invar $Fe_{86}B_{14}$ system in the $\mathbf{Q} \perp P$ configuration. The spin-flip scattering (solid circles) exhibits the usual spin wave excitations, while the non-spin-flip scattering (open circles) also has peaks near the spin wave energies (after ref.[48]).

These data initially suggested that there are indeed longitudinal propagating excitations unique to Invar systems, whose energies are close to the spin wave excitation energies. These are just the type of excitations which would be needed to explain the "Invar anomaly". However, additional polarized beam measurements were carried out on a non-Invar amorphous ferromagnet, $Fe_{40}Ni_{40}P_{14}B_6$.[48] The results show that there are also longitudinal excitations in this material as well, suggesting that these excitations might be related to the amorphous state rather than the Invar anomaly. Experiments were then carried out on a single crystal of Invar, $Fe_{65}Ni_{35}$,[49] as well as on single crystals of ordered and disordered $Fe_{72}Pt_{28}$,[50] another Invar-type material. No evidence was found for longitudinal propagating excitations in either of these systems, and thus the longitudinal excitations observed in $Fe_{86}B_{14}$ are likely not related to the Invar properties, but rather have the interesting interpretation that they are unique to the amorphous state. On the other hand, as the Curie temperature is approached from below, longitudinal spin fluctuations are observed but these fluctuations are diffusive in nature, *i.e.* they peak at $E = 0$. This is the behavior expected for simple spin diffusion. For $T \geq T_C$, the x, y, and z directions are equivalent in an isotropic ferromagnet by definition, and in the hydrodynamic regime only spin diffusion occurs.[13] Below T_C this symmetry is broken as we noted when we discussed Goldstone's theorem at the beginning of the section. The transverse susceptibility corresponds to the spin wave excitations, but in all the crystalline ferromagnets studied to date the longitudinal fluctuations remain diffusive below T_C. Hence this scattering appears not to be related to the Invar effect, but rather is the expected behavior for an isotropic ferromagnet. This still leaves the origin of the Invar anomaly as a mystery, and also implies that the longitudinal excitations observed in the amorphous ferromagnets are a new phenomenon. Both of these effects warrant further investigation.

3.3. Large q excitations

For amorphous solids the atomic structure can be conveniently modeled in terms of a dense random packing of hard spheres, with the density described by a radial distribution function which specifies the average number of atoms at distances between r and $r + dr$ from a typical origin. This behavior is reflected in the elastic scattering (or alternatively, the energy-integrated scattering) structure function $S(Q)$. $S(Q)$ has a delta function at $Q = 0$, corresponding to the reciprocal lattice vector $\tau = 0$, and then a relatively sharp peak in $S(Q)$ which typically occurs in these metallic glasses at $Q \sim 3.1$ Å^{-1}, and then smaller oscillations that converge to a constant at large Q. The first peak corresponds to the average nearest-neighbor separation, and since nearest-neighbor positions are well correlated this peak is relatively sharp, even allowing for the fact that the local environment in the amorphous case fluctuates from site to site. This can be contrasted with a periodic lattice where the atomic positions are perfectly defined and we have delta function Bragg peaks at all the "peaks" in $S(Q)$. For the glass system at larger Q's we find that the structure in $S(Q)$ becomes less pronounced and eventually $S(Q)$ becomes Q independent demonstrating that the atomic positions become uncorrelated at large separation. It is interesting to note that the first peak in $S(Q)$ is very close to the position of the first Bragg peak in elemental systems, such as for example fcc nickel (3.09 Å^{-1}) and bcc iron (3.10 Å^{-1}), demonstrating that the average nearest-neighbor separations (and densities) typically differ by only a few per cent between similarly constituted amorphous and crystalline materials.

For the magnetic structure we have a corresponding magnetic structure factor $S_m(Q)$ which relates the magnetic spin correlation function in the material. In the small-Q regime, near the $\tau = 0$ Bragg peak, we know that we have well-defined spin wave modes, but as we proceed to larger values of Q we expect the linewidths of the excitations to become large due to the randomness in the structure, and we should not really view the system in terms of plane-wave-like spin wave states, but rather in terms of a magnetic density of states. However, there is a sum rule for the magnetic susceptibility $\chi(Q)$ that relates the magnetic structure factor to the average energy

244

of the magnetic excitations at that Q, and this relation is given by [21]

$$\left\langle E^2(Q) \right\rangle = \frac{Q^2}{4m_e \chi(Q)} \tag{5.33}$$

where m_e is the electron mass. At small wave vectors we know that $\chi(Q) \propto Q^{-2}$ so that we recover the dispersion relation Eq. (5.27). We also obtain the more general result that the average energy will have a minimum when the magnetic susceptibility has a maximum. For a crystalline system this requires that the excitation energy approaches zero at every reciprocal lattice point (Goldstone mode), as it must for an isotropic ferromagnet. For an amorphous magnet we should expect that the average energy of the magnetic inelastic scattering will decrease substantially in the vicinity of the principal peak in the structure factor at around $Q \sim 3.1$ Å^{-1}.

Similar arguments hold for the lattice dynamical excitations in fluids and glasses, that is, the excitation spectrum should dip in energy when the structure factor has a maximum. Moreover, we noted previously that the strength of the phonon scattering is proportional to Q^2, so that at small Q we didn't have to worry about the phonon scattering as it has a negligible intensity. Around the first peak in $S(Q)$, on the other hand, both types of excitations will have significant strength, and they will be distributed over rather wide ranges of wave vector and energy, making polarized neutron measurements a necessity in order to distinguish the two types of contributions to the scattering. These types of measurements are exceedingly difficult, and there have been only a few studies so far.

The first measurements of this kind were on Co_4P [21] and $Fe_{75}P_{15}C_{10}$ [24], and these were followed by more detailed measurements on $Fe_{40}Ni_{40}B_{14}C_6$.[35] The measurements were taken with a time-of-flight spectrometer, a technique which is typically well suited to the studies of glasses and liquids since only the magnitude of Q is relevant and not the direction. However, generally time-of-flight measurements cannot be taken with polarization analysis. These particular measurements used a polarized incident beam, and essentially employed the ferromagnetic sample as the polarization analyzer, and there is some information that is lost in such a "half-polarized" experiment. The measurements were concentrated around the first peak in the structure factor $S(Q)$ where the scattering should exhibit a minimum in energy. Figure 5.9 shows a contour plot of the dynamic magnetic susceptibility for $Fe_{40}Ni_{40}B_{14}C_6$,[35] and shows that the scattering does indeed peak up around the first peak in the structure factor. Similar results, in terms of the enhanced scattering around the first peak in the structure factor, have been obtained with the triple-axis polarized beam technique on $Fe_{83}B_{17}$ [51, 52] and $Fe_{75}P_{15}C_{10}$.[25] However, there has been considerable debate about the proper interpretation of this scattering. Mook and collaborators [21, 24, 35] found that the time-of-flight data could be explained on the basis of a quadratic dispersion relation of the form of Eq. (5.27), but with excitations that were broadened in energy and displaced in energy by a large gap (12 meV for $Fe_{40}Ni_{40}B_{14}C_6$). Shirane et al.[25] found that the magnetic scattering extends to quite low energies, and con-

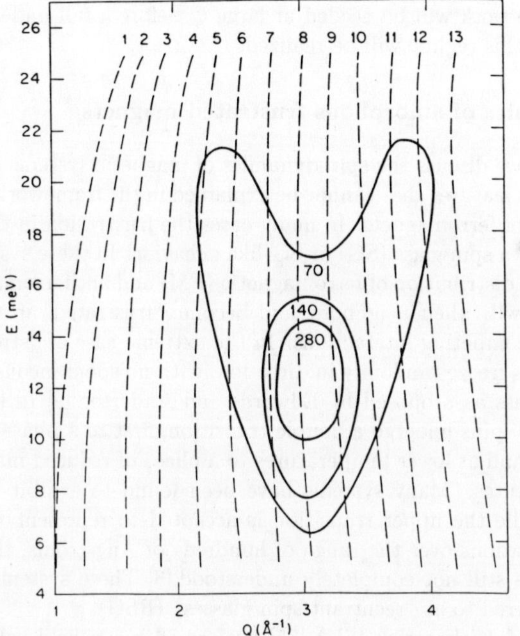

Fig. 5.9: Contour plot of the measured dynamic susceptibility, showing that the scattering is concentrated around the first peak in the magnetic structure factor (after ref. [35]).

cluded that there is no actually "gap" in the excitation spectrum, and hence there are no sharp spin-wave-like excitations in this regime. Rather, the low energy data could be interpreted in terms of a powder-averaged model, which predicts that the scattering should extend all the way to zero energy, and this was the interpretation for the $Fe_{83}B_{17}$ system as well.[51, 52] On the other hand, there appears to be considerably more structure in the scattering, as shown in Fig. 5.9, than can be accounted for in the powder-averaged model, and the powder-averaged model was also found to be inadequate in the more recent low frequency data.[53] These authors also found that there was considerable randomness to the atomic spin directions in these materials, giving rise to a non-collinear spin density locally with only very short correlations transverse to the magnetization direction. Separate freezing of the transverse spin components has also been observed in Mössbauer measurements.[54] This observation might help explain the longitudinal propagating spin excitations that have been observed in the amorphous ferromagnets.[48] It should be kept in mind that these are extremely difficult measurements to carry out experimentally, and it is clear that considerably more work will be needed at large Q before a full understanding of the spin dynamics in this regime will be realized.

4. Spin dynamics of amorphous frustrated magnets

In this section we discuss the spin dynamics of magnetic systems that due to spin frustration exhibit features that cannot be explained in the framework of conventional uniform Heisenberg ferromagnets. In many cases the frustration in these systems results in features of a spin glass (SG) or SG-like behavior. In these systems the atomic spins experience a distribution of ferromagnetic (FM) and antiferromagnetic (AF) exchange couplings with their neighbors, and become "frustrated" in their attempt to order with these competing interactions. In the extreme case of "strong" frustration the magnetic spins freeze randomly in direction, with no spontaneous magnetization, to form a spin glass as proposed by Edwards and Anderson.[9] In the intermediate case the magnetic spins undergo a double transition, first to a phase that appears to be ferromagnetic and at lower temperatures to a phase of reduced magnetization and other SG-like features. Many systems have been found to exhibit this "reentrant" behavior, and while the upper transition is accepted to represent the onset of ferromagnetic correlations over the range of hundreds of Ångströms, the nature of the lower transition is still not completely understood.[3] These systems are commonly (and loosely) referred to as "reentrant spin glasses" (RSG).

Sherrington and Kirkpatrick (SK) [55] performed mean field calculations of the magnetic phase diagram of a system of Ising spins with a Gaussian distribution of (infinitely ranged) exchange interactions, with the mean value of this Gaussian distribution being J_0 and its width δJ. These early calculations indicated that when $J_0 >> \delta J$ the system exhibits a transition from a paramagnetic (PM) state to a FM state, while when $J_0 = 0$ the system undergoes a transition from a PM state to a SG state. In the intermediate region the calculations of SK predict a sequence of transitions from PM to FM to SG. Gabay and Toulouse [56] extended the calculations of SK to a system of Heisenberg spins and predicted that in the intermediate region where $J_0 \approx \delta J$ there is a more complex sequence of transitions from PM to FM to two low temperature mixed phases. The first one is characterized by a coexistence of FM and SG phases, in which only the transverse components of the spins are frozen. The second phase is characterized by the onset of strong irreversibilities. More recent calculations by Thomson [57, 54], have shown that a system of frustrated Heisenberg spins with nearest-neighbor interactions can have two different order events. The first one corresponds to a conventional long-range ferromagnetic order, while a lower temperature event marks the freezing of the spins' transverse degrees of freedom without affecting the long-range order in the system.

In amorphous alloys a significant distribution of FM and AF exchange interactions can be obtained in diluted systems as they approach the percolation threshold, or in some iron-rich systems as they approach the pure amorphous iron structure.

In this section we will concentrate on two systems that illustrate these two cases, $(Fe_xNi_{1-x})_{75}P_{16}B_6Al_3$, and $Fe_{90-x}Ni_xZr_{10}$. Because these systems have a phase where conventional magnetism occurs they have been listed in tables 5.1 and 5.2. Their low temperature behavior, on the other hand, is not conventional as we discuss below.

4.1. Diluted magnets

One of the first amorphous magnetic systems to be extensively studied by neutron scattering techniques was $(Fe_xNi_{1-x})_{75}P_{16}B_6Al_3$.[30, 27, 28, 29, 31, 32] At high iron concentrations (*i.e.* $x \geq 0.5$) this system behaves as an ideal isotropic ferromagnet with a Curie temperature that ranges from $T_C = 482$ K for $x = 0.5$ to $T_C = 630$ K for $x = 1$ (see table 5.1). To a good approximation the nickel in this system carries no magnetic moment, and its main role is to dilute the magnetic iron atoms, making this an ideal system to study the effects of dilution. When decreasing the iron content the average exchange interaction J decreases at the same time that the width of the exchange distribution δJ increases until there is a substantial fraction of AF interactions in the vicinity of the (geometric) percolation concentration. This results in exchange frustration, reentrant magnetism (where the magnetization decreases drastically at low temperatures) and spin glass behavior (characterized by a irreversible component of magnetization at low temperatures).[58, 59, 60] Similar behavior has been found in the related system $(Fe_xT_{1-x})_{75}P_{16}B_6Al_3$, where T=Mn, Cr,[61, 59, 62] and in $(Fe_xCr_{1-x})_{75}P_{15}C_{10}$.[63] In these cases, however, both Mn and Cr are magnetic and favor antiferromagnetic exchange interactions, and their role in these compounds is considerably more complicated than just diluting magnetic Fe atoms, as suggested by the values of D/kT_C in table 5.1. Inelastic neutron experiments in all of these systems show that below their ordering temperature T_C (see table 5.1) they initially exhibit a long wavelength (small wave vector) spin dynamics similar to that described in section 3.1 of this chapter for conventional isotropic ferromagnets. In this regime their spin-wave excitations can be well described by the dispersion relation of Eq. (5.27) (with no quartic term in q), and the stiffness parameter D is temperature dependent. As explained earlier in this chapter the temperature dependence of the spin wave stiffness $D(T)$ should be qualitatively similar to that of the magnetization $M(T)$. Inelastic neutron scattering measurements on $(Fe_xNi_{1-x})_{75}P_{16}B_6Al_3$ for $x = 0.25, 0.30, 0.40$ show this behavior, where the spin wave energies first increase with decreasing temperature below T_C (as expected for a ferromagnet), but then they decrease upon further decreasing the temperature.[30, 29, 31, 32, 8] This reentrant behavior is illustrated in Fig. 5.10 where a plot of $D(T)/kT_C$ vs. T/T_C for these concentrations is shown.[32] In a simple mean-field-like dilution system (with no frustration) all the data would lie on a single curve (like that of Fig. 5.5), given roughly by the line for $x = 0.5$, also shown in this figure. In all cases the low-temperature softening of the spin-wave energies is accompanied by an increase of their intrinsic linewidths. This behavior is similar to that observed in amorphous $(Fe_xMn_{1-x})_{75}P_{16}B_6Al_3$,[37, 64]

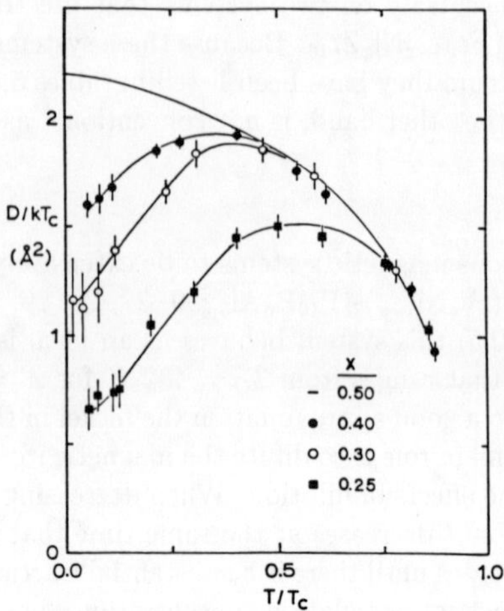

Fig. 5.10: Spin-wave stiffness $D(T)$ scaled by T_C vs. the scaled temperature T/T_C for $(Fe_xNi_{1-x})_{75}P_{16}B_6Al_3$ ($x = 0.25, 0.30, 0.40, 0.50$). In a simple mean-field-like dilution system all of the data would lie on the single curve, given roughly by the line for $x = 0.50$ (after ref [32]).

$(Fe_xCr_{1-x})_{75}P_{16}B_6Al_3$,[36], $(Fe_xCr_{1-x})_{75}P_{15}C_{10}$,[63] and in a variety of crystalline frustrated systems.[3] Hennion et al.[64] have reported that in the amorphous Fe-Mn system the spin-wave energies increase again at lower temperatures after exhibiting a softening similar to that of Fig. 5.10. The origin of such a low temperature increase remains unexplained to this date, but is clearly not associated with the return of the macroscopic magnetization.

An interesting feature in the low-temperature inelastic neutron scattering spectra of these frustrated systems is the development of a central "elastic" component, which coexists with the spin wave scattering and eventually dominates at the lowest temperatures. As mentioned in section (2.1) this central component provides, in principle a direct measurement of the Edwards–Anderson spin glass order parameter.[9] Strictly speaking this central component represents a true order parameter only if the intrinsic linewidth is strictly zero, while in practice if the energy resolution is γ then the elastic component is a measure of the number of spins that become static on a time scale of $\gtrsim \hbar/\gamma$. Figure 5.11 shows the development of the "elastic" component of the scattering for $(Fe_{0.30}Ni_{0.70})_{75}P_{16}B_6Al_3$, with has been measured with an energy resolution as good as $0.6\mu eV$.[65] This figure shows the development of spins that are "frozen" on a time scale of $> 10^{-9}$ s, which is five orders of magnitude greater than the characteristic time of the spin fluctuations in this system ($t = \hbar/(kT_C) \approx 10^{-14}$ s). In this figure data for a variety of wave vectors ($q = 0.04, 0.06, 0.08, 0.1$) Å^{-1} are seen to collapse

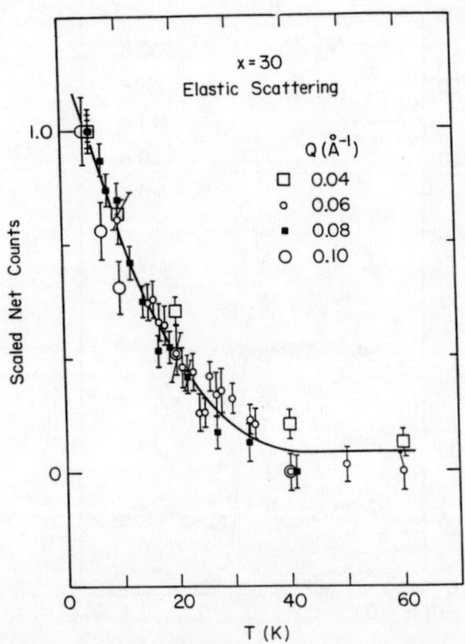

Fig. 5.11: Temperature dependence of the intensity of the elastic component of the scattering in $(Fe_{0.30}Ni_{0.70})_{75}P_{16}B_6Al_3$. The increase of this intensity at low temperatures can be associated with the development of spins that are "frozen" on a time scale of 10^{-9} s, and is a good experimental measurement of the Edwards–Anderson spin glass order parameter (after ref [32, 65]).

on to a single curve showing that the observed spin glass order parameter is not dependent on $\mathbf{Q}$. Figure 5.11 represents then a good experimental measurement of the spin glass order parameter in this system. The spin freezing occurs at a temperature T_{SG} that is lower than the temperature where the spin-wave energies soften, and below T_{SG} there is a coexistence of spin wave excitations and spin-freezing phenomena, consistent with the "transverse spin freezing" picture of Gabay and Toulouse.[56]

4.2. Iron-rich systems

As mentioned in section 1.1 of this chapter the structure of metallic glasses can be well described as a dense random packing of hard spheres, which provides a coordination number of about 12 and a local environment which is reminiscent of the fcc lattice. It is well known that the magnetic moments and exchange interactions of fcc iron are very sensitive to local environment effects, and it is not surprising that the distribution of nearest neighbor (nn) distances in the iron-rich amorphous alloys often leads to competing exchange interactions and spin frustration. Independent first-principle calculations predict that the magnetic moments and the spin correlations in fcc iron are strongly volume dependent, and that upon reducing the iron nearest neighbor (nn) distance below some value (between 2.5 and 2.6 Å) there

250

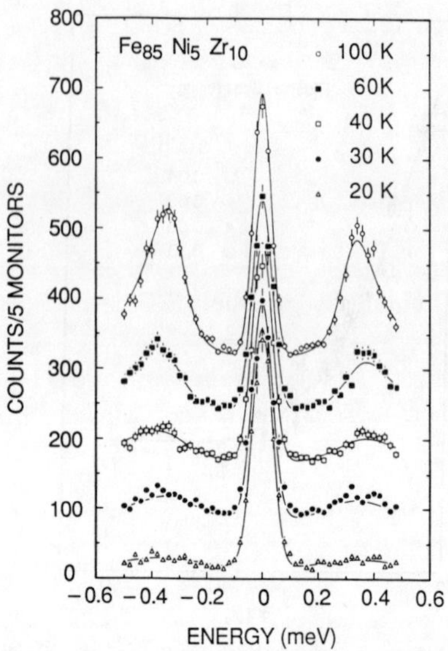

Fig. 5.12: Constant-q scans for $q = 0.08\text{Å}^{-1}$ for amorphous $Fe_{85}Ni_5Zr_{10}$ at various temperatures. The solid lines are the result of the least-squares fit described in the text (after ref [42]).

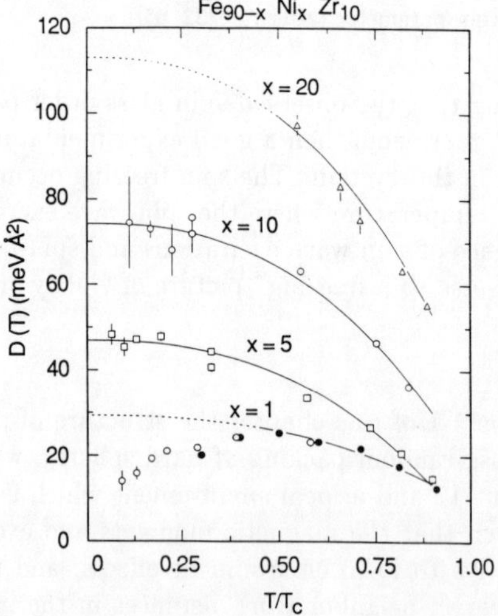

Fig. 5.13: $D(T)\,vs\,T/T_C$ for $Fe_{85}Ni_5Zr_{10}$ ($x = 1, 5, 10, 20$). The solid lines are the result of fits to the form of Eq. (5.30). The broken lines are extrapolations to low temperatures (after ref [42]).

is a transition from a high-moment ($\mu > 2\mu_B$) state to a low moment ($\mu < 1\mu_B$) state with the interaction between these moments changing from FM to AF.[66, 67] On the experimental side Xiao and Chien [68] have shown that the Fe-M (M=B,Si) amorphous alloys favor high Fe moment states, while similar Fe-ET (ET=Ti, Zr, Hf, Nb, Ta, Mo) alloys favor low Fe moment states. Thus, while the coupling of the magnetic moments in iron-rich Fe-M alloys is predominantly FM, the iron-rich Fe-ET alloys will see a distribution of FM and AF couplings and often exhibit spin frustration and reentrant magnetism. One such system is $Fe_{90-x}Ni_xZr_{10}$, where the distribution of FM and AF couplings can be altered continuously by varying the iron concentration. Thus for $x = 1$ (concentrated system) the iron moment is only $\mu \approx 1.4\mu_B$ and the system exhibits reentrant magnetism, while for $x = 20$ (diluted system) the system orders ferromagnetically with $\mu = 2.4\mu_B$ [42]. In this system Ni carries a relatively small moment ($\mu = 0.6\mu_B$) [69], and it is believed that its main role is just to dilute the iron moments and to increase the number of high-moment spins that are FM coupled. A similar enhancement of ferromagnetism by magnetic dilution has been observed in $Fe_{90}Zr_{10}$ by the partial substitution of Co for Fe,[70] and by hydrogenation.[71] Recent calculations by Krompiesky et al. [72] also support the idea that the enhancement of ferromagnetism in this system by hydrogenation is mainly a consequence of the strong volume dependence of the magnetic moments.

Inelastic neutron scattering experiments have been performed on $Fe_{90-x}Ni_xZr_{10}$ for $x = 1, 5, 10, 20$ [42, 18]. The specimens with $x = 5, 10, 20$ exhibit FM behavior while the specimen with $x = 1$ exhibits reentrant behavior. Figure 5.12 shows the inelastic scattering spectra for the specimen with x=5 at T=20, 30, 40, 60, 100 K for q=0.08Å^{-1}. As in Fig. (5.2) the peaks on the neutron energy gain ($E < 0$) and energy loss ($E > 0$) sides of the spectra correspond to the annihilation and creation of spin waves. As explained in section 2.3 the kinematic constraints in this type of experiment do not allow one to scan the full scattering lineshapes. The central peak in these spectra corresponds to "elastic" scattering originating from nuclear incoherent scattering from the sample and environment. The solid lines in this figure are the result of the least-squares fit to the cross section of Eq. (5.26) convoluted with the instrumental resolution. The spectral weight function used in this analysis was the double Lorentzian form of Eq. (5.10). In all cases the spin wave energies E_q can be well described by the dispersion relation of Eq. (5.27) with a small energy gap $\Delta E \approx 0.05meV$ (see figure 5.4), and no quartic term in q. For $x = 5, 10, 20$ the stiffness coefficient $D(T)$ follows a temperature dependence consistent with the two-magnon interaction form for ferromagnetic materials [Eq. (5.30)]. Figure 5.13 shows the plots of $D(T)$ vs. T/T_C for all the concentrations studied. Note in this figure that no low-temperature data are available for the specimen with $x = 20$, due to the fact that the spin-wave energies at these temperatures fall outside the accessible energy-transfer range [Eq. (5.25)]. The solid lines in this figure are the results of least-squares fits to Eq. (5.30) and the broken lines are extrapolations to low temperatures. The values of $D(0)$ obtained from the fits for all samples are listed in Table 5.2.

252

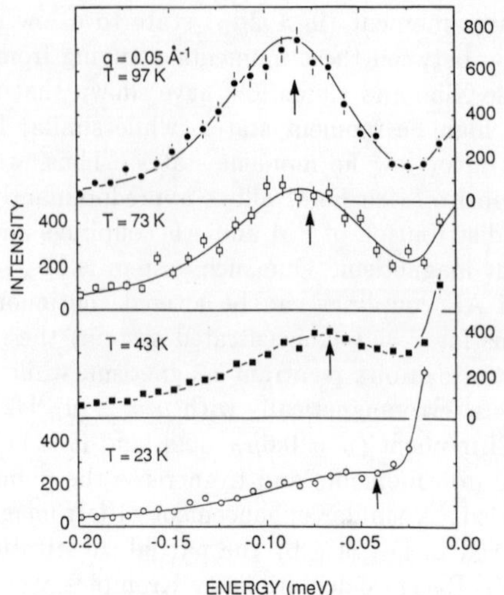

Fig. 5.14: Constant-q scans for $q = 0.05\text{Å}^{-1}$ for amorphous $Fe_{89}Ni_1Zr_{10}$. The arrows indicate the values of the spin-wave energies from the fits described in the text (after ref [42]).

While the results shown above seem to suggest that the systems for $x = 5, 10, 20$ are conventional ferromagnets, the decrease of the Curie temperature T_C and the spin-wave stiffness $D(T)$ with increasing Fe content is indicative of frustration effects that increase as the system approaches the "pure amorphous iron" structure. Also note (see table 5.2) that the ratio D/kT_C decreases considerably from 2.9Å^2 for $x = 20$ to 1.4Å^2 for $x = 1$. A typical range for this ratio for amorphous ferromagnets is between 2 and 3 Å^2. A detailed analysis of the spin-wave linewidth data for the systems with $x = 5, 10$ revealed that the q dependence of Γ_q can be adequately fit either to the q^4 form of Eq. [5.32] [20] or to the q^5 form predicted for topological disorder.[73] On the other hand, the temperature dependence of these linewidths is different from that expected for a conventional ferromagnet. At intermediate temperatures the linewidths were virtually temperature independent, and at lower temperatures the linewidths increased considerably. It is also observed that, in the same temperature range where the spin waves broaden, an apparent enhancement of the transverse susceptibility (coefficient A of Eq. (5.26)) similar to that found in RSG systems occurs.[42] For these systems (*i.e.* $x = 5, 10$), however, there is no evidence of any increase in the central component of the scattering (coefficient B of Eq. (5.26)), which could be associated with the development of the spin-glass order parameter, or a low-temperature spin-wave energy softening.

The results for the specimen with the highest Fe concentration, corresponding to

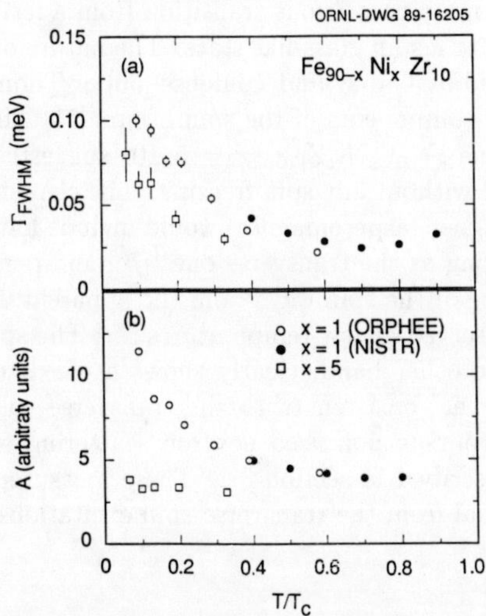

Fig. 5.15: Plots of the temperature dependence of (a) the intrinsic spin-wave linewidth Γ_q (FWHM) of the Lorentzian spectral weight function), and (b) the transverse susceptibility [parameter A from Eq. (5.26)] , for $Fe_{90-x}Ni_xZr_{10}$ ($x = 1, 5$, after ref [42]).

$x = 1$, show clear features of reentrant behavior. The spin-wave stiffness coefficient $D(T)$ shows a T dependence consistent with Eq. (5.30) only above $T = 0.4T_C$, and below this temperature the stiffness decreases (see Fig. 5.13). The softening of the spin-wave energies at low temperatures can be appreciated in Fig. 5.14, where the energy-gain side of the inelastic neutron spectra is shown for $T = 23, 43, 73$, and $97K$. The solid lines in this figure are the results of the least-squares fits explained above, and the arrows show the energy of the spin-wave excitations derived from these fits (the small shift in the spin-wave energies with respect to the peak maxima are due to resolution effects). The temperature dependence the spin-wave intrinsic linewidth Γ_q and the transverse part of the susceptibility (coefficient A of Eq. (5.26)) for $q = 0.05$ Å^{-1} are shown in Fig. 5.15. Also included for comparison in this figure are similar data for the specimen with $x = 5$. Note that both parameters increase below about $T/T_C \approx 0.6$. The elastic part of the scattering (coefficient B of Eq. (5.26)), on the other hand, shows a significant increase only below $T/T_C \approx 0.4$. This lower temperature, however, should not be interpreted as the temperature where "spin-freezing" occurs but as a temperature where the spins appear static on a time scale fixed by the finite experimental instrumental resolution.

The results presented above could perhaps be interpreted in the framework of the recent calculations of Lorenz and Hafner [74] who suggest that, with increasing density, amorphous iron shows a continuous transition from a ferromagnetic to a as-peromagnetic (spin canted) to a spin glass-like state. The nature of the spin glass-like state might be that proposed by Gabay and Toulouse [56] or Thomson,[57, 54] where a freezing of the transverse components of the spin occurs.[75] The low-temperature spin dynamics of amorphous $Fe_{90-x}Ni_xZr_{10}$ ($x = 5, 10$) suggests the existence of a noncollinear (canted) phase without any spin freezing. The elementary excitations of such a noncollinear ferromagnet (asperomagnet) would include longitudinal spin fluc-tuations that can be as strong as the transverse ones [76] and perhaps could explain the "anomalous" broadening of the spin waves and the apparent enhancement of the transverse susceptibility observed at low temperatures.[77] The spin dynamics of the system with $x = 1$, on the other hand, clearly shows a coexistence of spin waves and spin freezing phenomena, consistent with the transverse spin freezing picture. It would be interesting to pursue polarized neutron scattering experiments in this system, similar to those described in section 3.2. These measurements should allow separation of the longitudinal from the transverse spin excitations.

5. Future Directions

We hope that the reader has acquired some appreciation of the considerable progress that has been achieved in our understanding of the microscopic properties of amor-phous ferromagnets. Much of the experimental work conducted so far has concen-trated in the long wavelength spin fluctuation regime, thanks to the ready comparison of experimental results with hydrodynamic spin wave theory, and we now have a good overall understanding of the behavior of these materials on the long wavelength scale. Indeed many of these studies have gone past simply understanding amorphous mag-nets: These prototypical isotropic magnets have proved to be a potent testing ground for our ideas of how competing interactions and frustration should manifest them-selves in nature. These systems will certainly continue to play an important role in these types of studies. There are, though, unanswered questions about the nature of longitudinal excitations in isotropic ferromagnets in general, and amorphous mag-nets in particular, and we anticipate that the next decade should see these questions resolved.

One of the striking aspects of this present chapter, however, is the paucity of information regarding how the underlying randomness of the amorphous state affects the nature of the magnetic excitations. The excitations at long wavelength obey a quadratic dispersion relation, but there has been little information obtained about how the excitations evolve at larger q, either in terms of their mean energy or intrinsic linewidth. Likewise only limited information has been obtained so far about the scattering around the first peak in the magnetic structure factor, or indeed about the more general relationship between the fluctuation spectrum and S(Q). The reason is of course because these types of experiments are much more difficult and challenging to carry out. As experimental techniques improve we may anticipate progress on these topics, but the pace will be slow and the path slippery.

Acknowledgements

We would like to thank all of our colleagues who have collaborated with us in these studies. In particular we would like to thank R. W. Erwin, G. E. Fish, H. A. Mook, J. J. Rhyne, N. Rosov and S. C. Yu, for having assisted us in compiling this review. The work at ORNL was supported by the U. S. Department of Energy under contract No. DE-AC05-84OR21400 with Martin Marietta Energy Systems, Inc.

References

[1] For a review of tetrahedrally bonded glasses, see for example S. R. Elliott, *Physics of Amorphous Materials* (Wiley, New York 1993).

[2] J. W. Lynn and J. J. Rhyne, "Spin Dynamics of Amorphous Magnets", in *Spin Waves and Magnetic Excitations*, Part II Chapter 4, pg. 177, A. S. Borovik-Romanov and S. K. Sinha, editors (North Holland 1988).

[3] S. M. Shapiro, "Magnetic Excitations in Spin Glasses", in *Spin Waves and Magnetic Excitations*, Part II Chapter 5, pg. 219, A. S. Borovik-Romanov and S. K. Sinha, editors (North Holland 1988).

[4] S. W. Lovesey, *Theory of Neutron Scattering from Condensed Matter*, (Oxford, 1984).

[5] W. Gavin Williams, *Polarized Neutrons,* (Oxford, 1988).

[6] G. L. Squires, *Introduction to the Theory of Thermal Neutron Scattering*, (Cambridge, 1978).

[7] D. L. Price and K. Sköld in *Methods of Experimental Physics*, volume **23**-Neutron Scattering, part A, chapter 1; K. Sköld and D. L. Price, editors, (Academic Press, 1986).

[8] R. W. Erwin, J. W. Lynn, J. J. Rhyne and H. S. Chen, *J. Appl. Phys.* **57**, 3473 (1985).

[9] S. F. Edwards and P. W. Anderson, *J. Phys. F* **5**, 965 (1975).

[10] R. M. Moon, T. Riste and W. C. Koehler, *Phys. Rev.* **181**, 920 (1969).

[11] N. Rosov, J. W. Lynn, and R. W. Erwin, *Physica* B **180&181**, 1003 (1992).

[12] F. Keffer, *Handbuck der Physik* **18**, Part 2, p. 1, ed. by S. Flugge (Springer-Verlag, Berlin, 1966).

[13] B. I. Halperin and P. C. Hohenberg, *Phys. Rev.* **188**, 898 (1969); *ibid* **188**, 952 (1969). See also D. Forster, *Hydrodynamic Fluctuations, Broken Symmetry, and Correlation Functions* (Benjamin/Cummings, 1975).

[14] R. V. Lange, *Phys. Rev.* **146**, 301 (1966).

[15] J. A. Fernandez-Baca, J. W. Lynn, J. J. Rhyne and G. E. Fish, *Phys. Rev.* B **36**, 8497 (1987).

[16] J. A. Fernandez-Baca, J. W. Lynn, J. J. Rhyne and G. E. Fish, *J. Appl. Phys.* **57**, 3545 (1985); *Physica* **136B**, 53 (1986). J. J. Rhyne, G. E. Fish, and J. W. Lynn, *J. Appl. Phys.* **53**, 2316 (1982).

[17] S. C. Yu, J. W. Lynn, J. J. Rhyne and G. E. Fish, *J. Appl. Phys.* **63**, 4083 (1988); *J. Magn. Magn. Mater.* **97**, 286 (1991).

[18] J. A. Fernandez-Baca, J. W. Lynn, J. J. Rhyne and G. E. Fish, *J. Appl. Phys.* **61**, 3406 (1987).

[19] F. J. Dyson, *Phys. Rev.* **102**, 1217 (1956); *ibid* **102**, 1230 (1956).

[20] A. B. Harris *Phys. Rev.* **175**,674 (1968); *ibid* 606 (1969).

[21] H. A. Mook, N. Wakabayashi, and D. Pan, *Phys. Rev. Lett.* **34**, 1029 (1975).

[22] H. A. Mook, D. Pan, J. D. Axe and L. Passell, A.I.P. Conf. Proc. **24**, 112, ed. by C. D. Graham, Jr., G. H. Lander, and J. J. Rhyne (A.I.P., New York, 1975).

[23] J. D. Axe, L. Passell and C. C. Tsuei, A.I.P. Conf. Proc. **24**, 119, ed. by C. D. Graham, Jr., G. H. Lander, and J. J. Rhyne (A.I.P., New York, 1975).

[24] H. A. Mook and C. C. Tsuei, *Phys. Rev.* B **16**, 2184 (1977).

[25] G. Shirane, J. D. Axe, C. F. Majkrzak and T. Mizoguchi, *Phys. Rev.* B **26**, 2575 (1982).

[26] J. D. Axe, G. Shirane, T. Mizoguchi and K. Yamauchi, *Phys. Rev.* B **15**, 2763 (1977).

[27] R. J. Birgeneau, J. A. Tarvin, G. Shirane, E. M. Gyorgy, R. C. Sherwood, H. S. Chen and C. L. Chien, *Phys. Rev.* B **18**, 2192 (1978).

[28] J. A. Tarvin, G. Shirane, R. J. Birgeneau and H. S. Chen, *Phys. Rev.* B **17**, 241 (1978).

[29] J. W. Lynn, R. W. Erwin, J. J. Rhyne and H. S. Chen, *J. Appl. Phys.* **52**, 1738 (1981).

[30] J. W. Lynn, G. Shirane, R. J. Birgeneau and H. S. Chen, AIP Conference Proceedings # **34**, 313 (1976).

[31] J. W. Lynn, R. W. Erwin, H. S. Chen and J. J. Rhyne, *Solid State Commun.* **46**, 317 (1983); J. W. Lynn, R. W. Erwin, J. J. Rhyne and H. S. Chen, *J. Magn. Magn. Mater.* **34**, 1397 (1983).

[32] R. W. Erwin, J. W. Lynn, and H. Chen (to be published). R. W. Erwin, Ph.D. Thesis (University of Maryland, 1985).

[33] J. J. Rhyne, J. W. Lynn, F. E. Luborsky and J. L. Walter, *J. Appl. Phys.* **50**, 1583 (1979).

[34] Z. Xianyu, Y. Ishikawa, and S. Onodera, *J. Phys. Soc. Jpn* **51**, 1799 (1982).

[35] H. A. Mook and J. W. Lynn, *Phys. Rev.* B **29**, 4056 (1984).

[36] J. P. Wicksted, S. M. Shapiro and H. S. Chen, *J. Appl. Phys.* **55**, 1697 (1984).

[37] G. Aeppli, S. M. Shapiro, R. J. Birgeneau and H. S. Chen, *ibid* **25**, 4882 (1982); *Phys. Rev.* B **28**, 5160, (1983); *ibid* **29**, 2589 (1984).

[38] For a review of Invar systems see Y. Nakamura, IEEE Trans. Magn. **MAG-12**, 278 (1976). E. F. Wasserman "Invar: Moment-volume instabilities in transition metals and alloys", in *Ferromagnetic Materials*, volume 5 chapter 3, pg. 237, K. H. J. Buschow and E. P. Wohlfarth, editors (North Holland 1990).

[39] J. J. Rhyne, G. E. Fish, and J. W. Lynn, *J. Appl. Phys* **53**, 2316 (1982).

[40] Y. Ishikawa, K. Yamada, K. Tajima, and K. Fukamichi, *J. Phys. Soc. Jpn.* **50**, 1958 (1981).

[41] H. Grimsditch, A. Malozemoff, and A. Brunsch, *Phys. Rev. Lett.* **43**, 711 (1979).

[42] J. A. Fernandez-Baca, J. J, Rhyne, G. E. Fish, M. Hennion, and B. Hennion *J. Appl. Phys.* **67**, 5223 (1990); J. A. Fernandez-Baca, J. W. Lynn, J. J. Rhyne and G. E. Fish, *J. Appl. Phys.* **63**, 3749 (1988).

[43] S-C. Yu, J. W. Lynn, and G. E. Fish, *Jpn. J. Appl. Phys.* **32**, 67 (1993).

[44] J. A. Fernandez-Baca, J. J. Rhyne, and G. E. Fish, *J. Magn. Magn. Mater.* **54-57**, 289 (1986).

[45] W. Minor, B. Lebech, K. Clausen and W. Dmowski, *Rapidly Quenched Metals* (Elsevier, 1985), pg. 1149.

[46] See, for example, R. Raghavan and D. L. Huber, *Phys. Rev.* B **14**, 1185 (1976); J. K. Battacharjee, *ibid* **27**, 3058 (1983); S. V. Maleev, *Soc. Sci. Rev. A Phys.* **8**, 323 (1987). V. G. Vaks, A. I. Larkin, and S. A. Pikin, *Sov. Phys. JETP* **26** 647 (1968).

[47] J. W. Lynn, N. Rosov, C-H. Lee, Q. Lin, and G. Fish *Physica* B **180-181**, 253 (1992).

[48] J. W. Lynn, N. Rosov, and G. Fish, *J. Appl. Phys.* **73**, 5369 (1993).

[49] J. W. Lynn, N. Rosov, M. Acet, and H. Bach, *J. Appl. Phys.* **75**, 6069 (1994).

[50] N. Rosov, J. W. Lynn, J. Kastner, E. F. Wassermann, T. Chattopadhyay, and H. Bach, *J. Magn. Magn. Mater.* **140-144** (in press 1995). N. Rosov, J. W. Lynn, J. Kastner, E. F. Wassermann, and H. Bach, J. Appl. Phys. **75**, 6072 (1994).

[51] D. McK. Paul, R. A. Cowley, W. G. Stirling, N. Cowlam and H. A. Davies, 1982, *J. Phys.* F **12**, 2687 (1982).

[52] R. A. Cowley, D. McK. Paul, W. G. Stirling and N. Cowlam, *Physica* **120B**, 373 (1982).

[53] R. A. Cowley, N. Cowlam, and L. D. Cussens, J. de Physique **49, C8**, 1285 (1988); R. A. Cowley, N. Cowlam, P. K. Ivison, and J. Martinez, *J. Magn. Magn. Mater.* **104-107**, 159 (1992); R. A. Cowley, C. Patterson, N. Cowlam, P. K. Ivison, J. Martinez, and L. D. Cussen, *J. Phys.* F (in press).

[54] For a review on the subject see D. H. Ryan, "Exchange Frustration and Transverse Spin Freezing", in *Recent Progress in Random Magnets*, Chapter 1, pg. 1, (World Scientific, 1992).

[55] D. Sherrington and S. Kirkpatrick, *Phys. Rev. Lett.* **35**, 1792 (1975).

[56] M. Gabay and G. Toulouse, *Phys. Rev. Lett.* **47**, 201 (1981).

[57] J. R. Thomson, Hog Guo, D. H. Ryan, M. Zuckerman and Martin Grant, *Phys. Rev.* B**45**, 3129 (1992).

[58] S. M. Bhagat, J. A. Geohegan and H. S. Chen *Solid State Commun.* **36** 1 (1980).

[59] J. A. Geohegan and S. M. Bhagat, *J. Magn. Magn. Mater.* **25**, 17 (1981).

[60] P. Mazumdar, S. M. Bhagat and M. A. Manheimer, *J. App. Phys.* **57**, 3479 (1985).

[61] Y. M. Yeshurun, M. B. Salamon, N. V. Rao and H. S. Chen, *Phys. Rev. Lett.* **45**, 1366 (1980); *Solid State Commun.* **38**, 371 (1981).

[62] M. A. Manheimer, S. M. Bhagat and H. S. Chen, *J. Magn. Magn. Mater.* **38**, 147 (1983).

[63] Ph. Mangin, D. Bomazouza, B. George, J. J. Rhyne and R. W. Erwin, *Phys. Rev.* B **40**, 11123 (1989).

[64] B. Hennion, M. Hennion, I. Mirebeau and F. Hippert, Physica **136B**, 47 (1986); M. Hennion, B. Hennion, I. Mirebeau, S. Lequien, F. Hippert, *J. Appl. Phys.* **63**, 4071, (1988).

[65] R. W. Erwin, J. W. Lynn and A. Magerl, *J. Magn. Magn. Mater.* **54-57**, 101 (1986).

[66] F. J. Pinsky, J. Staunton, B. L. Gyorfly, D. D. Johnson and G. M. Stocks, *Phys. Rev. Lett.* **56**, 2096 (1986).

[67] C. S. Wang, B. M. Klein and H. Krakauer, *Phys. Rev. Lett.* **54**, 1852 (1985).

[68] G. Xiao and C. L. Chien, *Phys. Rev.* B **35**, 8763 (1987).

[69] Z. M. Stadnik, P. Griesbach, G. Dehe, P. Gülich and T. Miyazaki, *Phys. Rev.* B **35**, 430) (1987).

[70] K. Shirakawa, S. Ohnuma, M. Nose and T. Masumoto, *IEEE Trans. Magn.* **MAG-16**, 910 (1980).

[71] D. H. Ryan, J. M. D. Coey, E. Batalla, Z. Altounian and J. O. Strön Olsen, *Phys. Rev.* B **35**, 8630 (1987).

[72] S. Krompiesky, U. Krauss and U. Krey, *Phys. Rev.* B **39**, 2819 (1989).

[73] V. A. Singh and L. M. Roth, *J. Appl. Phys.* **49**, 1642 (1978); Kaneyoshi, *J. Phys. Soc. Jpn.* **45**, 1835 (1978).

[74] R. Lorenz and J. Hafner, *J. Magn. Magn. Mater.* **139**, 209 (1995).

[75] Kakehashi (see chapter one of this book) predicted that pure amorphous iron should have a transition from a ferromagnetic to a spin glass state. It must be pointed out that the transverse spin fluctuations are neglected in Kakehashi's treatment (see section 2.3.1 of chapter one) and thus spin canting or transverse spin freezing phenomena are not predicted.

[76] M. A. Continentino and N. Rivier, *J. Phys.* F **9**, L146 (1979); *J. Magn. Magn. Mater.* **15-18**, 1419 (1980).

[77] W. M. Saslow and R. Erwin, *Phys. Rev.* B **45**, 4759 (1992).

CHAPTER SIX

NUMERICAL STUDIES OF MAGNONS IN AMORPHOUS MAGNETS

D. L. Huber

Department of Physics
University of Wisconsin-Madison
Madison, Wisconsin, USA

1. Introduction

1.1. Overview

Understanding the nature of the low-lying excitations in amorphous magnets continues to be a challenging problem. Even in materials which order ferromagnetically, the structural disorder precludes a conventional analysis based on the translational symmetry of the Hamiltonian. For this reason, alternative analytic approaches utilizing various approximation schemes have been developed [1]. In recent years, the importance of numerical simulation studies of the magnetic excitations has begun to be recognized. This chapter is a brief overview of some of the numerical work in this field with examples taken from the areas of amorphous Heisenberg ferromagnets and ferromagnets with random-axis anisotropy.

In many areas of physics, simulation studies play an important role which is distinct from both experiment and analytic (traditional) theory. With the advent of supercomputers and high performance minicomputers and work stations, it has become possible to study the behavior of large numbers (N >> 100) of interacting spins, electrons, atoms, molecules, etc.. In many instances, the systems studied are sufficiently large that the calculated behavior is representative of the behavior in macroscopic arrays. In situations where the form of the underlying microscopic Hamiltonian is known, simulation studies can be used to estimate the parameters of the Hamiltonian by adjusting those parameters so as to obtain maximum agreement with experiment. A second role for simulation studies is to test approximate analytic theories under controlled conditions. Here the focus is on tractable model Hamiltonians incorporating the features which are thought to be essential in determining the physical phenomena of interest.

1.2. Magnons

The word "magnon" appearing in the title of this chapter refers to linear spin excitations which in periodic arrays are "spin waves". This can be made explicit by considering a Heisenberg system which has ferromagnetic ordering at T=0K. In applied field H, the Hamiltonian takes the form

$$\mathcal{H} = \Sigma_j g\mu_B H S_j^z - \Sigma J_{jk} S_j \cdot S_k , \tag{6.1}$$

where S_j^α (α=x,y,z) are the three components of the jth spin S_j, g is the g-factor, and μ_B denotes the Bohr magneton. The symbol J_{jk} denotes the exchange interaction between spins j and k, and the sum is over all (j,k) pairs. Since the system is assumed to order ferromagnetically, the dominant interactions are positive ($J_{jk}>0$).

At T=0K, the ground state of the system has $<S_j^z>_o=-S$ for all spins, where the angular brackets denote the thermal average. The linear excitations are obtained by making use of the Holstein-Primakoff transformation in lowest order. In this order, the spin raising and lowering operators S_j^+ ($=S_j^x+iS_j^y$) and S_j^- ($=S_j^x-iS_j^y$) are replaced by $(2S)^{1/2}a_j^\dagger$ and $(2S)^{1/2}a_j$, respectively, where $a_j^\dagger$ and a_j are boson creation and annihilation operators. With the z component of the jth being written as $-S+a_j^\dagger a_j$, the Hamiltonian becomes

$$\mathcal{H} = g\mu_B H \Sigma_j a_j^\dagger a_j + S\Sigma J_{jk}(a_j^\dagger a_j + a_k^\dagger a_k - a_j^\dagger a_k - a_k^\dagger a_j) + const.. \tag{6.2}$$

From Eq. (6.2), one can establish a equivalent dynamical matrix by making use of the equation of motion $ida_j/dt=[a_j,\mathcal{H}]$ ($\hbar$=1) and the commutation relations $[a_j,a_k^\dagger]=\delta_{jk}$. One has

$$ida_j/dt = (g\mu_B H + S\Sigma_k J_{jk})a_j - S\Sigma_k J_{jk}a_k. \tag{6.3}$$

Assuming a harmonic time dependence, $a_j \propto exp(-i\omega t)$, one obtains the matrix equation

$$\omega a_j = \Sigma_k U_{jk}a_k, \tag{6.4}$$

where the dynamical matrix U has the form

$$U_{jk} = (g\mu_B H + S\Sigma_l J_{jl})\delta_{jk} - (1-\delta_{jk})SJ_{jk}. \tag{6.5}$$

The eigenvalues of the dynamical matrix are the magnon energies ω_ν (ν=1,...,N, where N is the number of spins in the array). The eigenvectors $X_{j\nu}$ (j,ν=1,...,N) are the coefficients in the expansion of the eigenoperators c_ν and $c_\nu^\dagger$ in terms of the operators a_j and $a_j^\dagger$, viz.

$$c_v = \sum_j X_{jv}^* a_j, \tag{6.6a}$$

$$c_v^\dagger = \sum_j X_{jv} a_j^\dagger. \tag{6.6b}$$

Since the diagonalization is equivalent to a principal axis transformation [2], the eigenvectors form a unitary matrix, i.e. $XX^\dagger=1$, where $X_{\alpha\beta}^\dagger=X_{\beta\alpha}^*$. As a consequence, one has the inverse relations

$$a_j = \sum_v X_{jv} c_v, \tag{6.6c}$$

$$a_j^\dagger = \sum_v X_{jv}^* c_v^\dagger. \tag{6.6d}$$

Two further comments are in order. First, the same dynamical matrix would have been obtained had one started with the equations of motion for the spin operators S_j^- calculated with the full Hamiltonian, (6.1), and then linearized the equations by replacing S_j^z with its zero-temperature expectation value, -S. Secondly, in the case of a periodic array with nearest-neighbor interactions, the eigenvalues assume the familiar spin wave form

$$\omega_q = g\mu_B H + zJS(1 - \gamma_q), \tag{6.7}$$

where q is a vector in the Brillouin zone of the magnetic lattice and $\gamma_q=z^{-1}\sum_k{}' \exp(i q \cdot r_{jk})$, in which the sum is over the z nearest neighbors of the site j, and r_{jk} is the vector connecting them. The corresponding eigenvectors are expressed as

$$X_{jq} = N^{-\frac{1}{2}} \exp(-i q \cdot r_j), \tag{6.8}$$

where r_j is the position vector of the jth spin.

1.3. Eigenvalues and Eigenvectors

From the eigenvectors and eigenvalues of the dynamical matrix one can calculate a number of quantities of direct physical interest. The magnon contribution to the specific heat per spin, C_{mag}, is given by the standard expression for boson excitations:

$$C_{mag} = N^{-1}\sum_v (\omega_v^2/kT^2)\exp(\omega_v/kT)/(\exp(\omega_v/kT) - 1)^2. \tag{6.9}$$

Likewise, the leading correction to the zero-temperature magnetization per spin takes the form

$$M(T) = M(0) - N^{-1}g\mu_B\sum_j \langle a_j^\dagger a_j\rangle_T, \tag{6.10}$$

where, as before, the brackets denote a thermal average. Using (6.6c) and (6.6d) along with the unitary property of the matrix $\mathbf{X}$, Eq. (6.10) reduces to

$$M(T) = M(0) - N^{-1}g\mu_B\Sigma_v<c_v^\dagger c_v>_T. \qquad (6.11)$$

Since the product $c^\dagger c$ is the number operator for the magnons, the thermal average in (6.11) is the boson occupation number so that $M(T)$ can be written

$$M(T) = M(0) - N^{-1}g\mu_B\Sigma_v(\exp(\omega_v/kT) - 1)^{-1}. \qquad (6.12)$$

Another quantity of direct physical interest is the zero-temperature dynamic structure factor, $S_o(\mathbf{q},E)$, characterizing the one-magnon contribution to the inelastic neutron scattering. This function takes the form of a spatial and time Fourier transform of a spin-spin correlation function, viz.

$$S_o(\mathbf{q},E) = (2\pi N)^{-1}\int_{-\infty}^{+\infty}dt e^{-iEt}\Sigma\exp(i\mathbf{q}\cdot\mathbf{r}_{jk})<S_j^- e^{i\mathcal{H}t}S_k^+ e^{-i\mathcal{H}t}>_o, \qquad (6.13)$$

where the subscript on the angular brackets denotes a zero-temperature average, and the sum is over all j and k. Equation (6.13) can be evaluated by using the Holstein-Primakoff transformation and equations (6.6c) and (6.6d) to express the spin operators first in terms of the site boson operators a_j and $a_j^\dagger$ and then in terms of the normal mode operators c_v and $c_v^\dagger$. Since the latter have an exponential time dependence, the integral over t can be readily calculated with the result

$$S_o(\mathbf{q},E) = 2SN^{-1}\Sigma\exp(i\mathbf{q}\cdot\mathbf{r}_{jk})\Sigma_v X_{jv}X_{kv}^*\delta(E - \omega_v), \qquad (6.14)$$

showing the explicit dependence on the eigenvalues and eigenvectors.

Finally, information about the spatial extent of the eigenfunctions can be inferred from the localization indices, or inverse participation ratios, defined by [4]

$$L_v = \Sigma_j|X_{jv}|^4/(\Sigma_j|X_{jv}|^2)^2. \qquad (6.15)$$

For a perfectly periodic array, $L_v=N^{-1}$; more generally, $L_v^{-1}\approx$number of spins on which the mode has significant amplitude.

1.4. Magnetic Materials and Model Hamiltonians

Provided the localized spin picture is appropriate, the Hamiltonian (1) can be applied equally well to amorphous and crystalline ferromagnets. In the case of the crystalline systems, the exchange interactions depend only on the <u>relative</u> separation of the two spins. With amorphous systems, the interactions can, in principle, be different for all (j,k) pairs. In the calculations discussed in Sec. 2, the simplifying assumption is made that J_{jk} depends only on the distance <u>between</u> spins j and k. Since the separation between nearby spins will depend on the local configuration of magnetic atoms, the parameters of the Hamiltonian will vary throughout the material.

The other class of amorphous magnetic materials treated in this chapter are the intermetallic compounds containing rare earth ions with non-zero orbital angular momentum such as a-TbFe$_2$. Because the rare earth ions have orbital angular momentum, they interact with the crystalline electric fields arising from the neighboring atoms. In lowest order, this interaction can be represented by a uniaxial term of the form $-D_j(\mathbf{n}_j \cdot \mathbf{S}_j)^2$, where D_j is a measure of the strength of the anisotropy at spin site j, and the unit vector $\mathbf{n}_j$ is in the direction of the local anisotropy axis. With an interaction of this form, $\mathbf{S}$, strictly speaking, is a pseudospin characterizing the lowest crystal field multiplet of the rare earth ion. In the calculations discussed in Sec. 3, the spin Hamiltonian of the rare earth intermetallics is further simplified by placing the ions on a lattice and taking the strength of the anisotropy to be the same at all sites. Only the <u>direction</u> of the anisotropy axis is random.

2. Amorphous Ferromagnets

2.1. Overview

The applicability of the dynamical matrix formalism to the problem of determining the low-lying magnetic excitations in amorphous ferromagnets depends critically on the appropriateness of the local spin picture. In the case of amorphous ferromagnetic insulators and semiconductors, for which there are rather few examples, the isotropic Heisenberg Hamiltonian can be employed (with dipolar interactions added, if necessary) when the magnetic ions have zero orbital angular momentum, e.g. Fe^{3+}, Mn^{2+}, Gd^{3+}, and Eu^{2+}. In the case of amorphous ferromagnetic metals, which are far more common, the localized spin model represents a rather severe approximation, especially when the corresponding crystalline alloys are better described in terms of itinerant electrons. Nevertheless, the relative ease with which calculations of the magnon-related properties can be carried out makes the Heisenberg model a useful starting point for a quantitative analysis, although caution must exercised in interpreting the results.

Further complicating the analysis is a lack of detailed knowledge of the exchange interactions. In the work to be discussed below, this problem is dealt with by assuming that the interaction J_{jk} depends only on the separation between spins j and k and can be expressed in terms of a common function of that separation, i.e. $J_{jk}=J(r_{jk})$. Disorder then enters through the <u>configuration</u> of magnetic atoms. As discussed in Chapter 2, there exist a variety of methods for modeling the structure of amorphous materials, either in the form of finite clusters or in large "supercells" with periodic boundary conditions. In both approaches, radial distribution functions are obtained which are in good agreement with experiment.

To summarize this section, the application of numerical techniques to the problem of spin excitations in amorphous ferromagnetic metals depends on three approximations: the use of a local spin picture, the assumption that the exchange interactions can be described by a universal function of the distance between the spin pair, and a prescribed structural model. Of the three, it is fair to say that the structural model is the approximation which is under the greatest control in the sense that it can be systematically improved.

2.2. Calculations

It has been established for some time that the structure of the monatomic amorphous ferromagnets Fe, Ni, and Co are adequately modeled by the dense random packing of hard spheres. Furthermore, the same model works well for the metal atoms in "metglass" alloys such as $Co_{80}P_{20}$ since the metalloid atoms, e.g. P, can be viewed as occupying the voids between the metal atoms.

Calculations of the magnetic properties of dense random packing models of amorphous Heisenberg ferromagnets in zero field have been reported by Krey [4,5,6] and by Alben [7]. In both cases, algorithms were used which made possible direct calculations of the magnon density of states and the zero temperature dynamic structure factor for arrays of 10^3 spins without having to explicitly diagonalize the dynamical matrix. Krey employed a continued fraction approach, whereas Alben used equation-of-motion techniques. A brief summary of these two methods is given in the Appendix to this chapter.

Krey's work is based on the 888 atom relaxed random packing model of von Heimendahl [8] which reproduces the first three peaks in the radial distribution function including fine structure. The exchange interaction, $J(r)$, was assumed to be non-zero only when $0.75 \leq r/r_o \leq 1.25$ where r_o is the position of the first peak in the distribution function. Figure 6.1 shows the magnon density of states averaged over the full cluster (including surface sites) as a broken line, the density of states averaged over the 100 interior sites as a solid line, and the density of states of the fcc ferromagnet with nearest-neighbor interactions as a dotted line. It is evident that the density of states averaged over the interior sites, which is appropriate for comparison with the macroscopic density of states, resembles that of a "smoothed out" fcc lattice. In Ref. 5, the temperature dependence of the magnetization was calculated in lowest order (Eq. (6.12)) and compared with the experimental results of Chien and Hasegawa for various metglasses [9]. On the basis of this comparison, Krey concludes that a description based on localized spins and short ranged interactions should be much better for the amorphous metglasses than for their crystalline counterparts.

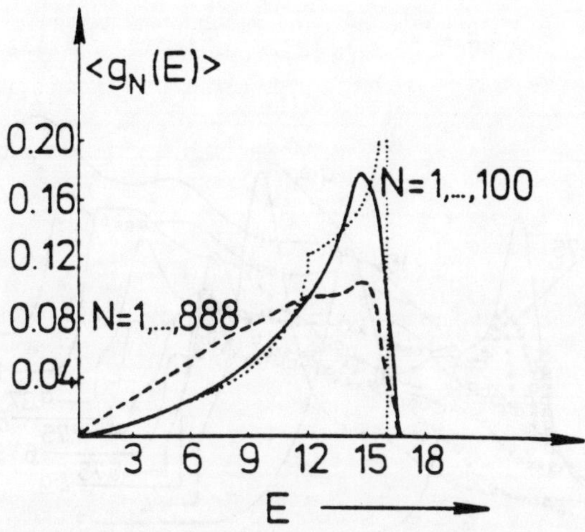

Fig. 6.1. The averaged density of states $<g_N(E)>$ (=g(E) in the text): -------- all 888 sites; ________ only the 100 interior sites; .... ideal fcc lattice. Energy in units of JS, where J is the value of the exchange interaction in the interval $0.75 \leq r/r_o \leq 1.25$ (nn interaction in the case of fcc). From Ref. 6.

Krey also used the continued fraction algorithm to calculate the dynamic structure factor (Fig. 6.2). For small q, he finds well defined spin wave (i.e. plane wave) peaks with a quadratic dispersion relation, i.e. $\omega_v \sim q^2$ for $qr_o \leq 5$. An additional low energy peak near $qr_o = 7.75$ was observed which might be connected with the "roton minimum" reported by Mook et al. [10]. However, the connection is far from certain.

Calculations of the dynamic structure factor and the magnon density of states were also carried out be Alben [7] for a model based on the dense random packing of hard (unrelaxed) spheres with results similar to those of Krey. Alben took the exchange interaction to be constant for $r < 1.25 \times$(sphere diameter) and zero otherwise. Like Krey, he obtained spin wave modes with quadratic dispersion for small q and some evidence of roton-like behavior at large q.

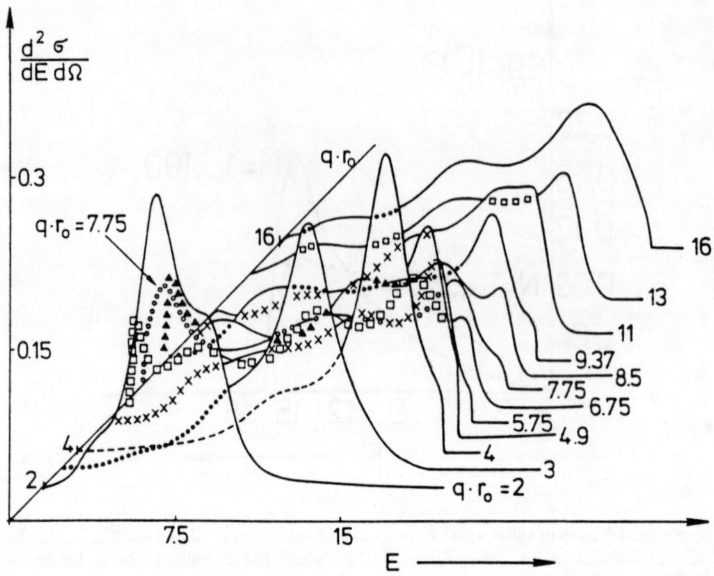

Fig. 6.2. Zero-temperature dynamic structure factor, $S_o(\mathbf{q},E)$, vs E. The numbers associated with the various curves correspond to different values of the product qr_o. Energy in units of JS, where J is the value of the exchange interaction in the interval $0.75 \leq r/r_o \leq 1.25$. From Ref. 6.

2.3. Discussion

The quadratic behavior of the peaks in the dynamic structure factor suggests that at least a fraction of the low-lying spin excitations are characterized by a dispersion relation similar to what is found in crystalline systems, $\omega_q = Dq^2$. Assuming <u>all</u> of the low-lying modes have this spin wave character, then the magnon density of states, g(E), has a characteristic $E^{\frac{1}{2}}$ behavior as $E \to 0$, with a coefficient determined by D, i.e.

$$g(E) = N^{-1}\Sigma_v \delta(E - \omega_v),$$

$$= (v_o/2\pi^2)\int q^2 dq \delta(E - Dq^2),$$

$$= (v_o/4\pi^2 D^{3/2})E^{\frac{1}{2}}, \tag{6.16}$$

where v_o is the volume per spin. Using (6.16) to evaluate Eq. (6.12), one finds

$$M(T) = M(0) - g\mu_{B0}\int^{\infty}dEg(E)/(e^{E/kT} - 1),$$

$$= M(0) - g\mu_{B}v_{o}(kT/4\pi D)^{3/2}\zeta(3/2), \tag{6.17}$$

where $\zeta(3/2)$ is a Riemann zeta function with the value 2.612....

Experimentally, it is found that the leading correction to the zero-temperature magnetization varies as $T^{3/2}$, consistent with the analysis outlined above [11]. Neutron scattering studies of isotropic amorphous magnets also confirm the existence of long wavelength excitations with quadratic dispersion. It should be kept in mind, however, that this agreement does not establish the validity of the local spin picture since similar results are obtained for itinerant ferromagnets in the long wavelength limit [12]. In general, the values of the dispersion parameter D inferred from neutron scattering agree with the values obtained from the $T^{3/2}$ coefficient of the magnetization, the notable, and not understood, exceptions being the Invar alloys [11].

3. Ferromagnets with Random Axis Anisotropy

3.1. Overview

The numerical studies of amorphous Heisenberg ferromagnets discussed in the preceding section were based on a realistic structure and Hamiltonian (within the local spin picture). Direct contact was made with experiment in order to establish the validity of the approximations for the exchange interaction. In contrast, the numerical studies of systems with random axis anisotropy start with a model Hamiltonian which is believed to incorporate the essential features of the spin Hamiltonian for amorphous rare earth intermetallic compounds. The focus of the effort is on elucidating the properties of the model system and making comparisons with analytic treatments.

In the model problem, the interest is in the effects of the random anisotropy on an otherwise isotropic system with translational symmetry. The Hamiltonian takes the form [13]

$$\mathcal{H} = g\mu_{B}H\Sigma_{j}S_{j}^{z} - J\Sigma'S_{j}\cdot S_{k} - D\Sigma_{j}(n_{j}\cdot S_{j})^{2}, \tag{6.18}$$

where the prime on the second summation indicates that it is limited to nearest-neighbor pairs on a lattice. The third term in (6.18) incorporates the randomness associated with the anisotropy in an amorphous material in the sense that the local anisotropy axis, n_{j}, is assumed to vary randomly throughout the system with no correlation between different sites (the strength of the anisotropy is assumed to be constant, however).

In zero and weak fields, the model Hamiltonian shows complex behavior with a highly degenerate spin glass-like ground state. In this case, little is known theoretically about the excitations in the random anisotropy model outside of the hydrodynamic (long wavelength) region [14,15]. The behavior in the high field limit is much better understood, in large part due to the interplay between theory and simulation. This will be discussed in the following section.

3.2. Calculations

In [16], it is shown that in the high field limit, $g\mu_B H \gg J,D$, the linearized magnon Hamiltonian takes the form

$$\mathcal{H} = g\mu_B H \Sigma_j a_j^\dagger a_j + JS\Sigma'(a_j^\dagger a_j + a_k^\dagger a_k - a_j^\dagger a_k - a_k^\dagger a_j)$$

$$+ 3D_R S \Sigma_j (\cos^2\theta_j - 1/3) a_j^\dagger a_j + \text{const.}, \tag{6.19}$$

where D_R $(=D(1-1/2S))$ is the renormalized anisotropy constant and θ_j is the angle between the applied field and $\mathbf{n}_j$. With the spins arrayed on a lattice, the first two terms in (6.19) have full translational symmetry; comparing the third term with the first, it is evident that the anisotropy term is equivalent to a random anisotropy field equal to $3D_R S(\cos^2\theta_j - 1/3)/g\mu_B$.

In so far as its excitations are concerned, the Hamiltonian (6.19) is formally equivalent to the random potential model for an electron in a disordered alloy. This equivalence has been exploited in Ref. 16 in developing a Coherent Anisotropy Field Approximation (CAFA) analogous to the Coherent Potential Approximation (CPA) for the electronic problem [17]. In the CAFA, the effective magnon energies that characterize the configuration average of the greens function take the form

$$\omega_{qeff}(E) = \omega_q + g\mu_B H_a^c(E), \tag{6.20}$$

where ω_q is given by Eq. (6.7) and the coherent anisotropy field is obtained as a solution to the equation

$$_{-1}\int^1 d(\cos\theta)[3D_R S(\cos^2\theta-1/3) - g\mu_B H_a^c(E)]/\{1-[3D_R S(\cos^2\theta-1/3) - g\mu_B H_a^c(E)]<G_o>\}$$

$$= 0, \tag{6.21}$$

where $<G_o>$ denotes the configuration average of a local greens function depending on E and $H_c^a(E)$. In deriving (6.21), is it assumed that the distribution of anisotropy axes is isotropic, i.e. $<\cos^2\theta>=1/3$.

In [18], results for the magnon density of states obtained by diagonalizing the dynamical matrix associated with Eq. (6.19) are compared with the predictions of the CAFA. Figure 6.3 shows the distribution of magnon energies calculated from 6 configurations of 8×8×8 arrays of spins on a simple cubic lattice with periodic boundary conditions. The calculations were carried out for various values of D_R taking $J=S=g\mu_B=1$. Figure 6.4 shows the corresponding results in the CAFA. From these figures it is evident that the CAFA accounts reasonably well for the modification of the density of states by the random anisotropy.

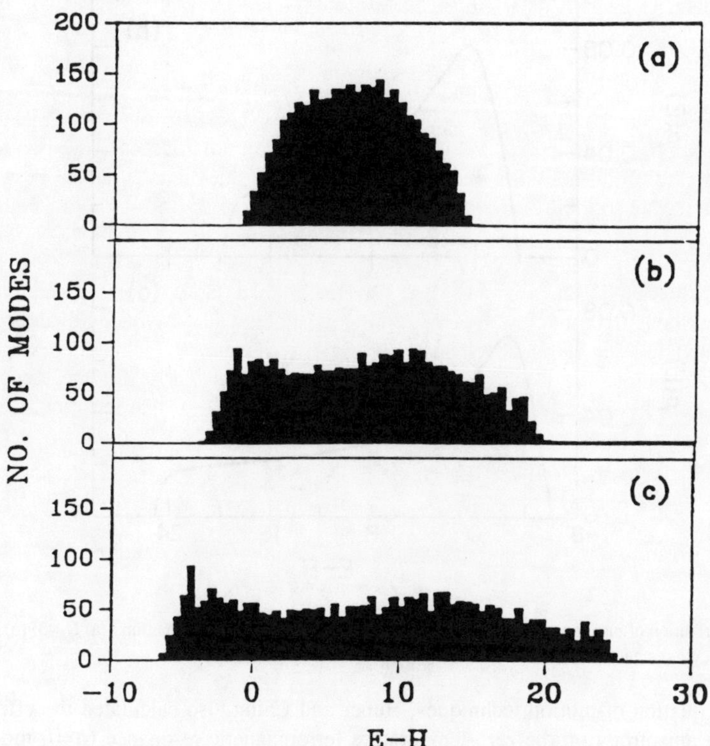

Fig. 6.3. Distribution of magnon modes in the high field limit for various values of D_R. (a) $D_R=3$; (b) $D_R=6$; (c) $D_R=9$. $J=S=g\mu_B=1$. From Ref. 18.

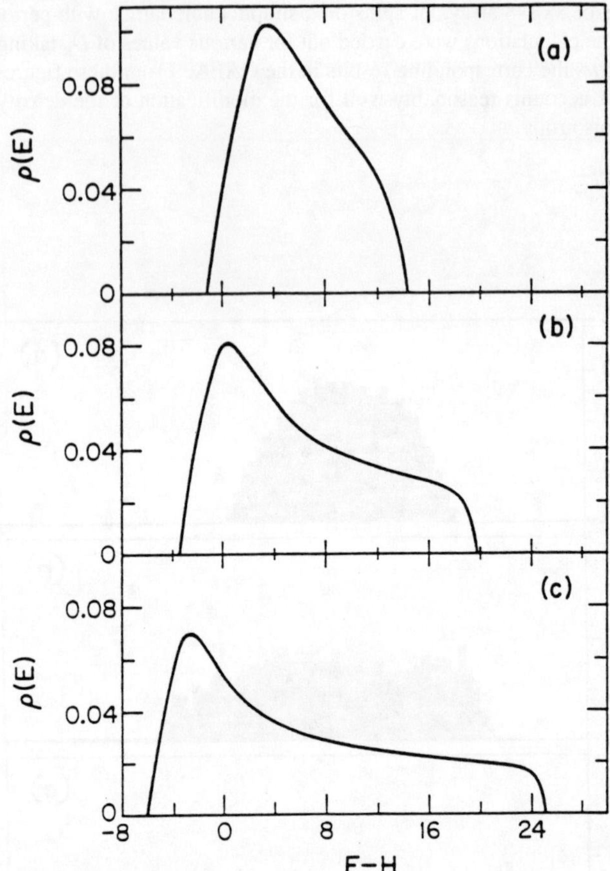

Fig. 6.4. Distribution of magnon modes in the Coherent Anisotropy Field Approximation. (a) $D_R=3$; (b) $D_R=6$; (c) $D_R=9$. All curves have the same area. $J=S=g\mu_B=1$. From Ref. 18.

Using equation-of-motion techniques, Huber and Ching also calculated the effect of the random anisotropy on the zero-temperature ferromagnetic resonance ($q=0$) mode in the high field limit. The results are shown in Fig. 6.5 where a comparison is made between the predictions of the CAFA and the numerical data from 12×12×12 and 16×16×16 arrays. From this figure, it is evident that the CAFA reproduces the shift in the mode to lower frequencies with increasing disorder. For small values of D_R relative to J, the shift can be calculated in lowest order perturbation theory [18], with the result

$$\omega_o - g\mu_B H = -0.20 D_R{}^2 S/J, \tag{6.22}$$

for a simple cubic lattice with nearest-neighbor interactions.

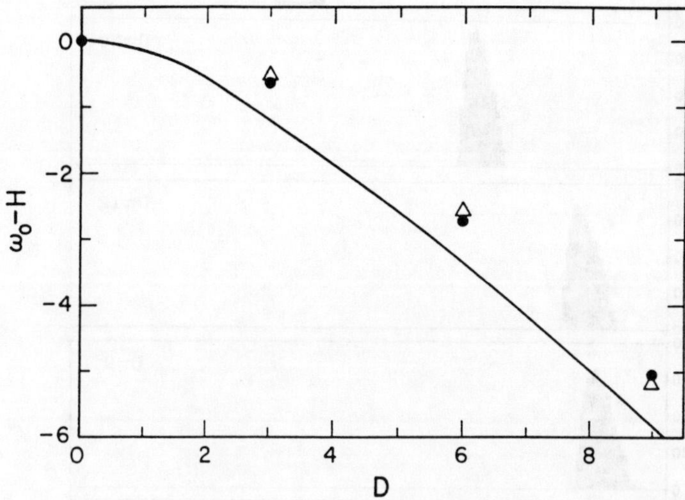

Fig. 6.5. Shift in frequency of the uniform mode in the high field limit vs the renormalized anisotropy constant (denoted here as D). Solid curve: Coherent Anisotropy Field Approximation; solid circles: results from 16×16×16 arrays; open triangles: results from 12×12×12 arrays. $J=S=g\mu_B=1$. From Ref. 18.

3.3. Discussion

The results outlined in Sec. 3.2 indicate that the magnon modes in the random anisotropy model are well understood in the high field limit, with the anisotropy term playing a role that is the counterpart of the role played by the potential in the random alloy problem. Such is not the case for the behavior in zero field. Figure 6.6 displays numerical results for the density of states in zero field obtained from three configurations of 256 spins arrayed on a fcc lattice [19]. With increasing disorder, the distribution shifts to higher energies becoming approximately centered about $2D_R S$, the Zeeman energy associated with the local anisotropy field. In this limit, the behavior reflects the fact that the equilibrium orientation of the spins is approximately along the local anisotropy axis. When this happens, it is the exchange interaction that is the effective source of the disorder since in the absence of exchange all modes have energy $2D_R S$. With finite J, the density of states has a width on the order of $z^{1/2}JS$, z being the number of nearest neighbors.

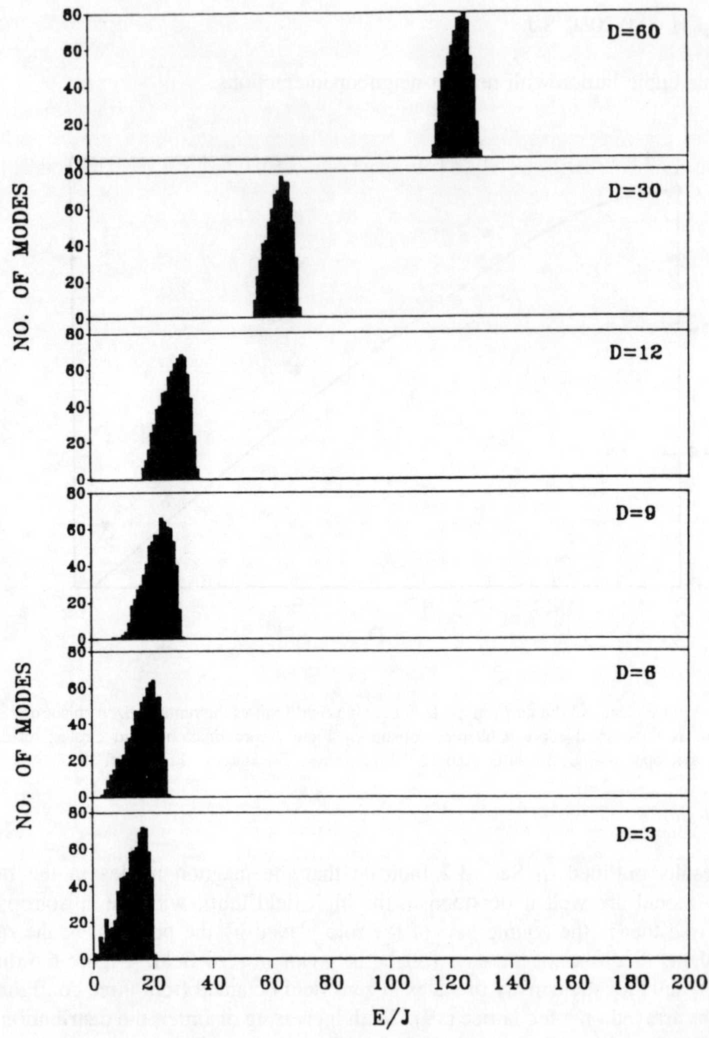

Fig. 6.6. Distribution of magnon modes in zero field for various values of the renormalized anisotropy constant (denoted here as D). In descending order, the panels correspond to D=60, 30, 12, 9, 6, and 3. J=S=1. From Ref. 19.

While the density of states in zero field behaves in an understandable fashion, the same can not be said for the spatial extent of the modes. A calculation of the localization indices (Eq.(6.15)) for arrays of 256 spins shows that for D_R as large as 60 (in units where J=S=1), the modes are localized across the band, i.e. $L \leq 10^{-2}$, with a small number of modes at the upper and lower band edges displaying a slightly greater degree of localization, i.e. $10^{-2} \leq L \leq 10^{-1}$ [19]. This behavior contrasts with what is found for the localization indices in saturating fields. As shown in Fig. 6.7, the modes become progressively more localized with increasing D_R, as is to be expected from the random potential analogy.

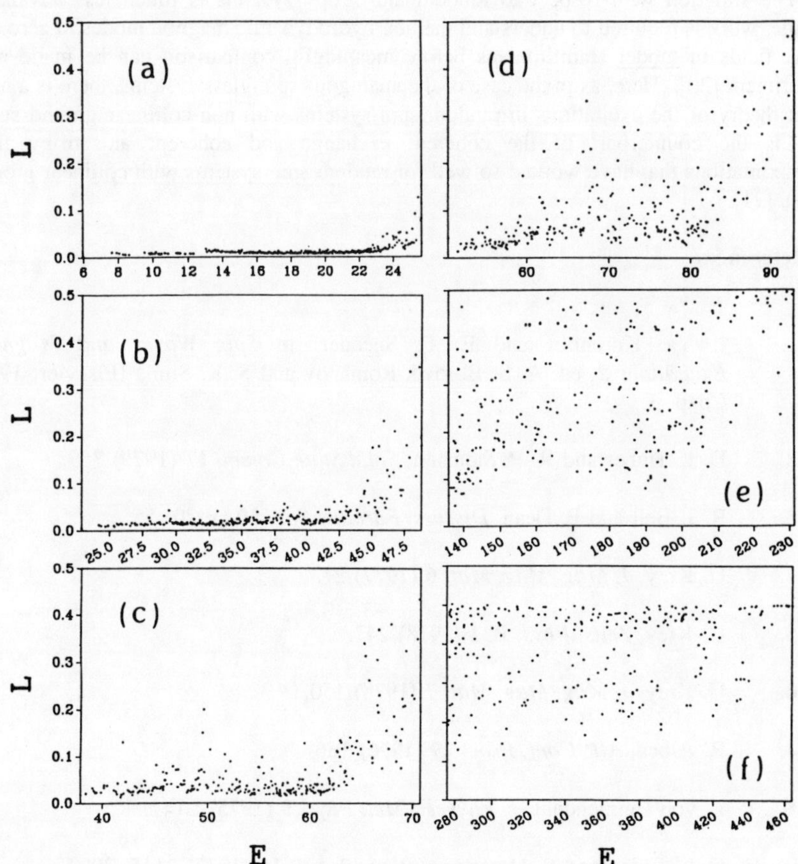

Fig. 6.7. Localization indices in saturating fields for various values of D_R. (a) D_R=3, H=6; (b) D_R=6, H=24; (c) D_R=9, H=40; (d) D_R=12, H=57; (e) D_R=30, H=160; (f) D_R=60, H=332. J=S=$g\mu_B$=1. From Ref. 19.

4. Summary and Conclusions

The purpose of this chapter has been to illustrate the role played by numerical simulation studies of spin excitations in amorphous magnets. In the case of isotropic amorphous ferromagnets, where quasi-realistic calculations can be carried out in the local spin approximation, further progress requires more accurate information about the exchange interactions. It seems likely that quantitative comparisons between experiment and theory, particularly in the case of the dynamic structure factor, may lead to a better understanding of J(r).

The situation with respect to random anisotropy systems is much less advanced. Further work is required to understand the non-hydrodynamic magnon modes in zero and weak fields in <u>model</u> Hamiltonians before meaningful comparison can be made with experiment [20]. Here, as in the case of the analogous spin glass systems, there is a need for a theory of the excitations in random spin systems with non-collinear ground states that is the counterpart of the coherent exchange and coherent anisotropy field approximations that have worked so well for random spin systems with collinear ground states [21].

5. References

1.   I. Ya. Korenblit and E. F. Shender, in *Spin Waves and Magnetic Excitations 2*, ed. A. S. Borovik-Romanov and S. K. Sinha (Elsevier, 1988) Chap. 3.

2.   D. L. Huber and R. P. Siemann, *Solid State Comm.* **17** (1975) 769.

3.   R. J. Bell and P. Dean, *Discuss. Faraday Soc.* **50** (1970) 55.

4.   U. Krey, *J. Mag. Mag. Mat.* **6** (1977) 27.

5.   U. Krey, *Zeits. Phys.* **B31** (1978) 247.

6.   U. Krey, *J. Mag. Mag. Mat.* **7** (1978) 150.

7.   R. Alben, *AIP Conf. Proc.* **29** (1976) 136.

8.   L. von Heimendahl, *J. Phys. F: Met. Phys.* **5** (1975) L141.

9.   C. L. Chien and R. Hasegawa, *Phys. Rev. B* **16** (1977) 2115, 3024.

10.   H. A. Mook, N. Wakabayashi, and D. Pan, *Phys. Rev. Lett.* **34** (1975) 1029.

11. J. W. Lynn and J. J. Rhyne, in *Spin Waves and Magnetic Excitations 2*, ed. A. S. Borovik-Romanov and S. K. Sinha (Elsevier, 1988) Chap. 4.

12. T. Izuyama, D.-J. Kim, and R. Kubo, *J. Phys. Soc. (Japan)* **18** (1963) 1025.

13. R. Harris, M. Plischke, and M. J. Zimmermann, *Phys. Rev. Lett.* **31** (1973) 160.

14. E. M. Chudnovsky, R. A. Serota, and W. M. Saslow, *Phys. Rev. B* **33** (1986) 251 and references therein. See also Chap. 3.

15. W. M. Saslow, *Phys. Rev. B* **35** (1987) 3454.

16. D. L. Huber, *Phys. Rev. B* **37** (1988) 3497.

17. R. J. Elliott, J. A. Krumhansl, and P. L. Leath, *Rev. Mod. Phys.* **46** (1974) 465.

18. D. L. Huber and W. Y. Ching, *Phys. Rev. B* **40** (1989) 2578.

19. D. L. Huber and W. Y. Ching, *Phys. Rev. B* **39** (1989) 4453.

20. D. L. Huber and W. Y. Ching, *J. Appl. Phys.* **64** (1988) 5627.

21. I. Avgin, D. L. Huber, and W. Y. Ching, *Phys. Rev. B* **46** (1992) 223.

22. R. Haydock, V. Heine, and M. Kelly, *J. Phys. C: Sol. State* **8** (1975) 2591.

23. R. Alben and M. F. Thorpe, *J. Phys. C: Sol. State* **8** (1975) L275.

6. Appendix

6.1. Overview

The purpose of the Appendix is to provide mathematical details on the two methods of calculating the magnon density of states and the dynamical structure factor that were mentioned in Sec. 2. The two methods, recursion relations and equation-of-motion, are discussed in the context of fully aligned (ferromagnetic) ground states but can be generalized to arbitrary ground state spin configurations.

6.2. Recursion Relations

The method of recursion relations allows one to evaluate diagonal matrix elements of

278

the resolvent operator, $(E+i\varepsilon - \mathcal{H})^{-1}$, by means of a continued fraction expansion with coefficients determined recursively [5,22]. This will be illustrated using states of the form $|j> = a_j^\dagger |0>$, where $|0>$ denotes the zero-magnon or vacuum state. Defining the matrix element of the resolvent operator by $R_{jj}(E)$, one has

$$R_{jj}(E) = <j|(E+i\varepsilon - \mathcal{H})^{-1}|j>,$$

$$= \cfrac{1}{z_1 - \cfrac{b_2^2}{z_2 - \cfrac{b_3^2}{z_3 - ...}}},$$

$$(6.23)$$

where $z_n = E+i\varepsilon+a_n$. The coefficients a_n and b_n are defined by [5]

$$a_n = \{n|\mathcal{H}|n\}, \tag{6.24}$$

$$b_n = \{\phi_n|\phi_n\}^{1/2}, \tag{6.25}$$

in which

$$|n\} = |\phi_n\}/b_n. \tag{6.26}$$

and the state vector $|\phi_n\}$ obeys the equation

$$|\phi_{n+1}\} = \mathcal{H}|n\} - a_n|n\} - b_n|n-1\}. \tag{6.27}$$

The recursive sequence is started at n=1 with $|1\}=|j>$, $|0\}=0$, and $b_1=0$. By evaluating the right hand side of (6.27) one obtains $|\phi_2\}$ and, thus, a_2 and b_2, etc.. Since the amorphous clusters contain finite numbers of atoms, the recursion relations must be terminated to avoid "contamination" by surface sites. In Ref. 5, this was done by taking $b_n=b_6$, for n>6 and $a_n=2b_6$ for n>5.

The quantity $(-1/\pi)\text{Im } R_{jj}(E)$, where Im denotes the imaginary part, is an approximation to the local density of states associated with spin j. One can obtain an approximation to the average density of states by using the recursion relations (6.24) - (6.27) and taking the starting state $|1\}$ to be a random superposition of the states $|j>$ with weights $\pm N^{-1/2}$ assigned at random. Likewise, the dynamical structure factor is given by $(-1/\pi)\text{Im } R_{qq}(E)$, where the resolvent is calculated with the starting state $|1\}$ given by the plane wave combination

$$N^{-1/2}\Sigma_j \exp(i\mathbf{q}\cdot\mathbf{r}_j)|j>.$$

6.3. *Equation-of-Motion Method*

The equation-of-motion method involves the calculation of the time-dependent correlation functions [7,23].

$$G_j(t,\alpha) = <0|a_j(t)\Sigma_k\alpha_k a_k^\dagger|0>, \tag{6.28}$$

where the coefficient α_k depends on the quantity to be calculated. These functions are solutions to a set of N first order linear differential equations associated with the N spins in the array under study

$$idG_j(t,\alpha)/dt = \Sigma_k U_{jk}G_k(t,\alpha), \qquad t \geq 0, \tag{6.29}$$

where U_{jk} is the dynamical matrix defined in Sec. 2. The corresponding initial conditions are $G_j(0+,\alpha)=\alpha_j$.

The dynamical structure factor and the density of states are calculated by integrating the equations of motion of the correlation functions and then taking a Fourier transform with appropriate weights. In the case of the structure factor, one has

$$S_o(\mathbf{q},E) = -(2S/\pi)\text{Im} \ _o\!\int^\infty dt e^{iEt}\Sigma_j\alpha_j^*G_j(t,\alpha), \tag{6.30}$$

with $\alpha_j=N^{-\frac{1}{2}}\exp(-i\mathbf{q}\cdot\tau_j)$. The (average) density of states is given by a similar expression, except that $\alpha_j=N^{-\frac{1}{2}}\exp(i\phi_j)$, where ϕ_j is a random number between 0 and 2π.

In applying the equation-of-motion method, one generally introduces a convergence factor, $\exp(-\lambda t)$, into the integrand in (6.30) and carries out the integration to a time $T >> 1/\lambda$. This convergence factor can also be looked upon as simulating the effects of a finite energy resolution on the order $\Delta E \approx \lambda$.

CHAPTER SEVEN

MAGNETISM AND MAGNETO-OPTICS OF RARE EARTH-TRANSITION METAL GLASSES AND MULTILAYERS

D. J. SELLMYER, R. D. KIRBY, and S. S. JASWAL

Behlen Laboratory of Physics

and

Center for Materials Research and Analysis

University of Nebraska, Lincoln, NE 68588-0113, USA

1.   Introduction and Scope

The goal of this chapter is to review recent advances in our understanding of the magnetic structure and magneto-optic properties of amorphous rare earth (RE) - transition metal (TM) alloys and multilayers. Early work in the field was reviewed by Cochrane, Harris and Zuckermann[1], Sellmyer and O'Shea[2], Sellmyer and Nafis[3], and in two books published in 1984 by Moorjani and Coey[4] and Kaneyoshi[5]. Very recent reviews on magnetic glasses include one on random anisotropy and phase transitions by Sellmyer and O'Shea[6] and on magnetic amorphous alloys and magneto-optics by Hansen[7]. Theoretical aspects of the random anisotropy problem are also covered in this book in chapter three by Chudnovsky. Regarding magneto-optic properties of RE-TM alloys and other materials, the proceedings of two recent conferences on this topic will serve to give the reader an excellent overview of the present status of the field[8,9]. In particular we refer to review articles by Gambino[10], Sellmyer, Shan, Shen and Kirby[11], Kryder[12], and Schoenes and Brandle.[13]

This contribution is not intended to provide a comprehensive review of all recent work in the field. The above-mentioned references should be consulted for this purpose. Rather, our aim is to focus on three *fundamental* aspects on which either (a) significant new understanding has emerged very recently, or (b) the level of understanding is so rudimentary that much additional work remains to be done. The organization of this chapter is as follows. In the second section we give a brief overview of the magnetic interactions and structures that are important for understanding amorphous RE-TM alloys. The roles that random magnetic anisotropy and exchange interactions play in determining the magnetic structure will be emphasized. The third section discusses the origins of perpendicular magnetic anisotropy (PMA) in films and multilayers, especially with regard to anisotropic pair correlations in films containing RE elements with orbital angular momentum. Recent developments in determining local atomic arrangements have led to a more complete understanding of the origins of PMA in some cases. In the fourth section we outline the complex problem of the relationship between the fundamental electronic structure and the resulting magneto-optic properties. The computational problems are especially severe because of the complexity of spin-orbit interactions and amorphous structures. In any magnetic material the magnetization reversal mechanisms and domain dynamics are of utmost importance for fundamental understanding and practical applications, and these topics are considered in section five. Particular emphasis is given to the discussion of physical models which attempt to explain the reversal process in these complex systems. Finally we conclude with a summary and give some suggestions for future work on unsolved problems

2.   Overview of Magnetic Interactions and Structures

We are interested in the magnetic properties of amorphous RE-TM alloys with some typical examples being a-$Tb_{20}Co_{80}$, a-$Gd_{20}Fe_{80}$ and a-$Nd_{40}Fe_{45}Co_{15}$. Alloys such as these are very complicated and to achieve an understanding of their properties it has been necessary to develop rather oversimplified, but useful, models. Because of the structural disorder it is important to account for the possibiity of fluctuations and randomness in both exchange interactions and directions of local single-ion anisotropy. Complicating matters further is the presence of both localized and itinerant electrons with their associated magnetic moments. We outline in the following the most important concepts used to rationalize the magnetic structures and properties of this class of materials.

In an amorphous material containing a RE ion with orbital-angular momentum, there are electric-field gradients which tend to produce an easy axis of magnetization for each RE ion. This axis is imagined to be random from site-to-site, and this led Harris *et al.* to the Hamiltonian[1]

$$H = -J_o \sum_{ij} S_i \cdot S_j - D \sum_i (\hat{n}_i \cdot S_i)^2 \tag{1}$$

in which the first term represents the exchange and the second term the random magnetic anisotropy (RMA). The local easy axes ($\hat{n}_i$) are taken as random in direction. If the exchange interactions are assumed ferromagnetic ($J_o > 0$) and the RMA interactions uniaxial ($D > 0$), then the effect of the first term is to *align* the spins while the RMA term tends to *scatter* them.

A somewhat more realistic situation than Eq. 1 can be described by a Hamiltonian which allows for exchange fluctuations (ΔJ_{ij}) as well as RMA, namely[2]

$$H = -\sum_{i,j} (J_o + \Delta J_{ij}) S_i \cdot S_j - D \sum_i (\hat{n}_i \cdot S_i)^2. \tag{2}$$

The exchange-fluctuation term can be quite important in the type of alloy we are considering in this review. Figure 1 illustrates the point in a schematic phase diagram in a three-dimensional space in which $t = k_B T / J_o$, $\alpha = D / J_o$ and $\delta = \langle \Delta J \rangle / J_o$.[3] In the $\alpha = 0$ plane there is a ferromagnetic phase (F) for small δ values and a spin-glass-like phase (SG) for large δ values; there also is the possibility of a mixed phase (M) in the intermediate region. On the other hand in the $\delta = 0$ case a speromagnetic (spin scattered) structure (SM) exists for large α values and a correlated speromagnetic structure (CSM) exists for small α values where there is short-range ferromagnetic order. It can be seen that in the presence of all three terms in the Hamiltonian there will be a very complicated three-dimensional phase space to be considered.

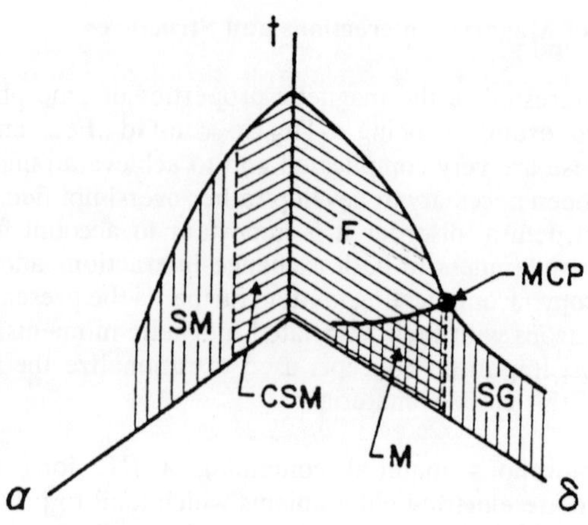

Fig. 1.   Schematic phase diagram of possible magnetic states in the presence of RMA and exchange fluctuations [after Ref. 3].

One of the important early theorems relevant to the magnetic structure in the presence of RMA is that of Imry and Ma.[5] These workers showed in the presence of a finite random anisotropy that the magnetic structure could not have long-range ferromagnetic order. This means that there is always a characteristic length over which the magnetization varies to minimize the energy of the finite D system. Of course, in the limit of very small D or equivalently very large J_0/D it is difficult to distinguish the resulting properties from the ferromagnetic case. The susceptibility approaches $(J_0/D)^4$ in this case.[6]

On the basis of the above considerations it has been shown that a fairly complicated set of magnetic structures can exist. Figure 2 shows some examples in which there are speromagnetic, asperomagnetic, and ferrimagnetic structures, for example, depending on the presence of RMA in the first two cases and its absence in the third case. In addition, when there are both rare-earth and transition metals such as iron or cobalt present, one can obtain even more complex structures as illustrated in the figure. An important point is that the heavy rare-earth moments tend to couple antiparallel whereas the light rare earths couple parallel to the transition-metal moments; this effect has been widely seen and confirmed in many amorphous rare earth-transition metal glasses. In this review we will distinguish spin-glass ordering from speromagnetic ordering on the basis of the fundamental origins of the scatter in the spins. In the spin-glass case there are fluctuations in exchange, including positive and negative signs, and this results in *frustration* which leads to the spin disorder. In the speromagnetic case it is the random anisotropy which leads to the spin disorder.

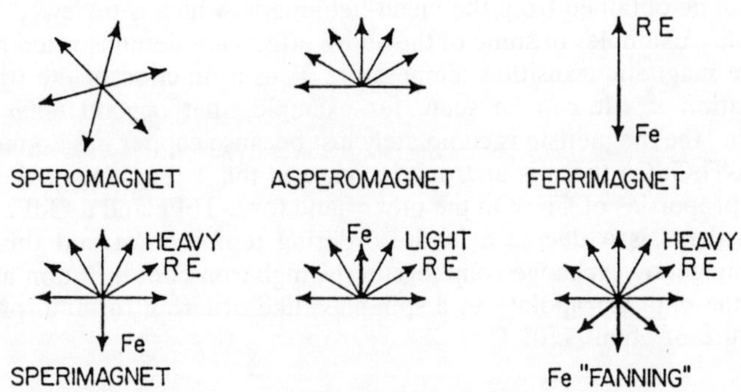

Fig. 2. Schematic diagrams and definitions of phases possible in one- and two-component RMA systems [after Ref. 2].

In attempting to apply Eq. 2 to understand RE-TM amorphous alloys there are several additional complicating features. One is that when considering the magnetic interactions between rare-earth ions in general, the relevant Hamiltonian has the form

$$H = -\sum_{ij} (J_{ij} S_i \cdot S_j) \qquad (3)$$

where in this case the exchange interaction is given by the Ruderman-Kittel interaction. This interaction can be expressed as

$$J_{ij} \sim J_{sf}^2 F(\xi_{ij}) e^{-r_{ij}/\lambda} \qquad (4)$$

where

$$F(\xi_{ij}) = \frac{\xi_{ij} \cos \xi_{ij} - \sin \xi_{ij}}{\xi_{ij}^4}. \qquad (5)$$

Here J_{sf} is the rare-earth conduction-electron interaction and λ is the mean free path. $\xi_{ij} = 2k_f r_{ij}$ where k_f is the Fermi momentum and r_{ij} is the distance between the two ions in question. Because of its oscillatory nature, this interaction generally will lead to a spin-glass-like behavior or at least a tendency in that direction. An additional complicating factor is that certain transition-metal moments, for example iron, can have local-environment effects which render the iron moment unstable depending on its nearest-neighbor configuration.[7] Given these difficulties it has been found that a zero-order model describing the magnetic structure of rare-earth transition-metal glasses can be obtained from the mean-field model which is reviewed in some detail by Hansen.[7] Examples of some of the above effects are demonstrated in Fig. 3 which shows the magnetic transition temperature T_c as a function of the transition-metal concentration x. It can be seen, for example, that beyond 60% of copper in a-$Gd_{1-x}Cu_x$ the magnetism is completely lost because copper has no moment. In the case of a-TbGdCo there is a strong increase in the T_c because of the strong local-moment properties of Co. On the other hand for a-TbFe and a-GdFe, for high iron contents, there is a decrease in the ordering temperature and this results from antiferromagnetic exchange couplings in the high iron-concentration alloys. That is, for x=1 the data extrapolate to a spin-glass-like ordering for amorphous iron at a temperature of about 150° C.

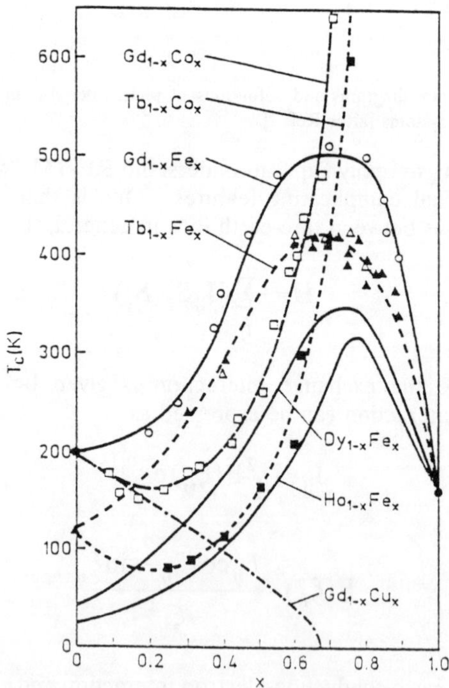

Fig. 3. Variation of Curie temperature with TM composition [after Ref. 7 and references therein].

In the Hamiltonians discussed above the itinerant character of the conduction electrons associated with transition metals has been ignored. A model that takes the d-electron character into account was developed by several workers and the Hamiltonian was written as[7]

$$H = \sum_{ij} t_{ij} c_{i\sigma}^{+} c_{j\sigma} + I \sum_{i} n_i^{\uparrow} n_i^{\downarrow} - J_{exch} \sum_{i} J_i \cdot \sigma_j - D \sum_{j} J_{z_j}^{2} . \tag{6}$$

In this equation the first term represents the kinetic energy of the d-band electrons, the second term is the Hubbard correlation; $c_{i\sigma}^{+}$ and $c_{j\sigma}$ are creation and annihilation operators for d-band electrons of spin σ at sites i and j, and $n_i^{\uparrow}$ and $n_i^{\downarrow}$ are particle number operators for spin-up and spin-down electrons. t_{ij} is the d-band transfer matrix and I is the Hubbard interaction. The third term represents the interaction between the rare-earth spins and the transition-metal atoms where J_{exch} is the coupling constant and σ_j is the d-band spin. The last term is the random anisotropy. In the molecular-field approximation this Hamiltonian transforms to

$$H = \sum_{k} \left(\epsilon_k - \frac{1}{2} \sigma \Delta \right) c_{k\sigma}^{+} c_{k\sigma} - \sum_{j} \left(J_{exch} \bar{\sigma}_z J_{zj} + D J_{zj}^2 \right) , \tag{7}$$

where ϵ_K is the Fourier transform of t_{ij}, $\Delta = (I\bar{\sigma}_z + 2J_{exch}\bar{J}_z/x)$ is the molecular field of the d electrons, and x is the TM concentration. $\bar{\sigma}_z$ is the polarization of the d-band electrons which are treated in the free-electron approximation. $\bar{\sigma}_z$ and the total number of d electrons, N, are determined by the equations

$$\bar{\sigma}_z = \sum_{k} \left[f(\epsilon_k - \Delta/2) - f(\epsilon_k + \Delta/2) \right] \tag{8}$$

$$N = \sum_{k} \left[f(\epsilon_k - \Delta/2) + f(\epsilon_k + \Delta/2) \right] , \tag{9}$$

where $f(\epsilon)$ is the Fermi function. Self-consistent values of $\bar{\sigma}_z$ and $\bar{J}_z$ are determined by an iterative procedure and the final result of the theory is that the net magnetization can be written as

$$M_s(T) = \mu_B \left[x\bar{\sigma}_z + (1-x) g\bar{J}_z \right] . \tag{10}$$

Figure 4 shows an example of magnetization data of a-RE-Co alloys by Jouve, et al.[14] The minima at temperatures between 100 and 200 K correspond to compensation temperatures where the rare-earth subnetwork cancels the cobalt subnetwork magnetization. Bhattacharjee et al.[15] have been able to obtain rather good fits to the data of Jouve et al.[14] by choosing appropriate values for the parameters $N(E_F)$, J_{exch}, D, and I. Perhaps the weakest aspect of this calculation is the free-electron approximation used for the amorphous metal d-band.

288

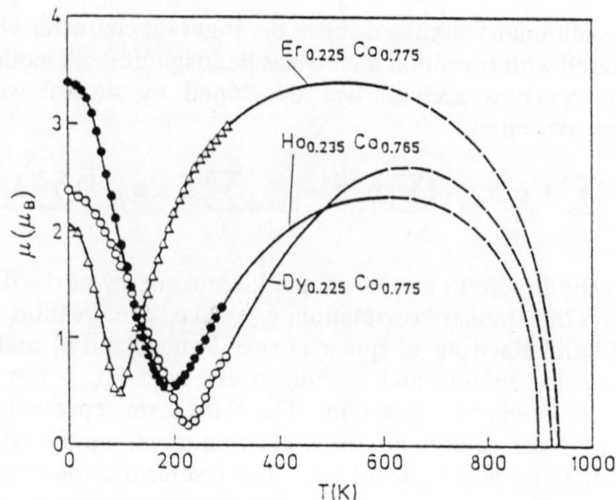

Fig. 4. Temperature dependence of average spontaneous moment per formula unit for a-RE-Co alloys [after Refs. 7 and 14].

In considering the magnetic phase transitions which exist in amorphous rare earth-transition metal alloys one must distinguish those situations in which RMA is of prime importance from those in which exchange fluctuations dominate. In those cases where to a good approximation the exchange fluctuations can be ignored it has been shown that there exist speromagnetic transitions which lead to a nonlinear susceptibility which diverges at the ordering temperature. Scaling analyses have been applied to such transitions for a number of a-RE-TM alloys, and these are reviewed in detail by Sellmyer and O'Shea.[6] In addition these latter authors discuss the interesting phenomenon of double transitions in anisotropic alloys in which as the temperature is lowered there is a transition from paramagnetism into a correlated speromagnetic state and then finally into another state characterized by speromagnetism. These phenomena can be rationalized on the basis of a phase diagram given by Chudnovsky.[16]

3.   Perpendicular Anisotropy in Films and Multilayers

It has been known since the early 1970s that sputtered amorphous RE-TM alloy films can exhibit a uniaxial anisotropy perpendicular to the film plane. That is, the alloys display a uniaxial anisotropy energy

$$E_a = K_u \sin^2\theta \tag{11}$$

where θ is the angle between the preferred axis and the magnetization, and the values of K_u range from 10^4 to 10^7 ergs/cm^3. This subject has been quite a controversial one and hundreds of papers have been written in an attempt to explain the origins of this perpendicular anisotropy. For a homogeneous glass there should be no anisotropy in the structure and therefore no magnetic anisotropy. Hansen has reviewed the status of this field up to about 1990 and points out that there have been several favored models for this perpendicular anisotropy.[7] Basically they have included: (1) a pair-ordering model or an atomic or nanoscale structural anisotropy, (2) a form of microstructural anisotropy having to do with columnar growth or other defect structures, and (3) magneto-elastic interactions involving strain due to a substrate plus inverse magnetostriction. For rare earths with orbital-angular momentum it is well known that the single-ion anisotropy caused by electric field gradients is the major source of anisotropy. Figure 5 shows the magnitude of the uniaxial anisotropy as a function of Tb content in some amorphous alloys based on GdFe or GdCo. It is clearly seen that the anisotropy increases linearly with the Tb content which is consistent with the interpretation of this anisotropy being caused by the single-ion anisotropy associated with Tb. Several workers have investigated the origins of perpendicular anisotropy in amorphous rare earth-transition metal alloys and have investigated specifically magnetostriction as well as the single-ion interactions plus anisotropic pair correlations. Figure 6 shows the uniaxial anisotropy constant for amorphous Gd-RE-Co alloys both before and after the alloys were removed from the substrate. The filled circles represent the data after the samples were removed from the substrate and therefore represent a possible contribution due to inverse magnetostriction. The lower panel of this figure shows the difference in the stress-free data and the K_u value for a-Gd-Co (where the single-ion contribution should be zero); thus ΔK_u as a function of the rare earth involved should represent the contribution due to the single-ion anisotropy plus anisotropic pair correlations. The full line in this graph was calculated on the basis of a formula due to Suzuki et al.[17] in which

$$K_u = -\frac{3}{10}\alpha_J\, J(J-\tfrac{1}{2})e^2q\frac{<r^2>}{a^3}\delta x, \tag{12}$$

where J is the total angular momentum, α_J is the Steven's factor, e,q,a and x are the electron charge, the valency of the neighboring ions, the average distance between next-nearest neighbors, and RE concentration, respectively. δ is an assumed local strain of the atomic environment and $<r^2>$ represents the average distribution of the 4f wave function. Clearly in this graph there is a correlation between the theoretical calculation based on single-ion interactions and the experimental data. The stress-induced contribution,

$$K_u^{\lambda} = -\frac{3}{2}\sigma\lambda_s \tag{13}$$

where σ is the stress and λ_s is the saturation magnetostriction constant can account for values ranging from 20 to 40% of the total perpendicular anisotropy.

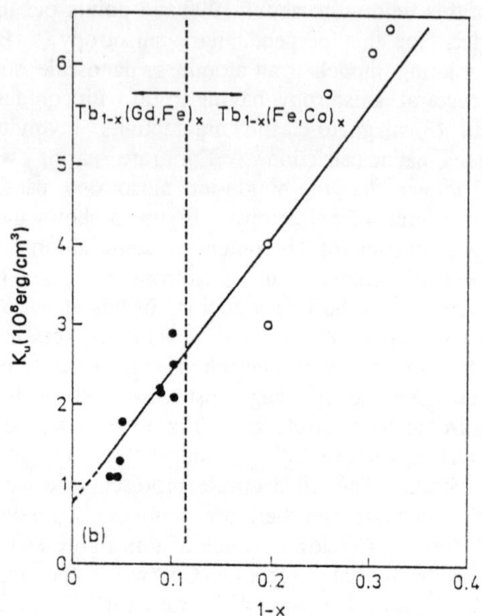

Fig. 5. Dependence of uniaxial anisotropy constant on Tb concentration in evaporated GdTbFe and TbFe alloys [after Ref. 7].

As discussed above, for about two decades the origins of uniaxial anisotropy in rare-earth transition metal amorphous alloys remained mysterious and no definitive evidence existed on such structural anisotropy. Recent studies by Shan and coworkers[18] have adopted a method of controlling the anisotropic pair correlations by producing compositionally modulated amorphous structures in a two-gun sputtering system. Alloys such as Dy/Co, Tb/Fe, Dy/Fe, and others were studied. Clearly in an idealized A/B multilayer, as the individual layer thicknesses approach one-to-two monolayers the relative number of AA and BB pairs in the plane of the film, and AB pairs perpendicular to the film, will be increased in comparison to the isotropic random structure. Thus one has in such a system control of the different types of pairs that exist. The situation in these RE/TM multilayers can be envisioned as an intermediate one between a completely isotropic homogeneous rare-earth transition-metal glass where $K_u = 0$ and a single-crystal RE-TM hexagonal or tetragonal thin film with the easy axis normal to the film plane where K_u is of the order of 10^7 ergs/cm^3.

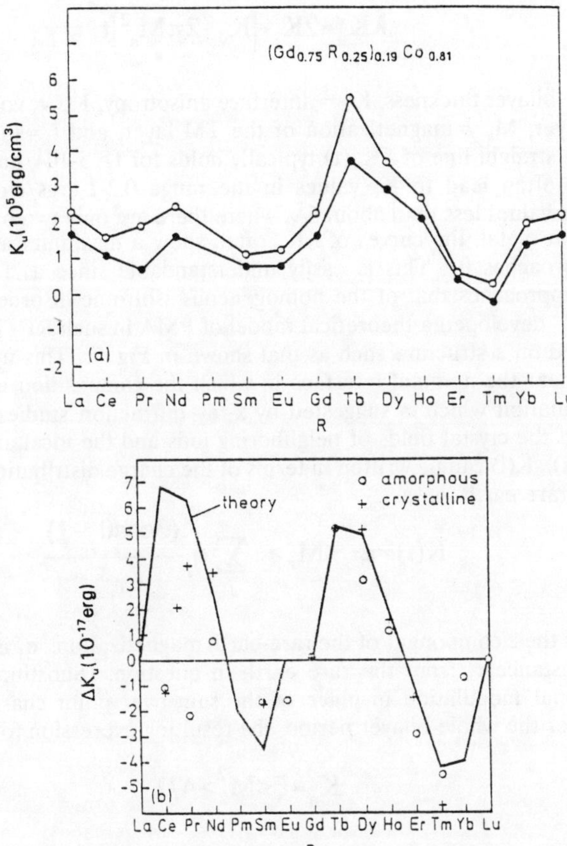

Fig. 6. (a) Uniaxial anisotropy constant for amorphous Gd-RE-Co alloys before (open circles), and after (closed circles) removal from substrate, versus rare-earth element,

(b) Estimate of the stress-free, single-ion contribution to the uniaxial anisotropy per RE atom (R). The crosses refer to crystalline compounds and the full line was calculated from the single-ion model of Ref. 17 [after Ref. 7].

The systematics of PMA in RE/TM multilayered films are similar to those of transition-metal/nonmagnetic films such as Co/Au or Fe/Cu. These have been reviewed by Sellmyer et $al.$[19] who show that both classes of films have PMA representable with the equation

$$\lambda K_u = 2K_i + \left[K_v - 2\pi M_s^2 \right] t, \tag{14}$$

where λ = bilayer thickness, K_i = interface anisotropy, K_v = volume anisotropy of the TM layer, M_s = magnetization of the TM layer, and t = thickness of the TM layer. The straight line of $\lambda K_u(t)$ typically holds for $t \geq$ 5-10Å and the intercepts of such plots often lead to K_i values in the range 0.2-1 ergs/cm^2. However, as t approaches values less than about 5Å, where there are only two or three monolayers of transition metal, the curves of λK_u often show a maximum and an approach to 0 as t approaches 0. This is easily understandable since as t approaches 0 the structure approaches that of the homogeneous isotropic disordered alloy or glass. Shan et $al.$[18] developed a theoretical model of PMA in such RE/TM multilayers and it was based on a structure such as that shown in Fig. 7. This figure shows a rare-earth ion near the nominal interface in which the composition actually has a sine-wave modulation which is suggested by x-ray diffraction studies. The calculation depends on the crystal fields of neighboring ions and the local anisotropy $K(i)$ at a position $z(i)$. $K(i)$ can be written in terms of the charge distribution q_j in the vicinity of this ith rare earth-ion as

$$K(i) \propto \alpha_J <M_z^2>_i \sum_j q_j \frac{(3\cos\theta_j - 1)}{r_j^3} \tag{15}$$

Here M_z is the z component of the rare-earth magnetization. q_j is the charge on the ion at a distance r_j from the rare earth in question. Substituting the sine-wave compositional modulation in place of the sum over point charges and taking an average over the whole bilayer period, the resulting expression for the multilayer is

$$K_u = \xi <M_z^2> A/\lambda \tag{16}$$

where ξ is a constant and A is the amplitude of the compositional modulation. Figure 8 shows an example of the mean-field fits for the composition and magnetization modulation for an 8Å Dy/8Å Co film. The theoretical fit of the theory via Eq. 16 to the K_u data for this sample is shown in Fig. 9. In this case it is seen that the fit is very good, and similar agreement is seen for other RE/TM multilayers. The maximum value of K_u is in the range of 10^6 ergs/cm^3 and this typically happens when both the TM and RE have nominal layer thicknesses of about 2 monolayers. This is consistent with the maximum degree of anisotropic pair correlations and maximum K_u occurring when there are roughly two monolayers of each constituent.

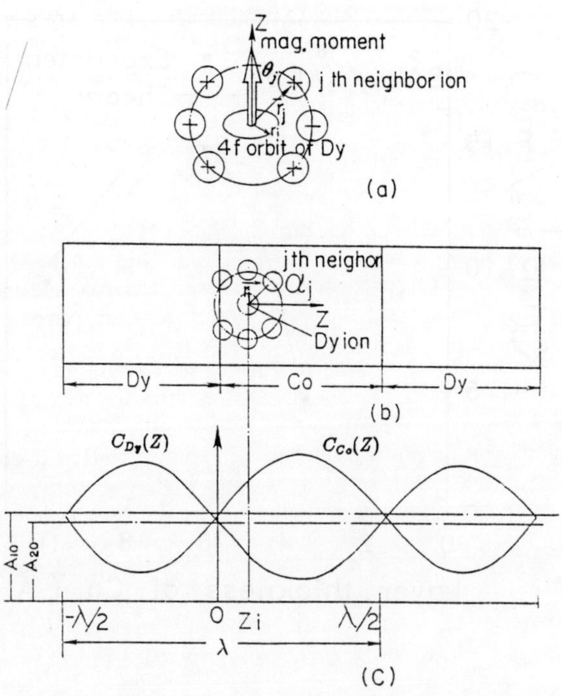

Fig. 7. Model for the crystal field(a), local anisotropy K(i) (b), and compositional modulation (c) for amorphous RE/TM multilayers [after Ref. 18].

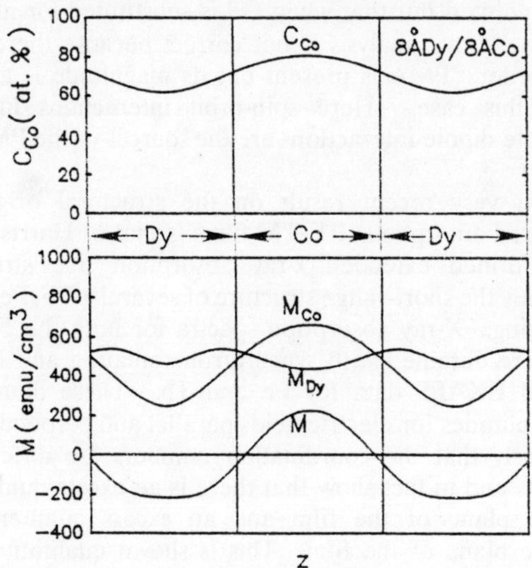

Fig. 8. The distribution of Co concentration and Co, Dy, and total magnetization for 8ÅDy/8ÅCo multilayers [after Ref. 18].

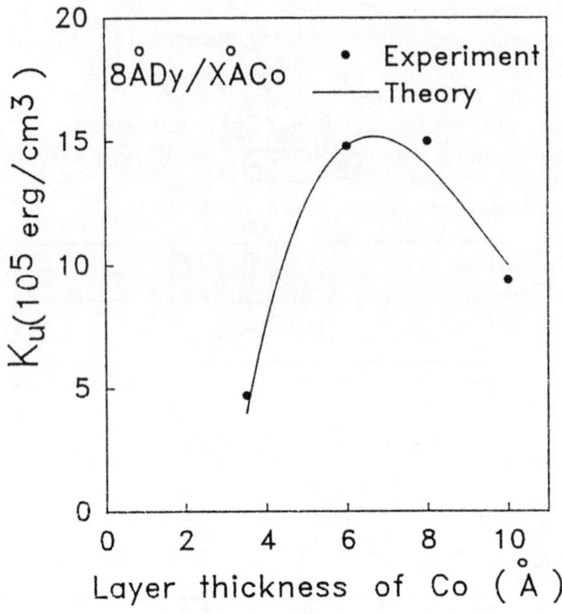

Fig. 9.   A comparison between the calculated and experimental anisotropy for 8Å Dy/XÅ Co [after Ref. 18].

It should be pointed out that when Gd is substituted for a heavy rare earth such as Dy or Tb the above analysis is not correct because there is no single-ion anisotropy to first order.   PMA is present but its magnitude is about an order of magnitude less in this case.   Here spin-orbit interactions involving the TM subnetwork and dipole-dipole interactions are the sources of the PMA when it exists.

An important very recent result on the structural origins of magnetic anisotropy in sputtered amorphous RE-TM films is that of Harris and coworkers.[20] These authors performed extended x-ray absorption fine structure (EXAFS) measurements to study the short-range structure of several a-Tb_xFe_{1-x} films prepared by ion-beam sputtering.   X-ray absorption spectra for both the Fe K and Tb LIII absorption edges were obtained with synchrotron radiation and Fig. 10 shows the Fourier transformed EXAFS data for Fe and Tb.   These data show a distinct difference in the amplitudes for electric fields parallel and perpendicular to the film. The data show clearly that the coordination numbers are different for the two different orientations and in fact show that there is an excess number of Fe/Fe and Tb/Tb pairs in the plane of the film and an excess number of Tb/Fe pairs perpendicular to the plane of the film.   This is shown qualitatively in the typical atomic arrangements shown in Fig. 11, where vertical and horizontal arrows correspond to normal and in-plane directions, respectively. These ideas are quite consistent with the above-mentioned anisotropy sources of Shan *et al.* based on single-ion anisotropy plus anisotropic pair correlations as induced by multilayering. Moreover, Harris *et al.* heat treated one of their samples $Tb_{0.26}Fe_{0.74}$ at 300°C for one hour and the resulting Fourier-transformed Fe EXAFS amplitude clearly shows that the structural anisotropy seen in Fig. 10 has been reduced to below the resolution of the experiment by the annealing. The authors thus suggest that the high temperature anneal has removed most of the anisotropic pair correlations and the remaining anisotropy which is perhaps 20% of the original is likely to be due to magneto-elastic interactions between the film and the substrate.

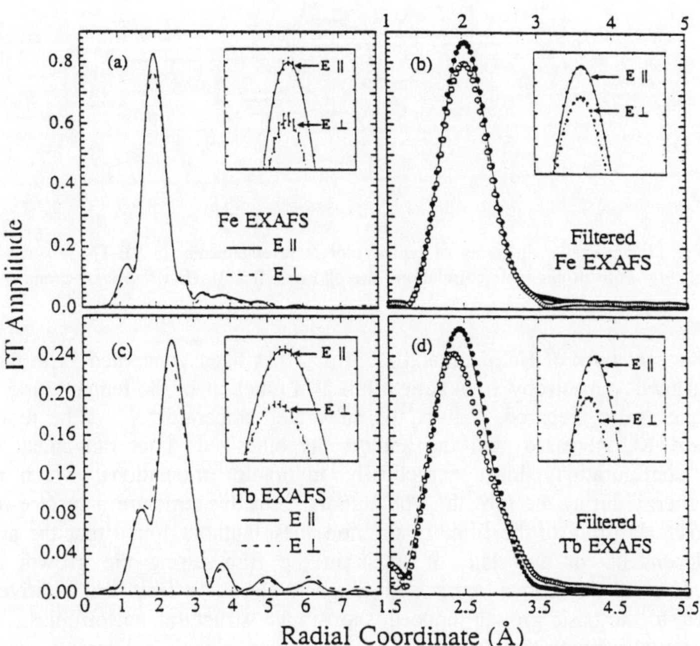

Fig. 10. Fourier-transformed EXAFS data for Fe(a) and Tb(c). (b) and (d) represent filtered EXAFS data compared to theoretical fits (filled circles, E∥; open circles, E⊥) [after Ref. 20].

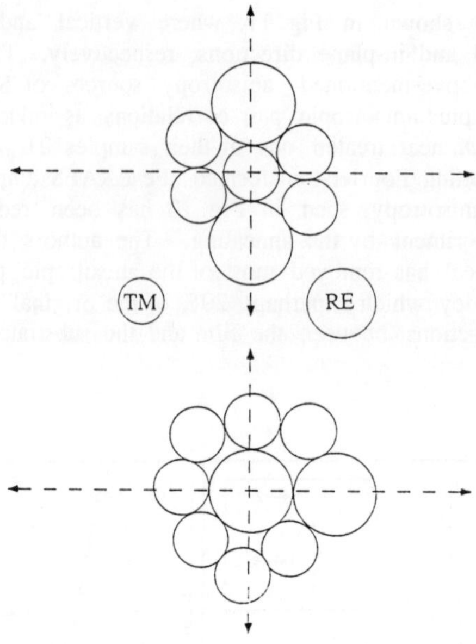

Fig. 11. Schematic diagrams of typical atomic arrangements in RE-TM amorphous alloys. Anisotropic pair correlations are shown (after V. Harris, private communication).

Recent work of Hellman and Georgy [21] has been concerned with the change in the magnetic anisotropy of a-TbFe films as a function of the temperature at which the samples were prepared, that is, the substrate temperature. As the temperature was raised K_u increased and the authors hypothesized a rearrangement of local adatom configurations into energetically favorable orientations which minimize surface energy during the growth. The authors term this structure a surface-mediated amorphous texturing of the film. In addition these authors found that the anisotropy was independent of the state of stress in the film during the growth and thus suggested as in the above work that magneto-elastic effects are relatively small compared to the basic growth-induced short-range structural anisotropies. Hellman and Georgy also argue that the model of bond-orientational anisotropy suggested by Yan and coworkers [22] is not likely to be a major source of magnetic anisotropy. This bond orientational anisotropy was postulated to result from an anelastic strain which would change sign if the stress at the surface of the film during growth changed from compressive to tensile. Hellman and Georgy found that this was not the case in their work. Finally, in these TbFe films the authors argued reasonably that the surface dipolar interactions considered by Fu et al.[23] are not an important source of anisotropy in a-RE-TM films containing heavy rare earths. Theoretical calculations of Jaswal also came to the same conclusion.[24]

Recently Mergel and coworkers have attempted a synthesis of much of the extant data on perpendicular anisotropy in amorphous rare earth-transition metal films.[25] They have proposed a "pseudocrystalline" or cluster model for the occurrence of perpendicular anisotropy. The basic idea is that during layer-by-layer growth small planar hexagonal units are formed which define on the average a preferential axis perpendicular to the film plane. The units are regarded as similar in structure to relaxed crystalline ones which comprise typically six rare-earth atoms and also take into account the known next-neighbor RE-TM and TM-TM coordination numbers in the amorphous state. Figure 12 shows planes of one such class of intermetallic compounds, here denoted as RT_5. The different shadowing indicates different planes and the atoms in the key planes are imagined to relax in the amorphous state with respect to the crystalline state as indicated by arrows. In Fig. 13 a plot is shown of K_u as a function of x in a variety of a-$R_{1-x}T_x$ alloys. The compositions of the corresponding crystalline alloys RT_2, RT_3 and RT_5 are shown at the top. It is concluded that the observed maxima of K_u for DyFe, DyCo, and NdCo are directly related to their composition and in the amorphous case a maximum is expected at compositions for which the maximum number of hexagonal units are formed during the thin-film growth. While there is no direct evidence in this work of such structural units it is argued that the model is able to explain known experimental results on the effects of composition, substrate temperature, annealing and bombardment during sputter deposition on the resulting magnetic anisotropy of this class of materials.

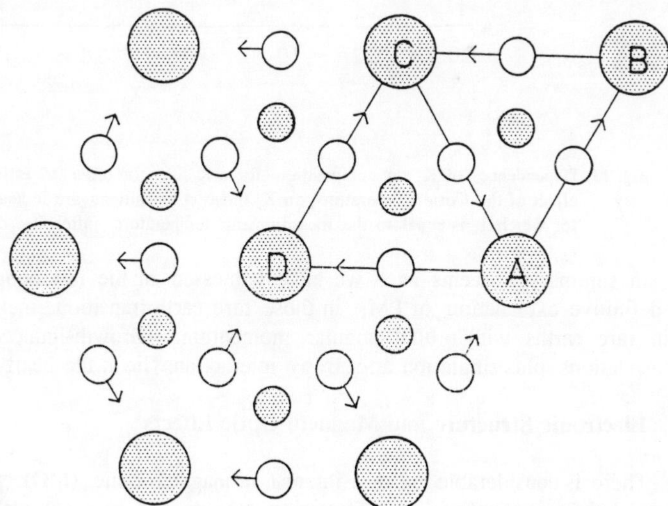

Fig. 12. Planes of RT_5 intermetallic compounds, with different shadowing representing different planes. The T-plane atoms are imagined to relax in the amorphous state relative to the crystalline state as indicated by arrows [after Ref. 25].

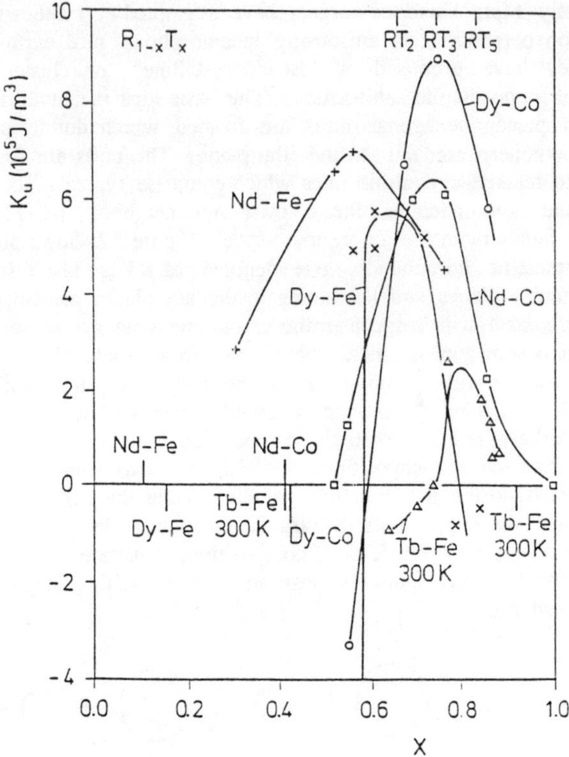

Fig. 13. Dependence of K_u on composition for a-$R_{1-x}T_x$. In order to estimate an indirect effect of the Curie temperature on K_u those compositions are indicated on the x-axis for which T_c is equal to the measurement temperature [after Ref. 25].

In summary it seems as if we have witnessed in the last couple of years a fairly definitive explanation of PMA in those rare earth-transition metal films which contain rare earths with orbital angular momentum. Growth-induced anisotropic pair correlations plus single-ion anisotropy interactions lie at the heart of the matter.

4. Electronic Structure and Magneto-Optic Effects

There is considerable current interest in magneto-optic (MO) effects in solids because of their applications in information storage and retrieval technology. The incident plane-polarized light is elliptically polarized upon reflection and refraction due to these effects. The former is known as the Kerr effect while the latter as the Faraday effect. The quantity of practical interest is the angle between the polarization vector of the incident light and major axis of the elliptically polarized light which is of the order of 0.3°. Often the materials with favorable magnetic properties are opaque so that one is interested in the Kerr effect where the angle mentioned above is known as the Kerr rotation.

The magneto-optic effects are produced when the indices of refraction for the right and left circularly polarized light are not the same.[26,27] This is possible in a ferromagnet with non-zero-spin-orbit interactions. Thus, one must carry out fully-relativistic self-consistent spin-polarized electronic-structure calculations to understand and explain magneto-optic effects.

In the present discussion we consider the polar Kerr effect in which the light-propagation and magnetization directions are along the z-axis perpendicular to the surface of the sample, the most favorable configuration for maximum Kerr rotation. The Kerr rotation θ_K is given by[26]

$$\theta_k = Re \left[\frac{N_r - N_l}{1 - N_r N_l} \right] \tag{17}$$

and the ellipticity (the ratio of minor to major axis)

$$\eta_k = Im \left[\frac{N_r - N_l}{1 - N_r N_l} \right] , \tag{18}$$

where Re and Im refer to the real and imaginary parts and N_r and N_l are the indices of refraction for right and left circularly polarized light respectively. N_r and N_l can be related to the conductivity tensor $\tilde{\sigma}$.

For a medium with three-fold or higher rotational symmetry about the z-axis, the conductivity tensor $\tilde{\sigma}$ has the form

$$\tilde{\sigma} = \begin{pmatrix} \sigma_{xx} & \sigma_{xy} & 0 \\ -\sigma_{xy} & \sigma_{xx} & 0 \\ 0 & 0 & \sigma_{zz} \end{pmatrix} . \tag{19}$$

Furthermore, the Onsager reciprocal relation $\sigma_{ij}(\vec{M}) = \sigma_{ji}(-\vec{M})$ gives $\sigma_{xy}(\vec{M}) = -\sigma_{xy}(-\vec{M})$ and $\sigma_{ii}(\vec{M}) = \sigma_{ii}(-\vec{M})$, where $\vec{M}$ is the magnetization. Thus σ_{xy} and σ_{ii} have odd-power and even-power dependence respectively on $\vec{M}$.

Using Maxwell's equations one can write N_r and N_l in terms of σ_{ij} as follows:

$$N_{r \atop l}^2 = 1 - i\frac{4\pi}{\omega}(\sigma_{xx} \pm i\sigma_{xy}) \tag{20}$$

300

where ω is the frequency of the light. With the approximation $N_r \sim N_1 = N \equiv (n+ik)$, we see that

$$(n+ik)^2 = (N_r^2 + N_1^2)/2 = 1 - i\frac{4\pi}{\omega}\sigma_{xx} \tag{21}$$

and

$$N_r - N_1 = \frac{4\pi}{\omega}\frac{\sigma_{xy}}{N} \tag{22}$$

so that

$$\theta_k = \frac{4\pi}{\omega} Re\left[\frac{\sigma_{xy}}{(n+i\kappa)[1-(n+i\kappa)^2]}\right]. \tag{23}$$

Thus, the Kerr rotation is determined by the off-diagonal component of the conductivity tensor and the indices of refraction of the medium.

σ_{xy} can be related to the electronic states by calculating the power P absorbed by the medium:

$$P = \frac{1}{2}\int Re\ \vec{J}^* \cdot \vec{E}\ d\vec{V} \tag{24}$$

where $\vec{J}, \vec{E}$ and $\vec{V}$ are the current density, electric field and volume of the sample respectively.

Simplifying Eq. 24 one gets for right and left circularly polarized light

$$P_{\overset{r}{1}} = \frac{1}{2}VE^2(\sigma_{1xx} \mp \sigma_{2xy}), \tag{25}$$

where subscripts 1 and 2 on σ_{ij} refer to the real and imaginary parts respectively.

Eq. 25 gives

$$\sigma_{1xx} = (P_r + P_1)/VE^2 \tag{26}$$

and

$$\sigma_{2xy} = (P_1 - P_r)/VE^2 . \tag{27}$$

From quantum mechanics, the power absorbed at frequency ω is given by

$$P = \hbar\omega \sum_{\alpha\beta} W_{\alpha\beta}\delta(\omega_{\alpha\beta} - \omega), \tag{28}$$

where $W_{\alpha\beta}$ is the transition probability per unit time between electronic states α and β. The perturbation Hamiltonian H' due to the electromagnetic radiation responsible for these transitions is

$$H' = \frac{e}{mc} \, \vec{p} \cdot \vec{A} \tag{29}$$

where $\vec{p}$ is the momentum operator for the electron and $\vec{A}$ is the vector potential for the radiation. We have ignored the term in Eq. 29 that leads to spin-flip transitions.

Using time-dependent perturbation theory with Eqs. 28 and 29 one gets

$$P_{r_1} - \frac{\pi e^2 E^2}{4m^2 \hbar \omega} \sum_{\alpha \beta} \left| < \beta \, | x \pm iy | \, \alpha > \right|^2 \delta(\omega_{\alpha \beta} - \omega). \tag{30}$$

Combining Equations 26, 27 and 30,

$$\sigma_{1xx} - \frac{\pi e^2}{4m^2 \hbar \omega V} \sum_{\alpha \beta} \left[\left| < \beta \, | x+iy | \, \alpha > \right|^2 + \left| < \beta \, | x-iy | \, \alpha > \right|^2 \right] \delta(\omega_{\alpha \beta} - \omega) \tag{31}$$

and

$$\sigma_{2xy} - \frac{\pi e^2}{4m^2 \hbar \omega V} \sum_{\alpha \beta} \left[\left| < \beta \, | x-iy | \, \alpha > \right|^2 - \left| < \beta \, | x+iy | \, \alpha > \right|^2 \right] \delta(\omega_{\alpha \beta} - \omega) \ . \tag{32}$$

Knowing the electronic states of the system one can calculate σ_{1xx} and σ_{2xy} and then use the Kramers-Kronig dispersion relations to find σ_{2xx} and σ_{1xy}.

Both the intraband and interband transitions contribute to σ_{ij}. Intraband transitions are normally estimated by the Drude formula. They are important at lower frequencies but their overall contribution appears to be small compared to interband transitions.

The indices of refraction n and κ can be calculated from σ_{xx} by using Eq. 21. It is clear from Eq. 23 that the small off-diagonal component of the conductivity tensor σ_{xy} is the most important quantity for the Kerr parameters. Therefore we look at Eq. 32 to determine the properties of a material necessary for non-zero Kerr parameters with interband transitions.

The selection rules for the dipolar transitions in Eq. 32 for an atom are $\Delta l = \pm 1$ and $\Delta m = \pm 1$, where l and m are the orbital and magnetic quantum numbers respectively. These rules do not apply to Bloch states constructed from atomic orbitals in a solid due to the near-neighbor overlap. However, they may be reasonable for qualitative explanations in a solid for fairly localized orbitals. In the absence of spin-orbit interactions, the first and second terms cancel each other exactly for every α and β. Thus there is no Kerr rotation without spin-orbit interactions. With spin-orbit interactions, σ_{2xy} is non-zero for a given pair of states. However, for a paramagnetic material, spin-up contributions to σ_{2xy} are exactly cancelled by the corresponding spin-down contributions. Thus a paramagnetic material does not produce any Kerr rotation. In the case of a ferromagnetic material with spin-orbit interactions, the above-mentioned cancellation is not complete leading to a non-zero Kerr rotation.

Since the spin-orbit interactions are normally very weak, the Kerr rotation is extremely small in most of the ferromagnetic materials and its calculation from first-principles poses a very challenging problem. A number of calculations on elemental metals and ordered compounds with different levels of sophistication have been carried out in recent years.[28,29,30,31,32]

A substantial body of experimental magneto-optic data on amorphous RE-TM systems exists in the literature today. Some of these results have been summarized in a 1988 review article by Buschow[33] and a 1990 review article by Reim and Schoenes.[34] Some electronic structure calculations have become available recently which have been reviewed by Kakehashi and Tanaka in the first chapter of this book. However, to our knowledge there are no first-principle calculations of the magneto-optic parameters in an amorphous system so far.

Since the off-diagonal component of the conductivity tensor (Eq. 32) is determined by the dipole matrix elements between the occupied and unoccupied electronic states of the same spin separated by the incident photon energy, one can qualitatively try to relate the features in the magneto-optic spectra to those in the spin-polarized density of states (DOS). This procedure amounts to making the simplistic assumption that the matrix elements in Eq. 32 are the same for all states and leads to the following equation:

$$\sigma_{2xy} \propto \left\{ \int_{\omega_f - \omega}^{\omega_f} \rho(\omega')\rho(\omega+\omega')d\omega' \right\}_{\uparrow\uparrow} + \left\{ \int_{\omega_f - \omega}^{\omega_f} \rho(\omega')\rho(\omega+\omega')d\omega' \right\}_{\downarrow\downarrow} \tag{33}$$

where $\rho(\omega')$ and $\rho(\omega+\omega')$ are the same-spin DOS for occupied and unoccupied states respectively and ω_f refers to the Fermi energy. As pointed out by Wang and Callaway,[35] it must be kept in mind that the constant-matrix-element assumption is inadequate for a quantitative comparison of such a weak effect.

Due to the lack of first-principle calculations, several authors have used the above qualitative arguments to correlate the structure in magneto-optic data with that in some available DOS of various RE-TM systems. The DOS information usually comes from photoemission for the occupied states and occasionally from inverse photoemission for the unoccupied states. These data do not give the spin-polarized PDOS needed in Eq. 31 and hence one simply uses a joint-DOS version for the occupied and unoccupied states as allowed by the available data.

Connell applied the approach mentioned above to analyze the magneto-optic data for a-FeTb.[36] He used the XPS and inverse XPS data for FeTb and Tb to estimate the partial DOS for Fe d-states and Tb d- and f- states. He concluded that Fe d-states dominate the magneto-optic properties of FeTb below 3 eV because the highly-localized main Tb f-peaks (occupied and unoccupied) in DOS are more than 3 eV away from the Fermi energy.

Weller *et al.* came to a similar conclusion from their work on RE-FeCo thin films when RE=Gd and Tb.[37] They find almost identical Kerr rotations for amorphous Tb-FeCo and Gd-FeCo up to 3 eV and spectral behavior similar to that of polycrystalline FeCo but reduced by about a factor of two (Figure 14). A small peak due to a Tb f-multiplet at about 2.5 eV below the Fermi energy does not produce any feature in the MO spectrum of Tb-FeCo perhaps due to small f-d matrix elements. They do observe a significant contribution to Kerr rotation due to Nd in Nd_xCo_{1-x} for $x=0.3$ when the photon energy is near 4 eV. This is due to the transitions from the main Nd 4f-band centered around 5 eV below the Fermi energy to the empty d-band. Some possible reasons for the overall reduction in θ_k with the addition of RE atoms are mentioned in the following.

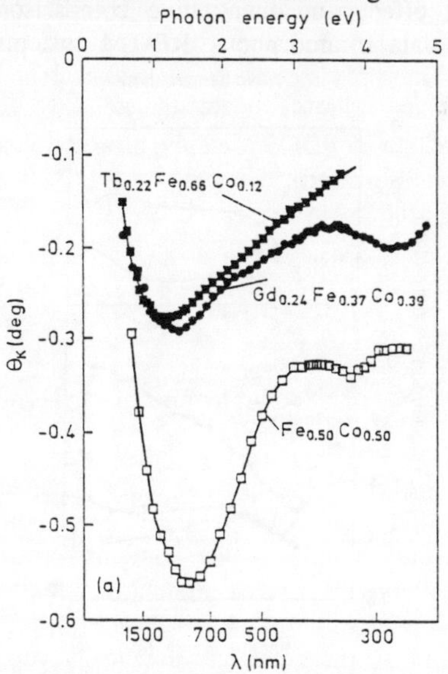

Fig. 14. Magneto-optical polar Kerr rotations as functions of photon energy for a-TbFeCo, a-GdFeCo, and a polycrystalline bcc $Fe_{50}Co_{50}$ film, [after Ref. 37].

304

The RE-TM systems are probably the most widely studied amorphous materials for their magneto-optic properties.[38,39,40] The qualitative agreement of the overall magneto-optic data with the available electronic structure information is similar to that described above. A summary of these results is as follows:

(a) For photon energies below about 2.5 eV RE f-bands are too far from the Fermi energy to contribute directly to the Kerr rotation.[41] Thus the Kerr rotation is dominated by the transition metals. Kerr rotation of RE-TM is lower than that of the pure TM system due to the dilution caused by the RE atoms and the ferrimagnetic coupling of their non-4f electrons. Of course RE atoms are added to achieve the perpendicular anisotropy and other characteristics essential for polar Kerr rotation.

(b) For shorter wavelengths or photon energies in the range of 2.5-5.0 eV, some of the RE 4f-states, especially the main 4f-band of the light RE, can be excited. Therefore there is some contribution to Kerr rotation from RE 4f-5d transitions. This is generally small due to the highly localized nature of 4f states giving small 4d-5f dipole matrix elements. So far Nd seems to make the largest contribution. Examples of this are the recent data of θ_K and η_K as functions of the wavelength for RE/TM multilayers shown in Fig. 15.[42] Films containing Nd have considerably higher θ_K than those with Tb in the 2-5 eV photon energy range. Homogeneous amorphous films give similar results.

First-principles calculations are needed for better understanding of the spin-orbit and matrix-element effects and quantitative comparisons of the magneto-optic and electronic structure data in amorphous RE-TM systems.

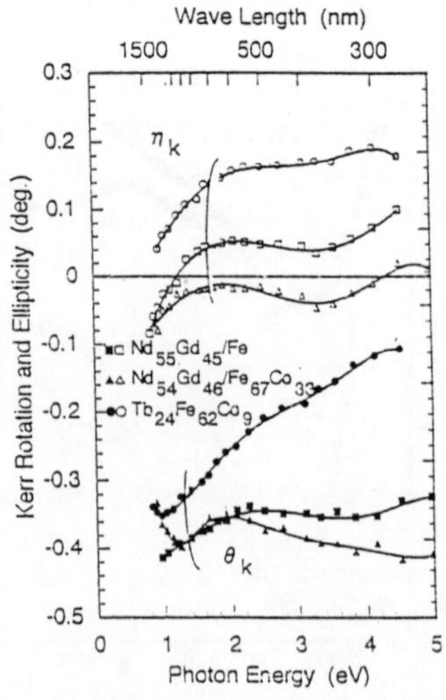

Fig. 15. Kerr spectra of NdGd/FeCo, NdGd/Fe and TbFeCo films as functions of the photon energy [after Ref. 42].

5. Domain Dynamics and Magnetization Reversal

The subject of magnetization reversal and coercivity in magnetic materials has been an active field of research for many decades. As would be expected, much of the early work dealt with crystalline or polycrystalline magnetic systems, and in most cases, defect structures (either magnetic, or structural) were found to play significant, and often dominant, roles in the reversal process. In this section, we consider the mechanisms by which magnetization reversal takes place in RE-TM amorphous alloys and multilayers, and how these mechanisms are related to the structural properties and the macroscopic magnetic parameters. We will limit the discussion to materials which have uniaxial perpendicular magnetic anisotropy (PMA), since such materials have been most thoroughly studied. Experimentally, the most common approach to reversal measurements is either to sweep the applied magnetic field and measure a hysteresis loop, or to first magnetically saturate the sample, then either reduce the field, or reverse the field to a constant value near the coercivity, and monitor the magnetization as a function of time. This latter measurement is commonly called a magnetic viscosity measurement, as the reversal occurs slowly over a long period of time. The most widely applied model to date has been the model of Street and Woolley,[43] who considered magnetic viscosity of polycrystalline permanent magnetic materials. This model presumes that reversal occurs by thermal activation over an energy barrier, and further that there is a wide distribution in barrier energies due to a distribution in single-domain particle sizes or variations in the magnetic parameters. The Street and Woolley model has been extended and used by a number of authors.[44-48] We will consider this model in more detail in Section 5.2

Much of the work on magnetization reversal in RE-TM materials has been stimulated by their potential application as magneto-optic storage media. Current magneto-optic storage devices use polar Kerr rotation to detect the *direction* of magnetization in the film. When plane polarized light is reflected from a magnetized material, the plane of polarization is typically rotated a small amount (Kerr rotation), with the sense of rotation depending on the direction of the magnetization. The reflected light may also have a significant ellipticity. In the polar Kerr effect, the light is incident at near normal, and the size of the rotation is typically a few tenths of a degree. For magnetic storage applications, small regions of the sample have their magnetizations pointing either into or out of the sample surface, resulting in positive or negative rotations of the plane of polarization which can be used as (say) bit 0 and bit 1, respectively, in data storage applications. Magneto-optic storage devices are currently commercially available, and most of them use a-TbFeCo as the storage medium.

In developing materials for magneto-optic storage applications several different and sometimes conflicting requirements must be considered, and the important physical ideas have been described in some detail by a number of authors.[7-9,36,49-52] We summarize their ideas here. For polar Kerr effect storage devices the material must have a strong perpendicular anisotropy so that the magnetization is perpendicular to the film plane. It must also have a square hysteresis loop and a large coercivity (a few kOe) at room temperature for stability of the magnetic domains. It must have a small coercivity at ~500 K (either close to T_C or significantly above the compensation temperature) if thermomagnetic writing is to be used. A large Kerr rotation and small media noise are most important. The current magneto-optic storage materials (a-TbFeCo) have a very strong perpendicular anisotropy and a reasonably large Kerr rotation (0.3°) in the near infrared, where infrared diode lasers ($\lambda = 800$ nm) can be used for both reading and writing. The amorphous nature of these materials leads to a very small media noise (because of the absence of grain boundaries) making them very suitable for storage applications. Current TbFeCo media have very high storage densities (0.3 Gb/in^2), corresponding to written single-domain bits about 1000 nm in diameter, and satisfactory carrier-to-noise ratios. The minimum bit size currently is determined by the size of the focussed laser spot used for reading and writing. It is expected that the next generation of magneto-optic storage media will use blue laser light ($\lambda = 400$ nm), which will increase the storage density by a factor of approximately four. Unfortunately, the magnitude of the Kerr rotation in TbFeCo falls off markedly in the blue, which means that new media materials will have to be developed. The smaller domain sizes will also have implications for domain stability, so that the nature of the magnetization reversal process will undoubtedly have to be well understood.

In this article, we limit our discussion to magnetization reversal and domain dynamics in a-RE-TM thin films, and we discuss how the reversal mechanisms are related to sample nanostructure and to the macroscopic magnetic parameters. In one sense, these materials are highly "defective", since there is no long-range order and since there will necessarily be local variations in the structural and magnetic properties. This argument suggests that perhaps a model similar to that of Street and Woolley could be useful. On the other hand, because of their amorphous nature they can be considered homogeneous and isotropic on a larger length scale, at least in the film plane. This spatial averaging could in principle lead only to a narrow distribution of activation energies, obviating the broad distribution of activation energies implicit in the Street and Woolley model. These kinds of considerations have been used to develop models of magnetization reversal which have been reasonably successful in describing the experimental results. In what follows, we summarize some of the experimental results which have been important in developing our understanding of magnetization reversal in a-RE-TM thin films, and we describe the models which have been used to interpret these results. It will be seen that a satisfactory microscopic picture of the reversal process is not yet available, but the models are getting closer to being predictive.

5.1   Experimental Results

5.1.1 Direct Observations of Magnetic Domains

A number of investigations of magnetization reversal have utilized direct observations of magnetic domains to study the reversal process. In most cases, the domains are observed by means of polarized light microscopy, with Kerr rotation being used to provide the contrast between domains, and in favorable cases this technique is capable of submicron resolution. Most samples investigated have been in thin film form, where it is presumed that the magnetic domains extend through the thickness of the film.

Magnetic domain observations in a-RE-TM thin films usually show that magnetization reversal occurs first by nucleation of a small section of the film. Once nucleated, the domain may expand either by dendritic growth into stripe domains, or by uniform expansion into roughly circular domains. Two limiting behaviors are observed in these materials. In some materials, nucleation dominates the reversal process, with the domains growing only slightly in size once nucleated. This limit is in general observed in materials with small uniaxial anisotropy and/or large saturation magnetization. In the other limit, usually observed in materials with large uniaxial anisotropy and/or small saturation magnetization, nucleation is observed to be a relatively rare event. But once nucleation occurs, the domain expands rapidly. It is this diverse behavior which has proved troublesome to interpret in terms of simple models of magnetization reversal. (It should be noted that a considerable literature on reversal in "bubble" domain materials exists. See for example the article by Cape and Lehman.[53]) In what follows, we discuss some polarized light observations of the reversal process in a-RE-TM materials.

Ohashi et $al.$[54] used polarized light microscopy and the polar Kerr effect to study magnetization reversal and remanence decay in a-Tb_xFe_{1-x} near compensation ($x = 0.21$ at 300 K). Their $remanence$ $decay$ observations showed that reversed domains first nucleated and then expanded by dendritic growth from the nucleation sites. Repeated observations showed that the nucleation sites were not always the same, but they did preferentially occur at certain sites. They also found that the density of reversed nuclei (n_n) depended strongly on the magnitude of the initial saturating magnetic field H_S as: $\log(n_n) = \log(n_0) - CH_s$, where C is a constant. This suggests that either extrinsic defects or intrinsic random fluctuations in the local anisotropy result in hard pinning centers which cannot be reversed by small saturating fields. Their results, shown in Fig. 16, suggest that these nuclei have a

308

broad distribution of activation energies. These data were obtained by first saturating the magnetization with a premagnetizing field, then reversing the field to a constant value and counting the number of reversed nuclei for each value of the premagnetizing field.

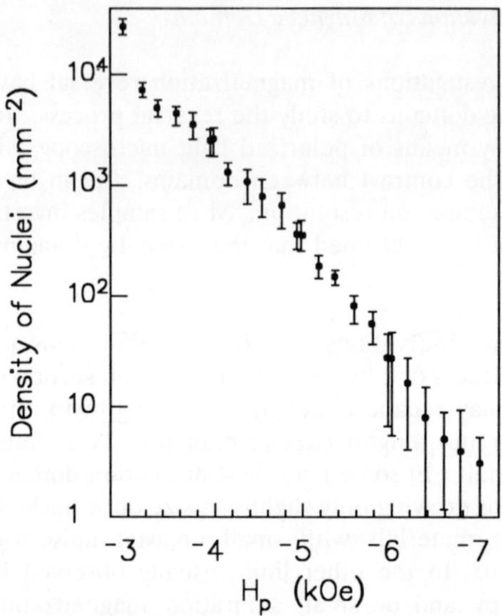

Fig. 16. Density of nucleation sites in TbFe as a function of premagnetizing field. [After Ref. 54.]

Ohashi *et al.*[54] also studied magnetization reversal by first saturating the film and then *reversing* the field to a value less than the coercivity and observing domain growth. They concluded that reversal occurs in three stages: (1) nucleation of reversed domains; (2) growth of domains from the nuclei, which first involves domain tip propagation, ultimately leading to reversed stripe domains; and (3) widening of the stripe domains by parallel wall displacement. This appears to be the first report of direct observation of domain-wall motion in semi-hard magnetic materials. In a later report on similar Tb-Fe materials, Ohashi *et al.*[55] observed roughly circular expansion of domains during reversal, and they were able to measure the domain wall velocity v_w as a function of reversing field H_r. We will discuss these latter results in Section 5.1.2.

Connell and Allen[56] also studied magnetization reversal in a-Tb-Fe near compensation. Their observations were generally in accord with those of Ohashi *et al.*[55] in that they found that reversal occurred by nucleation at isolated sites followed by roughly uniform expansion of domains. Connell and Allen interpreted their results in terms of a model originally developed by Fatuzzo[57] for polarization reversal in ferroelectrics. This model will be discussed in more detail in Section 5.2.3.

Mizoguchi and Kronmüller[58] studied rf sputtered $Fe_{63}Tb_{37}$ 1 μm thick films. Their samples had PMA with a room-temperature coercivity of 500 Oe. Their samples had considerably less than 100 % remanence, and they could not be completely saturated with applied fields up to 9 kOe. In their observations of magnetic domains, they nearly saturated the film and then reduced or reversed the applied field. They found a well-defined critical field of 280 Oe for nucleation of reversed domains, with the critical field being independent of the saturating field for saturating fields above 2500 Oe. After nucleation, they observed two distinct demagnetization processes. First, for positive fields ($0 < H < 280$ Oe), the domain length was noted to grow inhomogeneously. For field demagnetized films, they found a maze-like structure. For negative applied fields, reversal occured by a more or less homogeneous increase in the domain width.

Winkler *et al.*[59] studied 114 nm thick films of TbFeCo near compensation. In their experiments, they first saturated the film, and then reduced the field in small steps until the detector signal changed by 10%. The field was then reduced to freeze the domain pattern. After this, the field was increased to restart the after-effect detection of the domain structure. Presumably, the first phase of this experiment nucleated some domains, while no new domains were nucleated in the second phase -- only domain expansion occured. Domain pictures for the Fe-rich samples (near 20 at.% Tb) showed irregular growth of domains from the nuclei, with some evidence of stripe expansion. The domain patterns for the Tb rich samples (near 30 at.% Tb) showed round nuclei which initially ran out into stripes. The stripes grew partly together with increasing time, finally forming an equilibrium maze-like pattern. They found that no new nuclei were formed during this reversal process.

Merchant and Kryder[60] carried out a comprehensive study of coercive squareness in TbFeCo films with a range of compositions both RE and TM rich. They used Kerr microscopy in conjunction with advanced video processing techniques to investigate the reversal process in samples having widely varying values of M_s and K_u. Their results were in general agreement with those of Ohashi *et al.*[54,55] in that reversal in some samples occured largely by nucleation, whereas in other samples wall motion (either dendritic or regular) was dominant. They measured the width ΔH_c of the $-M_s$ to $+M_s$ transition and found that it was largest for nucleation-dominated samples and smallest for samples showing regular domain expansion. They also found a striking dependence of ΔH_c on M_s, as shown in Fig. 17, with a fairly well-defined boundary between regular and irregular domain expansion. Figure 17 shows that ΔH_c increases monotonically with M_s, as would be expected from simple models of uniaxial anisotropy. Furthermore, regular domain wall growth is only observed for samples which have small ΔH_c and small M_s. This result suggests that the magnetization reversal properties in real films are largely determined by the macroscopic magnetic parameters. Merchant and Kryder also found that ΔH_c depended on the sputtering conditions under which the samples were prepared, but their conclusion was that the variations in ΔH_c could be explained by variations in M_s and K_u. They also noted that films exhibiting nucleation-dominated reversal showed nucleation even if the initial saturating field was as high as 50 kOe. This suggests that the spread in local anisotropies for films with small K_u facilitates the nucleation process.

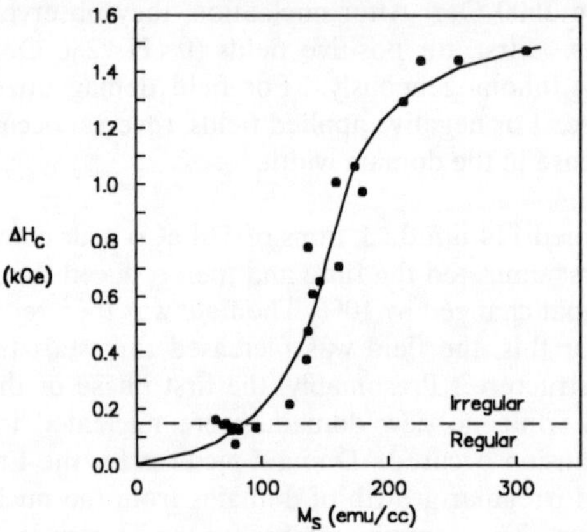

Fig. 17. Width of the switching transition as a function of saturation magnetization for TbFe. For narrow switching transitions reversal is regular, occuring by uniform expansion of nucleated domains. [After Ref. 60]

Merchant and Kryder were also able to correlate the type of reversal with the ratio $R = H_k/H_c$, where $H_k = 2K_u/M_s$ is the anisotropy field. They found that for $R < 7$, reversal was predominantly by nucleation, while for $R > 8$, reversal proceeded by dendritic domain expansion. This result emphasizes the conclusion that magnetization reversal can be largely understood in terms of the macroscopic magnetic parameters, without the need for including defects.

Merchant and Kryder interpreted the dependence of ΔH_c on M_s and K_u in terms of the wall stiffness model of Thiele[61]. Thiele considered the stability of circular domains to elliptical deformations and was able to discuss the stability of the domain in terms of the domain wall "stiffness". He found that the wall energy (due largely to exchange and anisotropy) increases the wall stiffness, and hence tends to maintain a circular domain, while the demagnetizing energy tends to decrease the wall stiffness and leads to deformation from a circular shape.

Forkl *et al.*[62] studied 1000 nm thick films of $Fe_{63}Tb_{37}$ with a 100 nm SiO_x overcoat. These samples showed no compensation point down to 4 K. They observed domains using an optical polarization microscope at temperatures between 298 K and 375 K in fields up to 6 kOe. They recorded the domain patterns and motion using a low light level video camera (with background subtraction). After saturation at fields greater than 2 kOe, they reduced the field to below 1 kOe (field still in same direction) and observed demagnetization. The hysteresis loop of this sample has only about 70% remanence. They find general agreement with earlier publications, and in particular they noted that domains nucleated at certain sites which were randomly distributed, and further that nucleation did not always take place at the same positions. Starting from these nuclei, the domains grew asymmetrically in two, sometimes three, directions with a characteristic velocity even if the applied field was kept constant. This velocity seemed to fluctuate around a mean value and sometimes a branch would develop In negative applied fields, they observed the stripe width to increase, so that the separation between reversed domains decreased. At large enough reversed fields, all stripes merged together so that the film was again saturated.

5.1.2 Time Dependence of Remanence Decay and Magnetization Reversal

The most common type of magnetization reversal measurement is the hysteresis loop, where sample magnetization (or alternately Kerr rotation) is measured as a function of external magnetic field. The coercivity H_c and the width ΔH_c of the switching transition are two indicators of the magnetization reversal process. One of the problems with using hysteresis loops to study magnetization reversal is that H_c is often quite strongly dependent on the magnetic field sweep rate (see Ohashi *et al.*[54] and Wolniansky *et al.*[63]). More illuminating measurements for most samples are the time dependence of remanence decay and magnetization reversal with a constant applied magnetic field. These measurements are schematically indicated in Fig. 18. The sample is first saturated by applying a large magnetic field in the negative direction. In a remanence decay measurement, the field is then reduced to zero and the magnetization is monitored as a function of time. In a magnetization reversal measurement, after saturation the magnetic field is changed to a constant positive value and the sample magnetization is measured as a function of time. For these measurements, the constant field usually has to be near H_c. In many cases however it is possible to observe substantial (or even complete) magnetization reversal for applied fields well below H_c. In these cases, the reversal process is undoubtedly dominated by thermal activation.

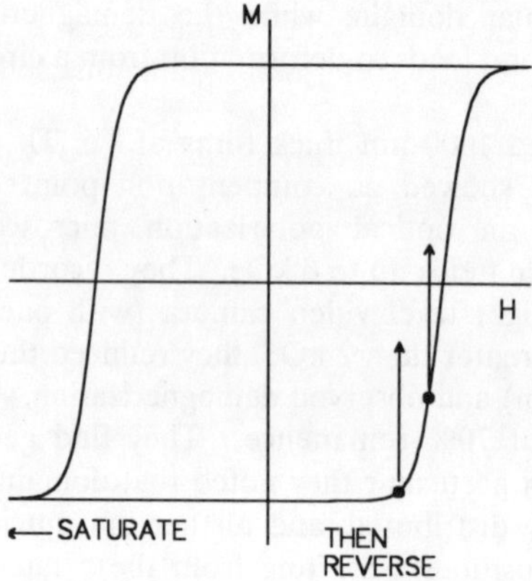

Fig. 18. To make a magnetization reversal measurement, first magnetically saturate the sample with a large negative magnetic field. Then reverse the field to near the coercive field and monitor the magnetization as a function of time.

Magnetization reversal measurements of the type described above generally lead to two distinctly different reversal behaviors. Ohashi *et al.*[55] studied $Tb_{22}Fe_{78}$ and found the reversal curves shown in Fig. 19.

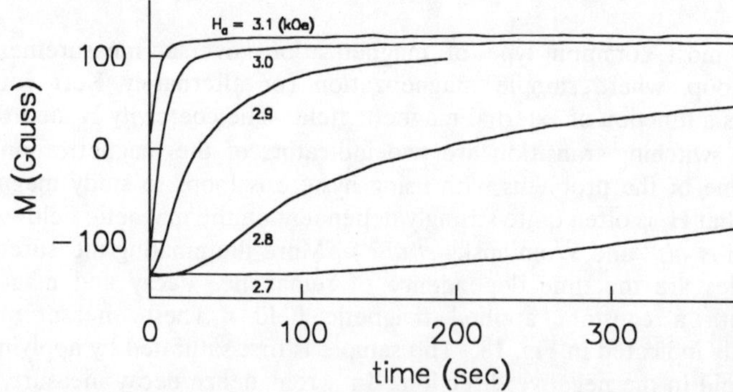

Fig. 19. Magnetization reversal curves in $Tb_{22}Fe_{78}$ for several values of the applied magnetic field. [After Ref. 55]

Note in particular that the reversal process occurs on a time scale of seconds to hundreds of seconds for the values of applied field used. The domain observations on this sample showed that it reversed by slow nucleation (indicated by the small initial slope of the M vs. t curves) followed by relatively rapid uniform domain expansion. Reversal curves similar to those of Ohashi *et al.* have been reported for other RE-TM systems by Connell and Allen[56], Labrune *et al.*[64], and Kirby *et al.*[65] in compositionally-modulated Dy/Fe amorphous thin films. Labrune *et al.* showed explicitly that the shape of the magnetization reversal curves are often correlated with the types of domain patterns observed during reversal. Fig. 20 shows two reversal curves obtained by Labrune *et al.*

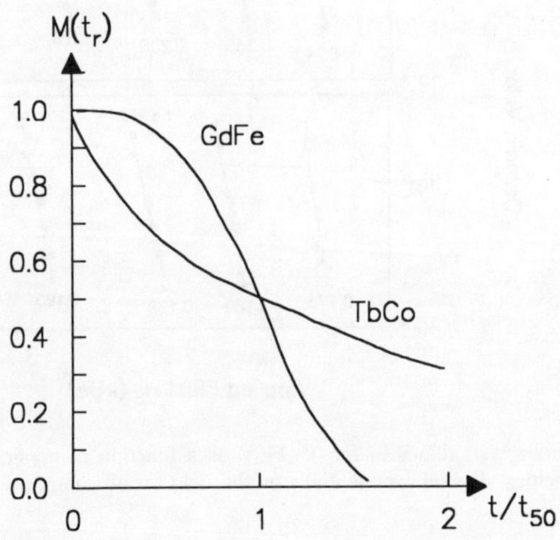

Fig. 20. Magnetization reversal curves for GdFe and TbCo samples. The TbCo sample reverses primarily by nucleation, while the GdFe sample reverses primarily by domain wall motion. [After Ref. 64]

Domain observations on these same samples by Labrune *et al.* showed that GdFe sample in Fig. 20 reverses by slow nucleation followed by rapid domain expansion, while the TbCo sample reverses primarily by nucleation.

Ohashi *et al.*[55] were able to measure the domain growth velocities as a function of applied field for a range of sample compositions, as shown in Fig. 21. A similar exponential dependence of the growth velocity on applied field has been reported by other authors.[58,62] Labrune *et al.*[64] found a similar result in their measurements on GdTbFe. They showed that the time for reversal, as measured by the time required for the magnetization to reach the $M=0$ state, varied exponentially with the applied field. Their results are shown in Fig. 22.

314

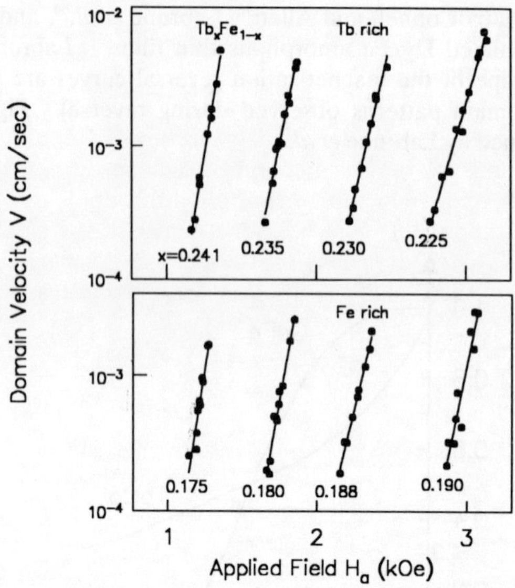

Fig. 21. Domain wall velocities for Tb_xFe_{1-x} as a function of applied magnetic field. The velocities depend exponentially on the field for all compositions studied [After Ref. 55]

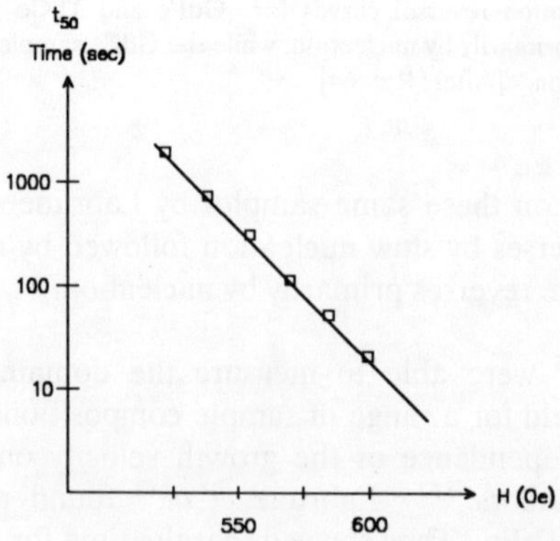

Fig. 22. Magnetic field dependence of the half reversal time in GdFeCo. Note that the half reversal time depends exponentially on the magnitude of the field. [After Ref. 64]

5.2   Models of Magnetization Reversal

5.2.1   Street and Woolley Model

Street and Woolley[43] considered magnetization reversal in polycrystalline magnetic materials consisting of non-interacting magnetic grains. They assumed that the magnetic material, upon reduction or reversal of the appllied magnetic field, was in a metastable state, and that it could relax from that state by thermally-activated processes. Letting $N = f(E)dE$ represent the number of magnetic domains with activation energies between E and $E + dE$, they wrote the rate of change of N as:

$$\frac{dN}{dt} = -\lambda N = -C\exp(-E / k_B T) f(E) dE \tag{34}$$

where $\lambda = C\exp(-E/k_B T)$ is a decay constant which depends on the activation energy and the temperature in the usual fashion, and C is a constant representing an attempt frequency. This equation is obviously satisfied by $f = f_0 \exp(-\lambda t)$, with f_0 being the value of f at time $t = 0$.

Street and Woolley assume that each activation process contributes the same amount $\overline{m}$ to the change in magnetization. Then if dN domains are activated in time dt, the change in the magnetization of the sample can be written:

$$dM = \overline{m} C f_0 \exp(-\lambda t)\exp(-E / k_B T) dE dt \tag{35}$$

Integrating over all activation energies then gives the time rate of change of the magnetization:

$$\frac{dM}{dt} = \overline{m} C \int_{E_{MIN}}^{E_{MAX}} f_0 \exp(-\lambda t)\exp(-E / k_B T) dE \tag{36}$$

The distribution of activation energies has its origin in distributions of the sizes and orientations of the magnetic particles, the size and nature of magnetic and structural defects, and it may additionally have contributions from domain wall motion within a magnetic particle. Making the assumption that there is a uniform distribution of activation energies, so that $f_0 = p$ (independent of E), they find a result which has been extensively applied to magnetization reversal in a variety of magnetic systems:

$$\Delta M = \overline{m} p k_B T \cdot \log(t) + \text{constant} \quad (Ct \gg 1) \tag{37}$$

316

Thus their prediction is that the change in magnetization (with constant applied field) will vary as log(t) at long times. Such behavior has been observed in a variety of magnetic materials, and it is often assumed that such behavior is *prima facie* evidence of a broad distribution of activation energies. As will be discussed later, this is not in general true, and some other physical effects must be considered. The reader is also referred to a thoughtful discussion of log(t) behavior by Aharoni.[66]

The Street and Woolley model would at first sight seem to be appropriate for a-RE-TM thin films. These materials are believed to have large random variations in the local anisotropy and structural properties due to their amorphous structures. These intrinsic variations might be expected to lead to a broad distribution of activation energies for reversal, either by nucleation or domain wall motion. However, the experimental situation is not so clear. Some materials do show at least roughly log(t) behavior during a magnetic viscosity measurement, which is suggestive of a broad distribution of activation energies. An example is shown in Fig. 23.

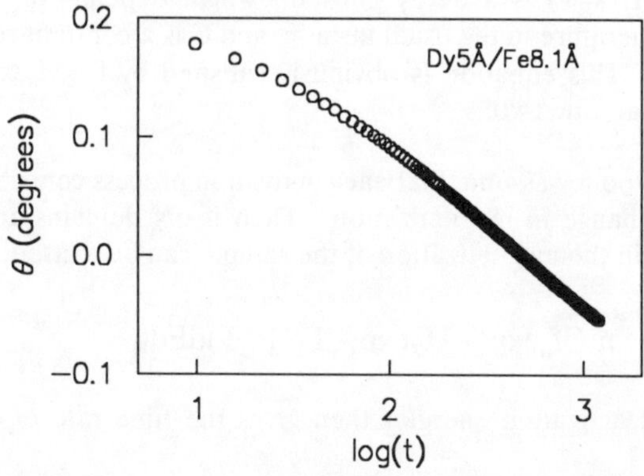

Fig. 23. Kerr rotation angle versus log(t) for Dy5A/Fe8.1A multilayer.

Samples which exhibit quasi-log(t) behavior largely appear to reverse by a nucleation process, with domain-wall motion only playing a small role. Other samples however show a slow rate of reversed domain nucleation, with the bulk of the sample reversing by expansion of the nucleated domains. Samples which show these widely disparate behaviors are thought to be structurally similar. This suggests that the effects of random anisotropy fluctuations on the reversal process may in fact be quite subtle.

Reversal in crystalline or polycrystalline materials is often couched in similar terms, with small particles reversing by nucleation, and larger particles reversing by nucleation followed by domain-wall motion. The limiting factor in domain-wall motion is generally thought to be the strength of wall-pinning mechanisms, and the reversal process has been discussed by a number of authors.[67-69] Intrinsic domain-wall pinning can occur in even perfect single crystals because of the atomic nature of the crystals. Egami[68] has shown that wall motion in this case is largely a diffusive process. In the case of polycrystalline materials, grain boundaries and other structural defects determine the strength of the pinning, and there can be a wide range of pinning strengths, even within one particular sample.

Domain-wall pinning in amorphous materials may in some ways be an easier process to understand. The pinning is expected to occur largely through fluctuations in anisotropy, which to first-order may average to a nearly constant value over dimensions on the order of domain wall thicknesses.

5.2.2 *Single Barrier Height Approach*

A number of researchers have analyzed magnetization reversal behavior in a-RE-TM systems assuming reversal occurs by thermal activation over barriers with a very narrow height distribution, especially during the wall expansion process. Once a reversed domain has nucleated, the domain wall is presumed to move by thermal activation over a barrier which arises due to local fluctuations in anisotropy and nanostructure. In this case, two different limits are generally considered. In the weak pinning limit, the wall is assumed to be pinned by a large number of weak pinning sites which are distributed randomly.[70-72] In this case, the barrier height for thermally-activated wall motion is written as $E_B = E_0 - 2VM_sH$, where E_0 is the zero-field barrier height and V is the volume associated with the discrete wall motion over the barrier. Note that the barrier height is directly proportional to the applied magnetic field. In the strong pinning limit, the wall is pinned by relatively widely-spaced pinning centers.[73,74] As the applied field is increased the wall remains locally constrained, but it can "bow" to reduce the magnetic energy. Finally, the wall breaks away from the pinning center and moves irreversibly until its motion is stopped by new pinning centers. The field dependence of the activation energy in this case is usually written in the form: $E_B = C/M_sH$, where C is a constant and the barrier height is now inversely proportional to the applied magnetic field. In either pinning limit, once nucleation has occured the wall is assumed to move by thermal activation, which leads to a wall velocity v_w given by

$$v_w = Rd \cdot \exp(-E_B / k_B T) \tag{38}$$

where R is an attempt frequency, typically chosen to be on the order of 10^9 s^{-1} and d is the distance moved by the wall. Thus in the weak pinning limit the wall velocity depends on H according to $\ln(v_w) \propto H$, while in the strong pinning limit $\ln(v_w) \propto 1/H$. Ohashi *et al.*[55] measured wall velocities as a function of field during uniform (roughly circular) domain expansion in a-TbFe samples near compensation. The

318

range of velocities which they could measure was between 10^{-4} and 10^{-2} cm/s, which limited the range of applied fields which could be investigated. Because of this limited range of H, they were unable to distinguish whether $\ln(v_w)$ was proportional to H or to 1/H.

Forkl *et al.*[62] were able to measure wall velocities over a large enough range to determine that $\ln(v_w) \propto H$ in a-$Tb_{63}Fe_{37}$. They measured the velocity of the *length* of the reversed domains for constant applied field (they had to do this during the early stages of reversal (<15%), as the dipolar interaction appreciably influenced the velocity.) Their results are shown in Fig. 24, and it is evident that $\ln(v_w) \propto H$, which strongly suggests that this film is in the weak pinning limit.

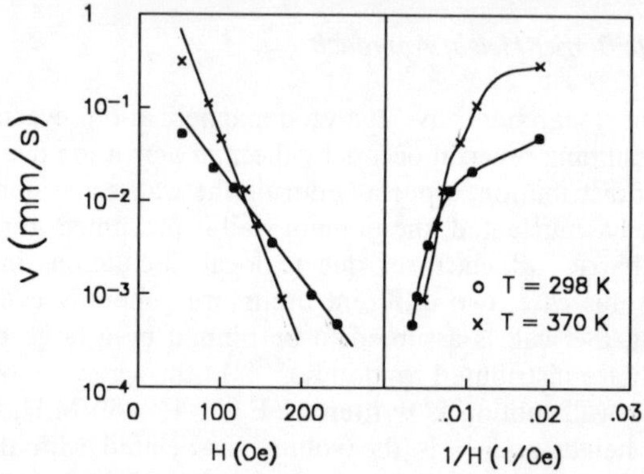

Fig. 24. Logarithm of the domain wall velocities as a function of magnetic field (left hand side) and inverse magnetic field (right hand side) for $Fe_{63}Tb_{37}$. [After Ref. 62]

Forkl *et al.* further claim that the activation volume increases with increasing temperature, from 4.2 x 10^{-18} cm^3 at 298 K to 21.4 x 10^{-18} cm^3 at 370 K. They found no time dependence for reversal by increase in domain width. They also did experiments where they saturated and then reduced or reversed the applied field and then monitored the magnetization as a function of time. They find approximately log(t) behavior of ΔM (over only 1 decade in time), which they discussed in terms of the irreversible susceptibility.

Winkler *et al.*[59] used the polar Kerr effect to study domain nucleation and wall motion in TbFeCo for Tb concentrations near 20 at.% and near 30 at.%. The magnetization reversal curves for the samples near 20% Tb showed behavior intermediate between pure nucleation (log(t)) and slow nucleation followed by rapid wall motion. They assumed that the activation energy for this could be written as $E = E_0 - 2VHM_s$, and that $dM/dt \propto \exp(-E/k_BT)$. They then measured the slope of the M vs t curves at M=0 and plotted the log of this slope vs. 1/T. From these results they found the surprisingly large value $E_0 \sim 100$ eV and $V \sim 10^{-16}$ cm^3. Note that their temperature range for these measurements was only ~ 3K.

Thompson et al.[75,76] have taken a slightly different approach to magnetization reversal in a-RE-TM alloys. They have used the magnetic equation of state proposed by Street et al.[77] for such materials to relate the activation volume during reversal to the field dependence of the reversal rates. This approach leads to the equation:

$$\frac{dH}{d(\ln \dot{M})} = \frac{k_B T}{V M_s} \tag{39}$$

where H is the applied magnetic field and $\dot{M}$ is the time rate of change of magnetization. It should be noted that this same equation can be obtained by taking the thermal activation energy barrier as $E_B = E_0 - V M_s H$ and writing

$$\frac{dM}{dt} = C \cdot \exp(-E_B / k_B T) \tag{40}$$

where C is the product of an attempt frequency times a magnetization. Taking the natural log of both sides of the equation and then differentiating H with respect to $\ln(dM/dt)$ results in Eq. 39. In the approach of Thomson et al., measuring the field dependence of the reversal curves at constant M allows a determination of the activation volume V. This analysis for $Tb_{20}Fe_{72}Co_8$ leads to field-dependent activation volumes on the order of 10^{-18} cm^3. The activation volume determined in this manner goes through a broad minimum for fields in the vicinity of H_c as shown in Fig. 25.

Thompson et al. carried out the above analysis for magnetization decay measurements made using both magnetic measurements and Kerr rotation measurements. The two curves in fig. 25 have similar shapes, but the minimum in the activation volume occurs at a smaller field for the magnetic measurements. Thomson et al. attribute this to a combination of a slower magnetic field sweep rate in the magnetic measurements (which will lead to a smaller measured coercivity) and to the surface sensitivity of the Kerr measurements. They do not give a physical interpretation of the field dependence of the activation volume, but it is likely that the initial decrease in activation volume with increasing field is due to the fact that for small fields, the minimum stable volume of a cylindrical domain during nucleation decreases with increasing field. The subsequent increase in activation

320

volume is probably due to the rapid field-induced expansion of the domain after nucleation takes place. The size of this latter effect will depend critically on the strength of the domain-wall pinning.

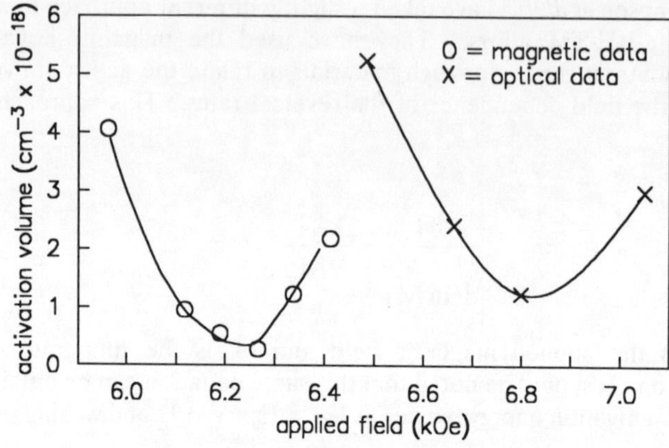

Fig. 25.   Activation volumes as a function of applied field for Tb$_{20}$Fe$_{72}$Co$_8$. [After Ref. 75]

5.2.3   The Fatuzzo Model

Fatuzzo[57] considered polarization reversal in ferroelectrics, and his model has subsequently been used to describe magnetization reversal in a-RE-TM films. Fatuzzo assumed that reversal occurred first by nucleation of reversed domains, followed by uniform (circular) expansion at constant velocity. In his model, the fractional area reversed as a function of time is written:

$$A(t) = 1 - \exp\left\{-2k^2\left[1 - (Rt + k^{-1}) + \frac{1}{2}(Rt + k^{-1})^2 - e^{-Rt}(1 - k^{-1}) - \frac{1}{2}k^{-2}(1 - Rt)\right]\right\}$$

(41)

where R is the rate at which reversed nuclei are created and k is a parameter related to the radial velocity. In the limit of $k = 0$, reversal occurs by nucleation only (with no expansion of domains after nucleation), whereas large values of k correspond to slow nucleation followed by rapid domain wall motion. Connell and Allen[56] found that the time dependence of the magnetization reversal could be described very well by this equation over a range of reversing fields and temperature. The value of k≈5 seemed to be satisfactory for the experimental conditions they used. Figure 26 shows their reversal curves for several different values of the applied field and sample temperature. The solid line is the prediction of the Fatuzzo model for k = 5.

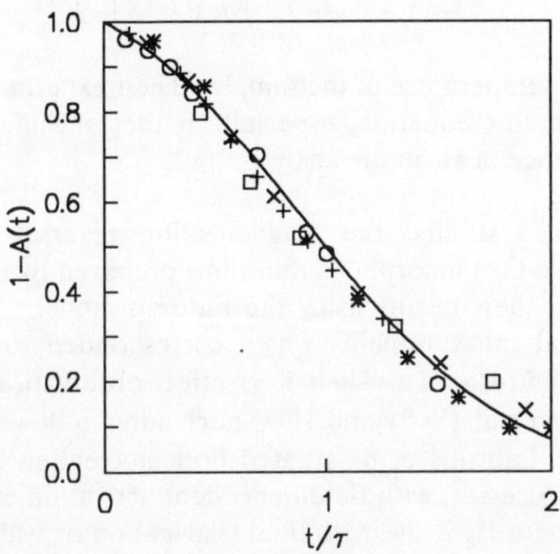

Fig. 26. Magnetization reversal curves for TbFe for several different applied fields and temperatures. [After Ref. 56]

In order to relate the Fatuzzo model to the physical properties of the sample being investigated, it is necessary to consider in more detail the mechanisms by which magnetization reversal proceeds. That is, what magnetic or structural parameters determine the nucleation rate R and the domain-wall velocity? (It should be noted that the room temperature experimental evidence to date indicates that the domain wall moves in small localized steps by a thermally-activated process.) Generally, two possibilities which differ in the magnitude of the domain wall pinning forces are considered. In the weak pinning limit,[70,74,78] the thermal activation energy decreases linearly with applied field, whereas in the strong pinning limit,[73,74] the thermal activation energy is proportional to $1/H$.

Connell and Allen plotted the time τ required for reversal to proceed halfway, and found that $\log(\tau) \propto 1/H$, where H is the reversing field. (Over the limited range of reversing fields used, it is possible that $\log(\tau) \propto H$ would also provide a reasonable fit to experiment). Thus they interpret their results in terms of a strong pinning model. They considered the free energy of a nucleating domain and incorporated the exchange stiffness constant A, the uniaxial anisotropy K_u and

322

the anisotropy field H_k to arrive at an expression for the temperature and field dependence of τ:

$$\ln\tau = \ln\tau_0 - \frac{E_0}{k_B}\left(\frac{1}{T} - \frac{1}{T_c}\right) + \frac{E_0}{k_B}\left(\frac{1}{T} - \frac{1}{T_c}\right)\frac{H_k}{H} \tag{42}$$

where T_c is the Curie temperature of the sample. Their experimental results were in good agreement with this equation, especially if they included the experimental temperature dependence of H_k in the analysis.

Labrune et al.[64] studied the magnetization reversal process in Gd-Fe, GdTbFe and TbCo(Fe-Gd) amorphous thin films prepared by e-beam evaporation. They also interpreted their results using the Fatuzzo model. This comprehensive study showed reversal measurements which corresponded to both limits of the Fatuzzo model ($k\approx0$ and $k>>1$). Their Kerr effect observations showed examples of nucleation-only reversal ($k\approx0$) and slow nucleation followed by rapid domain expansion ($k\approx1000$). Labrune et al. treated both nucleation and wall motion as thermally-activated processes, with field-dependent activation energies of the form: $E_A = E_0 - 2HM_sV$, where E_0 is the zero-field (nucleation or wall-motion) activation energy and V is the volume associated with the particular reversal process. Thus Labrune et al. treated the domain wall motion in the weak pinning approximation. By comparing the predictions of the Fatuzzo model with their experimental reversal curves, they were able to estimate $E_0 = 3$ eV for the nucleation process.

Kirby et al.[65] applied the Fatuzzo model to magnetization reversal in amorphous Dy/Fe multilayers and found results qualitatively similar to those found in amorphous alloys, suggesting that the multilayering process does not materially affect the reversal mechanisms. Multilayering does however allow quite precise control of the macroscopic magnetic parameters, as has been shown by Shan et al.[18] Kirby et al. found that the type of reversal behavior (nucleation only, or wall-motion dominated) depended strongly on the individual layer thicknesses. For example Dy5Å/Fe4.5Å reversed by nucleation, while Dy5Å/Fe5Å reversed by rapid wall motion, as indicated in Fig. 27. In this latter case, the Fatuzzo model provided fairly good fits to the experimental reversal curves. Subsequent measurements on additional samples showed that this sensitivity to multilayer thickness held true. As found by Merchant and Kryder,[60] there is a strong correlation between the width of the switching transition and the type of reversal behavior that occurs. The results of Kirby et al. for a fairly wide range of individual layer thicknesses are summarized in Table I. The results in Table I can also be correlated with the values of K_u, which is strongly dependent on the individual layer thicknesses.

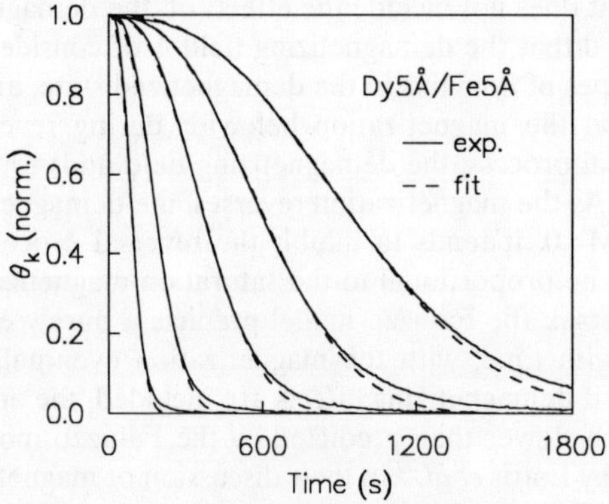

Fig. 27. Magnetization reversal in Dy/Fe multilayers for several values of the reversing field. The solid lines are the experimental measurements, and the dashed lines are fits to Eq. 39. [After Ref. 65]

Dy/Fe (Å)	Behavior	$\Delta H_C/H_C$
4.5/4.5	I	0.17
5.0/4.5	N	0.90
5.0/5.0	W	0.10
5.0/5.5	W	0.13
5.5/5.5	W	0.10
5.5/5.5	W	0.18
5.0/6.0	N	1.00
6.0/6.0	I	0.22
5.0/6.5	N	0.41
5.0/7.0	N	0.79

Table I. Reversal behavior as a function of switching transition width in Dy/Fe multilayers. Here, N = nucleation dominated reversal, W = wall motion dominated reversal, I = intermediate behavior between N and W.

The Fatuzzo model offers a simple picture of the magnetization reversal process which is very useful in interpreting experimental results. It does however suffer from some shortcomings. It presumes that the expansion of reversal domains is radial, while many systems show reversal by domain tip propagation and/or expansion of stripe domains. The Fatuzzo model cannot describe the reversal process in such systems. A more fundamental objection to the Fatuzzo model in its present form is that it does not include the effects of the demagnetizing field. It has long been understood that the demagnetizing field is of considerable importance in determining the shapes of domains in the demagnetized state, and it can also have a significant impact on the magnetization behavior during reversal. In the initial phases of the reversal process, the demagnetizing field adds to the external field in systems with PMA. As the magnetization reverses, the demagnetizing field becomes smaller, and after $M = 0$, it tends to inhibit the reversal process. The size of the effect will obviously be proportional to the saturation magnetization. In the case of nucleation only reversal, the Fatuzzo model predicts a purely exponential decay of the magnetization with time, with the magnetization eventually being completely reversed. However, if demagnetizing effects are included, the approach to complete reversal can be much slower than predicted by the Fatuzzo model. This point has been demonstrated by Lottis *et al.*[79] in their discussion of magnetization reversal in a particulate system. They consider a collection of single-domain magnetic particles with strong uniaxial anisotropy in the mean-field approximation. This case is equivalent to the $k = 0$ regime of the Fatuzzo model. Lottis *et al.* showed that reversal proceeded much more slowly than would be expected from the Fatuzzo model, and in fact that the magnetization varied approximately as $\log(t)$ at long times. This quasi-$\log(t)$ behavior is a consequence of the fact that the driving force for reversal is decreasing with time. This effect is partially mitigated, in the case of slow nucleation followed by rapid wall motion, by the fact that the local demagnetizing field at the edge of a large (micron size) but isolated domain will be near zero. If the magnetization is fairly large, the effects of the demagnetizing field can be extremely important. On the other hand, it is often difficult to separate demagnetizing field effects from other effects such as sample inhomogeneity. If a sample is sufficiently inhomogeneous, then magnetization reversal may not proceed to completion for fields smaller than the saturation field, and in fact the magnetization will vary as $\log(t)$ at long times for reversing fields smaller than the saturation field.

5.2.4 Micromagnetic Calculations and Simulations

A major effort has been made in recent years to carry out micromagnetic calculations and simulations of magnetization reversal in a-RE-TM materials. The starting point for most micromagnetic calculations has been the Landau-Lifshiftz equation, which relates the time rate of change of angular momentum of the moments in the sample to the applied torque resulting from external magnetic fields and internal interactions (demagnetizing fields, anisotropy, and exchange). Usually a viscous damping term is also included. The micromagnetics approach is deterministic, in that coupled equations of motion are solved to follow the moment orientations as a function of time. While a detailed discussion of these techniques is beyond the scope of this review, we will summarize some of the results which have been obtained. It is important to note that such calculations require considerable computational power if they are to be carried out on a large scale, especially if long-range magnetic dipole interactions are to be properly included. For more details of the calculations, the reader is referred to the review article by Bertram and Zhu[80] and a series of papers by Manisuripur and co-workers.[81-84] It should be noted that these calculations are almost always equilibrium calculations carried out for $T=0$, and thus will not be directly comparable to experimental results at finite temperatures, where thermal activation can play an important role in the reversal process.

Mansuripur and McDaniel[83] solved the Landau-Lifshitz-Gilbert equations for a two-dimensional square lattice of dipoles. They divided the thin film into 10 A x 10 A cells and chose values of M_s and K_u appropriate for an a-RE-TM magneto optic material. Assuming a perfectly anisotropic crystal (every cell has the same magnitude and direction for K_u), they start their simulations with an abrupt switch from up to down magnetization. They find that in zero applied field, the narrow wall relaxes to a finite width wall in times on the order of nanoseconds. When they apply an external field, they find that the wall moves with constant speed. To make the calculation more realistic they chose the *direction* of the local anisotropy in different cells to be random distributed. Their simulations for this case showed that Bloch lines appeared in the domain wall, and the presence of the random axis anisotropy led to a non-zero wall coercivity. Thus this approach may lead to detailed pictures of the nucleation energy and domain wall pinning mechanisms responsible for the coercivity at $T=0$.

In a more recent calculation, Mansuripur and co-workers[84,85] used simulations to investigate domain-wall coercivity on the submicron scale in the presence of a variety of simulated defects and inhomogeneities. They found that in homogeneous samples the experimental nucleation fields were smaller than those necessary in the simulations, but if they introduced reversed magnetized seeds in regions of large local anisotropy, the experimental and simulated nucleation fields were brought much closer together. They mainly addressed the problem of wall coercivity in these materials, where the problem is that the simulated wall coercivity for homogenous samples was smaller than the experimental values for homogeneous samples. In this case, they find that the inclusion of spatial fluctuations in the anisotropy and other structural and magnetic defects could *increase* the wall coercivity in the simulations. Thus it is suggested that the fluctuations in the local magnetic and structural parameters are responsible for many of the macroscopic magnetic properties of a-RE-TM materials.

In a paper dealing with magnetization reversal in particulate systems, Lyberatos *et al.*[86] considered a thin film consisting of identical single domain

326

particles. These particles were assumed to have a strong uniaxial anisotropy and their anisotropy axes were all aligned perpendicular to the film direction. They wrote the energy of a single particle as:

$$E = K_u V \cdot \sin^2\theta - M_s V \cdot (H - DM) \cdot \cos\theta \qquad (43)$$

where K_u is the value of the anisotropy constant, V is the volume of the particle, H is the external field (applied perpendicular to the film), and DM is the local demagnetizing field at the site of the particle due to the presence of the other particles. This energy leads to an energy barrier

$$E_B = K_u V \cdot \left[1 \pm \frac{M_s \cdot (H - DM)}{2K_u} \right]^2 \qquad (44)$$

where the $\pm$ signs indicate the direction of the transition. Lyberatos *et al.* then assumed that reversal occurred by thermal activation over the anisotropy barrier and carried out a time-dependent Monte-Carlo simulation of the reversal process in the presence of a constant magnetic field. Lottis *et al.*[79] did a similar calculation in the mean field approximation, where they were able to calculate magnetization as a function of time by numerically solving a differential equation. Both groups showed that the magnetic viscosity was a strong function of temperature. Lottis *et al.* further showed that the demagnetizing field could lead to quasi-log(t) behavior in the remanent magnetization decay even in homogenous samples, without the need for introducing a distribution of particle volumes. The results of one calculation done by Lottis *et al.* are shown in Fig. 28. In the initial phases of the remanent decay, the demagnetizing field assists the thermal activation process by reducing the height of the energy barrier. As reversal proceeds however, the demagnetizing field reduces, and eventually reverses, so that the height of the energy barrier increases with time. Thus, in these calculations, there is a distribution of barrier heights (which leads to the quasi-log(t) behavior), but only as a function of time.

Recently Kirby *et al.*[87] introduced a simple model of magnetization reversal in a-RE-TM thin films which used the calculations of Lyberatos *et al.*[86] and Lottis *et al.*[79] as a starting point. In their calculation, Kirby *et al.* considered a thin-film ferromagnet which has strong perpendicular anisotropy and assumed that Eq. 43 described the energy of a small volume of the film (This volume was assumed to extend through the entire thickness of the film). The Individual atomic moments in this volume (cell) were assumed to be fluctuating coherently due to thermal processes, and the magnetization in each cell was assumed to be uniform perpendicular to the film. They then added a "domain wall" energy term to Eq. 43,

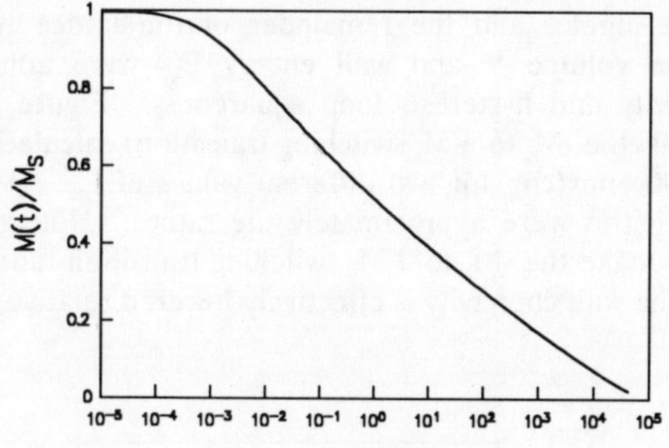

Fig. 28. Simulated remanent magnetization decay for $K_u = 5 \times 10^5$ erg/cm^3, $M_s = 200$ emu/cm^3, particle volume $= 2 \times 10^{-18}$ cm^3 at 300 K. [After Ref. 79].

so that the energy of cell i could be written:

$$E_i = K_u V \cdot \sin^2\theta - M_s V \cdot (H - DM) \cdot \cos\theta + \sum_j S_{ij} E_w \cos\theta \qquad (45)$$

The magnitude and sign of the last term on the right depends on the direction of magnetization in the nearest neighbor cells. If all the nearest neighbor cells have their magnetizations in the same direction as cell (ij), then the sum of S_{ij} over nearest neighbors is $+1$. If the nearest neighbor magnetizations are all antiparallel to that of cell (ij), then this sum is -1. This approach then favors reversal of a cell if its neighbors have already reversed. Qualitatively, it will tend to make the wall coercivity smaller than the nucleation coercivity, which seems to be true for most a-RE-TM films. Mansuripur[81] included a similar wall energy term in his *equilibrium* simulations of magnetization reversal.

In this model magnetization reversal occurs by thermal activation over the barrier height associated with Eq. 45. This barrier height can be written:

$$E_B = K_u V \left[1 \pm \frac{M_s V \cdot (H - DM) + \sum_j S_{ij} E_w}{2 K_u V} \right]^2 \qquad (46)$$

Kirby *et al.* have carried out time-dependent Monte Carlo calculations using this model, and some of their results are summarized in Figs. 29-31. In these calculations the demagnetizing field was calculated by treating neighboring cells out to third

328

neighbors as point dipoles and the remainder of the lattice in the mean-field approximation. The volume V and wall energy E_W were adjusted to give an appropriate coercivity and hysteresis loop squareness. Figure 29 shows partial hysteresis loops (only the $-M_s$ to $+M_s$ switching transition) calculated for $K_u = 1 \times 10^6$ erg/cm^3 and $M_s = 90$ emu/cm^3 for two different values of E_w. (V was adjusted so that the two coercivities were approximately the same). Note that the effect of increasing E_w is to make the $-M_s$ to $+M_s$ switching transition more rapid, as would be expected since the wall coercivity is effectively lowered relative to the nucleation coercivity.

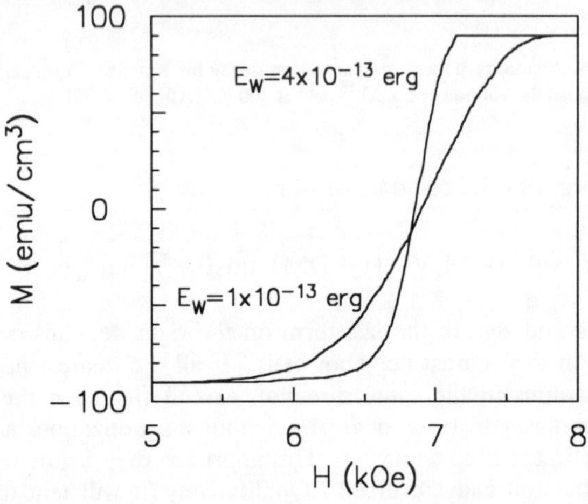

Fig. 29. Simulated hysteresis loop switching transitions for two different values of the wall energy term in Eq. 45. [After Ref.87]

The time dependence of the magnetization decay curves for the same values of the parameters is shown in Fig. 30. The reversal curve for $E_w = 1 \times 10^{-13}$ ergs shows quasi-log(t) behavior at long times. The reader should note that these reversal curves are qualitatively similar to the two reversal curves in Fig. 20, which are illustrative of reversal dominated by nucleation and reversal dominated by wall motion. The simulation results are in general agreement with these two kinds of behavior, as is illustrated in Fig. 31. This figure shows the state of the simulated samples at ~40% reversal. Clearly a larger value of E_w enhances the growth by wall motion. The simulation for small wall energy shows that reversal has occurred largely by nucleation, whereas for large wall energy, nucleation is a relatively rare event, but once nucleated, the domain expands roughly uniformly. These qualitative results are consistent with the experimental results of Labrune et al.[64]

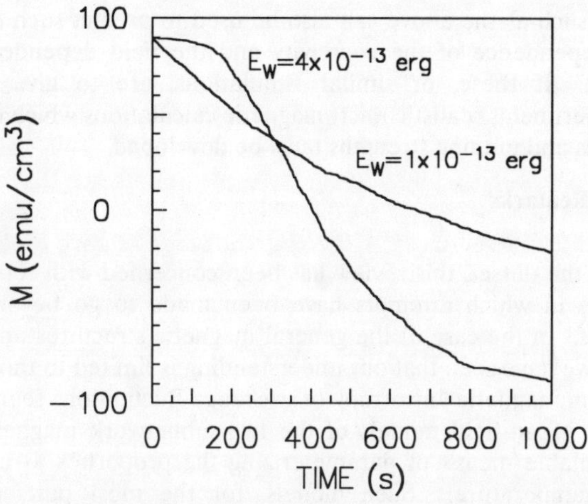

Fig. 30. Simulated magnetization reversal curves for two different values of the wall energy term of Eq. 45 . [After Ref. 87]

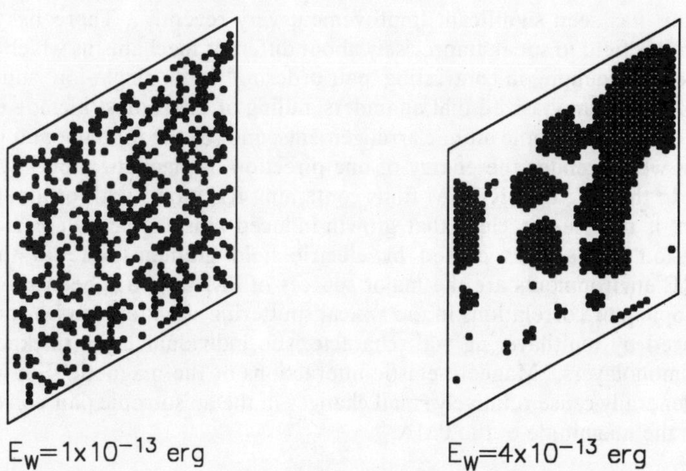

Fig. 31. Simulated domain patterns for K_u = 1 x 10^6 erg/cm^3 and M_s = 90 emu/cm^3 [After Ref.87].

Simulations such as the above can also be used to predict such quantities as the temperature dependence of the coercivity and the field dependencies of the reversal behaviors. If these, or similar simulations, are to give quantitative agreement with experiment, realistic micromagnetic calculations which can estimate the reversal volumes and pinning strengths must be developed.

6.   Concluding Remarks

As stated at the outset, this review has been concerned with recent research on a-RE-TM alloys in which attempts have been made to go beyond the early rudimentary models. In the case of the general magnetic structures and properties of these materials, we have seen that our understanding is limited to those situations where either random magnetic anisotropy or exchange fluctuations seem to operate nearly exclusively. Mean-field models of the two-subnetwork magnetic alloy are about the only available means of parameterizing the properties as a function of composition and temperature. Such models, for the most part, ignore local environment effects which are known to be important in certain cases where the TM moments are unstable. Basic aspects of the phase transitions in the presence of *both* random anisotropy and exchange fluctuations are largely unexplored.

Our understanding of the origins of perpendicular magnetic anisotropy in a-RE-TM films has seen significant improvement very recently. There has been a tendency in this field to speak imprecisely about different mechanisms which are not comparable, for example, in contrasting "pair ordering" with "single-ion" sources. It should be strongly emphasized that an understanding of PMA must include *both* (a) the nature of the anisotropic atomic arrangements, and (b) the fundamental physical interactions which render the energy of one direction of magnetization lower than the other. In the case of a-RE-TM films containing RE atoms with orbital angular momentum, it now seems clear that growth-induced anisotropic pair correlations plus single-ion interactions caused by electric field gradients due to the non-isotropic RE environments are the major sources of PMA. It has been shown that the anisotropic pair correlations in the typical sputtering situation can be controlled and enhanced by multilayering with characteristic individual layer thicknesses of about two monolayers. Magneto-elastic interactions of the magnetic film with the substrate generally cause relatively small changes in the anisotropic pair correlations and thus in the magnitude of the PMA.

For those a-RE-TM films containing Gd or other S-state ions, usually the magnitude of the PMA is rather small if present at all. In such cases spin-orbit interactions involving the TM-sub-network and dipole-dipole interactions are the sources of the anisotropy. In certain cases of Gd-TM films where the value of K_u

has been measured to be fairly large, the sources of both the anisotropic structure and the interactions are not entirely clear. This remains a topic for further research.

The subject of the electronic structure of a-RE-TM alloys and its relationship to magneto-optic properties has been shown to be in a rudimentary state. Electronic structure calculations in amorphous materials, in general, are very difficult although reasonably good approximate calculations have been performed for inter transition-metal glasses. TM glasses containing RE elements are very difficult for local-density calculations because of the problems of handling the nearly localized 4f levels. These latter difficulties, even for *ordered* RE intermetallic compounds, have impeded progress considerably, and there are essentially no satisfactory self-consistent calculations for RE-TM alloys containing RE elements other than Gd. When these problems are added to the complications inherent in a fundamental calculation of the magneto-optic parameters, it can be seen that there is room for considerable work in this area.

On the experimental side, it has been shown that the maximum value of θ_k obtainable from a-RE-TM alloys is about $0.4°$ from the TM component, and about an additional $0.2°$ from the RE. This latter value can occur only for light RE's such as Nd for which the 4f level is close enough to the Fermi level to be involved in the optical transitions at, say, $\lambda \approx 400$ nm. It has been shown that by multilayering a TM with a light RE, it is possible to obtain both the needed PMA and a relatively large Kerr rotation value at the desired wavelength (400 nm) for the next generation of magneto-optic storage devices. It remains to be seen whether other competitive materials such as fine-grained MnBi-based alloys or the Bi-doped crystalline garnets will be able to provide the needed carrier-to-noise ratios at 400 nm, and thus supersede the a-RE-TM films.

We have given an overview in this contribution of the status of our understanding of domain dynamics and magnetization reversal in a-RE-TM films. Early models such as those of Street and Woolley and Fatuzzo have been shown to have severe limitations. Recent simulational studies have shown how semiquantitative understanding can be obtained based on fundamental magnetic parameters such as M_s and K_u, plus a small number of other parameters which must be determined by fitting to experimental data. These parameters, including cell volume and a domain-wall energy, must in future work be related to the fundamental atomic properties such as magnetic moments, local anisotropy, structural defects, and fluctuations in these quantities. The work has shown the importance of thermally-activated processes in the magnetization decay and reversal processes.

Acknowledgments

We are grateful to our colleagues J.X. Shen, Z.S. Shan, and R.J. Hardy for much experimental assistance and many helpful discussions. For enlightening conversations and correspondence, we thank Drs. V. Harris, T. Suzuki, M. Mansuripur, D. Mergel and H. Kronmüller. We are indebted for financial support to the National Science Foundation under grant DMR-8918889, to the Research Corporation, and to the Center for Materials Research and Analysis at the University of Nebraska.

References

1. R.W. Cochrane, R. Harris and M.J. Zuckermann, *Physics Reports* **48** (1978) 1.
2. D.J. Sellmyer and M.J. O'Shea, *J. Less. Comm. Metals* **94** (1983) 59.
3. D.J. Sellmyer and S. Nafis, *J. Appl. Phys.* **57** (1985) 3584.
4. K. Moorjani and J.M.D. Coey, *Magnetic Glasses* (Elsevier, New York, 1984).
5. T. Kaneyoshi, *Amorphous Magnetism* (CRC Press, Boca Raton, 1984).
6. D.J. Sellmyer and M.J. O'Shea in *Recent Progress in Random Magnets*, ed. D.H. Ryan (World Scientific, Singapore, 1992), p. 71.
7. P. Hansen in *Handbook of Magnetic Materials*, ed. K.H.J. Buschow (North-Holland, Amsterdam, 1991), p. 289.
8. Proc. of Magneto-Optical Recording International Symposium '91, *J. Mag. Soc. Jpn.* Vol. **15**, Suppl. S1 (1991).
9. Proc. of Magneto-Optical Recording International Symposium '92, *J. Mag. Soc. Jpn.* (in press).
10. R.J. Gambino, in Ref. 8, p. 1.
11. D.J. Sellmyer, Z.S. Shan, J.X. Shen and R.D. Kirby, in Ref. 8, p. 9.
12. M.H. Kryder, in Ref. 8, p. 139.
13. J. Schoenes and H. Brandle, in Ref. 8, p. 213.
14. H. Jouve, J.P. Rebouillat and R. Meyer, AIP Conf. Proc. **29** (1976) 97.
15. A.K. Bhattacharjee, R. Julien, B. Cogblin and M.J. Zuckermann, *Physica B* **91** (1977) 179.
16. E.M. Chudnovsky, *J. Appl. Phys.* **64** (1988) 5770.
17. Y. Suzuki, S. Takayama, F. Kirino and N. Ohta, *IEEE Trans. Mag.* **23** (1987) 2275.
18. Z.S. Shan, D.J. Sellmyer, S.S. Jaswal, Y.J. Wang and J.X. Shen, *Phys. Rev. Lett.* **63** (1989) 449; *Phys. Rev. B* **42** (1990) 10,433; *ibid.* **42** (1990) 10,446.
19. D.J. Sellmyer, Z.S. Shan and S.S. Jaswal, *Matls. Sci. Eng.* **B6** (1990) 137.
20. V.G. Harris, K.D. Aylesworth, B.N. Dos, W.T. Elam and N.C. Koon, *Phys. Rev. Lett.* **69** (1992) 1939; *IEEE Trans. Mag.* **28** (1992) 2958.
21. F. Hellman and E.M. Gyorgy, *Phys. Rev. Lett.* **68** (1992) 1391.
22. X. Yan, M. Hirscher, T. Egami and E.E. Marinero, *Phys. Rev. B* **43** (1991) 9300.

23. H. Fu, M. Mansuripur and P. Meystre, *Phys. Rev. Lett.* **66** (1991) 1086; *ibid* **68** (1992) 1441.

24. S.S. Jaswal, *Phys. Rev. Lett* **68** (1992) 1440.

25. D. Mergel, H. Heitmann and P. Hansen, *Phys. Rev. B* (1993) 1 Jan.

26. P.N. Argyres, *Phys. Rev.* **97** (1955) 334.

27. J.L. Erskine and E.A. Stern, *Phys. Rev. B* **8** (1973) 1239.

28. D.K. Misemer, *J. Magn. Magn. Mat.* **72** (1988) 267.

29. G.H.O. Daalderop, F.M. Mueller, R.C. Albers and A.M. Boring, *J. Magn. Magn. Mat.* **74** (1988) 211.

30. H. Ebert, *Physica B* **161** (1989) 175.

31. S.V. Halilov and E.T. Kulatov, *J. Phys. Condens. Matter* **3** (1991) 6363.

32. P.M. Oppeneer, T. Maurer, J. Sticht, and J. Kübler, *Phys. Rev. B* **45** (1992) 10924.

33. K.H.J. Buschow in *Ferromagnetic Materials*, Vol. 4, edited by E.P. Wolfarth and K.H.J. Buschow (Elsevier, New York, 1988), p. 493.

34. W. Reim and J. Shoenes in *Ferromagnetic Materials*, Vol. 5, edited by K.H.J. Buschow and E.P. Wohlfarth (Elsevier, New York, 1990), p. 133.

35. C.S. Wang and J. Callaway, *Phys. Rev. B* **9** (1974) 4897.

36. G.A.N. Connell, *J. Magn. Magn. Mat.* **54-57** (1985) 1561.

37. D. Weller, W. Reim and P. Schrijner, *IEEE Trans. Mag. MAG-24* (1988) 2554.

38. R.J. Gambino and T.R. McGuire, *J. Magn. Magn. Mat.* **54-57** (1986) 1365.

39. P. Hansen, M. Hartmann and K. Witter, Proc. Int. Symp. Magneto-Optics, *J. Magn. Soc. Jpn.* **11**, Suppl. **S1** (1987) 257.

40. Y.J. Choe, S. Tsunashima, T. Katayama, and S. Uchiyama, *Proc. Int. Symp. Magneto-Optics,* Suppl. **S1** (1987) 273.

41. J.K. Lang, Y. Baer and P.A. Cox, *J. Phys. F: Metal Phys.* **11** (1981) 121.

42. X.Y. Yu, H. Watabe, S. Iwata, S. Tsunashima and S. Uchiyama, *Proc. Magneto-Optic Recording Int. Symp., J. Magn. Soc. Jpn.*, 1992 (In Press).

43. R. Street and J. C. Woolley, *Proc. Phys. Soc. A* **62** (1949) 562.

44. S. H. Charap, *J. Appl. Phys.* **63** (1988) 2054.

45. M. P. Sharrock and J. T. McKinney, *IEEE Trans. Magn.* **MAG-17** (1981) 302.

46. M.P. Sharrock, *IEEE Trans. Magn.* **26** (1990) 193.

47. R. W. Chantrell, J. Popplewell, and S. W. Charles, *J. Magn. Magn. Mater.* **15-18** (1980) 1123.

48. R. W. Chantrell, *J. Magn. Magn. Mater.* **95** (1991) 365.

49. P. Hansen and H. Heitmann, *IEEE Trans. Magn.* **25** (1989) 4390.

50. M. H. Kryder, *Mat. Res. Soc. Symp. Proc. Vol.* **150** (1990) 3.

51. M. H. Kryder, *J. Magn. Magn. Mater.* **83** (1990) 1.

52. P. Hansen, *J. Magn. Magn. Mater.* **83** (1990) 6.

53. J. A. Cape and G. W. Lehman, *J. Appl. Phys.* **42** (1971) 5732.

54. K. Ohashi, H. Takagi, S. Tsunashima, S. Uchiyama, and T. Fujii, *J. Appl. Phys.* **50** (1979) 1611.

55. K. Ohashi, J. Tsuji, S. Tsunashima, and S. Uchiyama, *Japan. J. Appl. Phys.* **19** (1980) 1333.

56. G. A. N. Connell and R. Allen, *Proc. 4th Intern. Conf. on Rapidly Quenched Metals*, Sendai, (1980) p. 981.

57. E. Fatuzzo, *Phys. Rev.* **127** (1962) 1999.

58. T. Mizoguchi and H. Kronmüller, *IEEE Trans. Magn.* **MAG-25** (1989) 3347.

59. S. Winkler, W. Reim, and K. Schuster, *Thin Solid Films* **175** (1989) 265.

60. A. A. Merchant and M. H. Kryder, *IEEE Trans. Magn.* **27** (1991) 3690.

61. A. A. Thiele, *Bell Syst. Tech. J.* **48** (1969) 3287.

62. A. Forkl, M. Hirscher, T. Mizoguchi, and H. Kronmüller, *J. Magn. Magn. Mater.* **93** (1991) 261.

63. P. Wolniansky, S. Chase, R. Rosenvold, M. Ruane, and M. Mansuripur, *J. Appl. Phys.* **60** (1986) 346.

64. M. Labrune, S. Andrieu, F. Rio, and P. Bernstein, *J. Magn. Magn. Mat.* **80** (1989) 211.

65. R. D. Kirby, J. X. Shen, Z. S. Shan, and D. J. Sellmyer, *J. Appl. Phys.* **70** (1991) 6200.

66. A. Aharoni, *Phys. Rev. B* **46** (1992) 5434.

67. W. F. Brown, *Rev. Mod. Phys.* **17** (1945) 15.

68. T. Egami, *Phys. Stat. Sol. (a)* **19** (1973) 747.

69. H. Kronmüller, *Phys. Stat. Sol. (b)* **144** (1987) 385.

70. F. D. Stacey, *Austr. J. Phys.* **13** (1960) 599.

71. J. A. Baldwin and F. Milstein, *J. Appl. Phys.* **45** (1974) 4006.

72. P. Gaunt, *J. Appl. Phys.* **48** (1977) 3470.

73. B. Barbara, J. Magnin, and H. Jouve, *Appl. Phys. Lett.* **31** (1979) 133.

74. P. Gaunt, *Phil. Mag. B* **48** (1983) 261.

75. T. Thomson, K. O'Grady, S. Brown, P.W. Haycock, and E.W. Williams, *IEEE Trans. Magn.* **28** (1992) 2515.

76. T. Thomson, K. O'Grady, C. M. Perlov, and R. W. Chantrell, *IEEE Trans. Magn.* **28** (1992) 2518.

77. R. Street, P. G. McCormick, and L. Folks, *J. Magn. Magn. Mater.* **104-107** (1992) 368.

78. P. Gaunt, *Phil. Mag.* **34** (1976) 775 and 781.

79. D. K. Lottis, R. White, and E. D. Dahlberg, *Phys. Rev. Letters* **67** (1991) 362.

80. H. N. Bertram and J. G. Zhu, *Solid State Phys.* **46** (1992) 271.

81. M. Mansuripur, *J. Appl. Phys.* **61** (1987) 1580.

82. M. Mansuripur, *J. Appl. Phys.* **61** (1987) 3334.

83. M. Mansuripur and T. W. McDaniel, *J. Appl. Phys.* **63** (1988) 3831.

84. M. Mansuripur, R. C. Giles, and G. Patterson, *J. Mag. Soc. Japan* **15** (1991) 17.

85. R. Giles and M. Mansuripur, *Computers in Physics* **4** (1991) 204.

86. A. Lyberatos, R. W. Chantrell, and A. Hoare, *IEEE Trans. Magn.* **26** (1990) 222.

87. R. D. Kirby, J. X. Shen, R. J. Hardy, and D. J. Sellmyer, *to be published.*